Hans Bisswanger
Enzyme Kinetics

Related Titles

Bisswanger, H.
Practical Enzymology
2004
ISBN: 978-3-527-30444-8

Aehle, W. (Ed.)
Enzymes in Industry
Production and Applications
2007
ISBN: 978-3-527-31689-2

Reymond, J.-L. (Ed.)
Enzyme Assays
High-throughput Screening, Genetic Selection and Fingerprinting
2006
ISBN: 978-3-527-31095-1

Breslow, R. (Ed.)
Artificial Enzymes
2005
ISBN: 978-3-527-31165-1

Buchholz, K., Kasche, V., Bornscheuer, U.T.
Biocatalysts and Enzyme Technology
2005
ISBN: 978-3-527-30497-4

Hans Bisswanger

Enzyme Kinetics

Principles and Methods

Second, Revised and Updated Edition

WILEY-VCH Verlag GmbH & Co. KGaA

The Author

Prof. Dr. Hans Bisswanger
Interfakultäres Institut für Biochemie
Hoppe-Seyler-Str. 4
72076 Tübingen
Germany

1st edition translated by Leonie Bubenheim.

■ All books published by Wiley-VCH are carefully produced. Nevertheless, authors, editors, and publisher do not warrant the information contained in these books, including this book, to be free of errors. Readers are advised to keep in mind that statements, data, illustrations, procedural details or other items may inadvertently be inaccurate.

Library of Congress Card No.: applied for

British Library Cataloguing-in-Publication Data
A catalogue record for this book is available from the British Library.

Bibliographic information published by the Deutsche Nationalbibliothek
The Deutsche Nationalbibliothek lists this publication in the Deutsche Nationalbibliografie; detailed bibliographic data are available on the Internet at http://dnb.d-nb.de.

© 2008 WILEY-VCH Verlag GmbH & Co. KGaA, Weinheim, Germany

All rights reserved (including those of translation into other languages). No part of this book may be reproduced in any form – by photoprinting, microfilm, or any other means – nor transmitted or translated into a machine language without written permission from the publishers. Registered names, trademarks, etc. used in this book, even when not specifically marked as such, are not to be considered unprotected by law.

Composition K+V Fotosatz GmbH, Beerfelden
Printing Betz-Druck GmbH, Darmstadt
Bookbinding Litges & Dopf GmbH, Heppenheim

Printed in the Federal Republic of Germany
Printed on acid-free paper

ISBN: 978-3-527-31957-2

For Anna and Michael

Preface to the Second English Edition

Die Zeit, innerhalb welcher eine bestimmte Menge Substrat verändert wird, also das Maß der Reaktionsbeschleunigung durch den Katalysator, hängt in erster Linie von seiner Menge ab. In sehr vielen Fällen ist sie sogar direkt proportional der wirksamen Menge des Fermentes, in anderen Fällen bestehen kompliziertere Beziehungen, die man in den sogenannten „Fermentgesetzen" hat ausdrücken wollen, die aber zum großen Teil sehr mangelhaft fundiert sind.

The time needed for a distinct amount of substrate to be changed, hence the degree of acceleration of the reaction by the catalyst, depends primarily on its amount. In a great number of cases it is even directly proportional to the efficient amount of the ferment. In other cases more complicated relationships exist. It was attempted to formulate these in the so-called "ferment laws". However, to a large extent, they are very insufficiently substantiated.

Carl Oppenheimer (1919) *Biochemie*
Georg-Thieme-Verlag Leipzig

Upon preparing a new edition of a book inevitably the question arises: what should be changed? A textbook is bound to present the fundamental topics, which cannot easily be modified without disturbing the whole concept and, since enzyme kinetics is not a very expanding area of biochemistry, most topics can be regarded as fundamental. One challenging area not satisfactorily covered by classical enzyme kinetics concerns enzymes located at boundary layers, especially at membranes. Actually, about half of the enzymes in a cell are in more or less intense contact with membranes. Such enzymes and their reactions are different to reactions in aqueous solutions and cannot be treated in the same way because of profound differences in the immediate environment, the access of substrates and the release of products. The situation is complicated by the fact that there exists no general mode of membrane association of enzymes. While various enzymes are completely integrated into the membrane, others are more or less loosely associated. Further, it must be considered that (non-enzymatic) transport processes through membranes also exhibit similarities with enzyme reactions, translocation from one site to another in the membrane can be compared with the conversion of substrate to product. This opens a broad field for the extension of the rules of enzyme kinetics. However, since such systems cannot satisfactorily be described by one or a few general rules, special mechanisms have to be developed for each individual system; the discussion of this topic is restricted to prominent systems, immobilized enzymes, membrane integrated enzymes and transport systems, and enzymes on boundary layers between the aqueous solution of the cytosol and the membrane. A striking conformity of all these systems is the broad applicability of the basic equation of enzyme kinetics, the Michaelis-Menten law which, as a rule, can be used in a first approach, while the special features of the respective system may necessitate distinct modifications. The concept of allosteric enzymes is extended to membrane systems. Nevertheless, only suggestions can be given, while for distinct solutions for a certain system the special literature must be consulted.

In a separate section a comparison is drawn between enzyme kinetics and pharmacokinetics, two closely related areas, both depending on enzyme reactions and thus sharing various similarities. However, there are also essential differences, enzyme kinetics treating defined enzyme reactions, while pharmacokinetics deals with more complex processes observing the fate of a drug during its journey through the organism. The different terminology employed in these fields is compared to facilitate mutual understanding.

The text was generally revised especially in order to simplify the understanding of the theoretical, often dry, matter. Sections not directly required for the continuous treatise are set in separate boxes.

I thank Mr. Zhougang Yang for updating and designing many of the figures.

Tübingen, January 2008 *Hans Bisswanger*

Preface to the First English Edition

The time about three decades ago may be regarded as the *Golden Age* of enzyme kinetics. Then it became obvious that many biological processes can be forced into terrifying formulas with which experts intimidate their colleagues from other fields. The subject has been treated in several competent textbooks, all published in the English language.

For students with English not being their mother tongue this did not provide a simple language problem, but rather confronted them simultaneously with a difficult matter *and* a foreign language. So the original intention to write a textbook in German was to minimise the fear of the difficult matter. Very difficult derivations were renounced realising the fact that most biochemists will never need or keep in mind every specialised formula. They rather require fundamentals and an understanding of the relationships between theoretical treatments and biological processes explained by such derivations as well as the knowledge which practical approaches are most suited to examine theoretical predictions. Therefore, the book is subdivided into three parts, only the central chapter dealing with classical enzyme kinetics. This is preceded by an introduction into the theory of binding equilibria and followed by a chapter about methods for both binding studies and enzyme kinetics including fast reactions.

Since the German edition is well introduced and the concept broadly accepted, the publication of an English edition appeared justified. This is supported by the fact that new editions in enzyme kinetics are rare, although a thorough understanding of the field as an essential branch of biochemistry is indispensable. The original principle of the former editions to present only fundamentals for a general understanding cannot consequently be maintained, as a specialist book on the subject must exceed the level of general textbooks and should assist the interested reader with comprehensive information to solve kinetic problems. Nevertheless, the main emphasis still is to mediate the understanding of the subject. The text is not limited to the derivation and presentation of formula, but much room is given for explanations of the treatments, their significance, applications, limits, and pitfalls. Special details and derivations turn to experts and may be skipped by students and generally interested readers.

The present English edition is a translation of the Third German edition including revisions to eliminate mistakes.

I would like to acknowledge many valuable suggestions especially from students from my enzyme kinetics courses as well as the support from WILEY-VCH, especially from Mrs. Karin Dembowsky. Her encouraging optimism was a permanent stimulus for this edition.

Tübingen, January 2002　　　　　　　　　　　　　　　　　　　Hans Bisswanger

Contents

Preface to the Second English Edition *VII*

Preface to the First English Edition *IX*

Symbols and Abbreviations *XVII*

Introduction and Definitions *1*
References *4*

1	**Multiple Equilibria** *7*	
1.1	Diffusion *8*	
1.2	Interaction between Macromolecules and Ligands *12*	
1.2.1	Binding Constants *12*	
1.2.2	Macromolecules with One Binding Site *13*	
1.3	Macromolecules with Identical Independent Binding Sites *14*	
1.3.1	General Binding Equation *14*	
1.3.2	Graphic Representations of the Binding Equation *20*	
1.3.2.1	Direct and Linear Diagrams *20*	
1.3.2.2	Analysis of Binding Data from Spectroscopic Titrations *22*	
1.3.3	Binding of Different Ligands, Competition *25*	
1.3.4	Non-competitive Binding *27*	
1.4	Macromolecules with Non-identical, Independent Binding Sites *29*	
1.5	Macromolecules with Identical, Interacting Binding Sites, Cooperativity *32*	
1.5.1	The Hill Equation *32*	
1.5.2	The Adair Equation *34*	
1.5.3	The Pauling Model *37*	
1.5.4	Allosteric Enzymes *38*	
1.5.5	The Symmetry or Concerted Model *39*	
1.5.6	The Sequential Model and Negative Cooperativity *44*	
1.5.7	Analysis of Cooperativity *48*	
1.5.8	Physiological Aspects of Cooperativity *50*	
1.5.9	Examples of Allosteric Enzymes *52*	

Enzyme Kinetics. Principles and Methods. 2nd Ed. Hans Bisswanger
Copyright © 2008 WILEY-VCH Verlag GmbH & Co. KGaA, Weinheim
ISBN: 978-3-527-31957-2

1.5.9.1	Hemoglobin 52
1.5.9.2	Aspartate Transcarbamoylase 53
1.5.9.3	Aspartokinase 54
1.5.9.4	Phosphofructokinase 55
1.5.9.5	Allosteric Regulation of the Glycogen Metabolism 55
1.5.9.6	Membrane Bound Enzymes and Receptors 55
1.6	Non-identical, Interacting Binding Sites 56
	References 57

2	**Enzyme Kinetics** 59
2.1	Reaction Order 59
2.1.1	First Order Reactions 60
2.1.2	Second Order Reactions 61
2.1.3	Zero Order Reactions 62
2.2	Steady-State Kinetics and the Michaelis-Menten Equation 63
2.2.1	Derivation of the Michaelis-Menten Equation 63
2.3	Analysis of Enzyme Kinetic Data 66
2.3.1	Graphical Representations of the Michaelis-Menten Equation 66
2.3.1.1	Direct and Semi-logarithmic Representations 66
2.3.1.2	Direct Linear Plots 73
2.3.1.3	Linearization Methods 75
2.3.2	Analysis of Progress Curves 77
2.3.2.1	Integrated Michaelis-Menten Equation 78
2.3.2.2	Determination of Reaction Rates 80
2.3.2.3	Graphic Methods for Rate Determination 82
2.3.2.4	Graphic Determination of True Initial Rates 84
2.4	Reversible Enzyme Reactions 85
2.4.1	Rate Equation for Reversible Enzyme Reactions 85
2.4.2	The Haldane Relationship 87
2.4.3	Product Inhibition 88
2.5	Enzyme Inhibition 91
2.5.1	Unspecific Enzyme Inhibition 91
2.5.2	Irreversible Enzyme Inhibition 92
2.5.2.1	General Features of Irreversible Enzyme Inhibition 92
2.5.2.2	Suicide Substrates 93
2.5.2.3	Transition State Analogs 95
2.5.2.4	Analysis of Irreversible Inhibitions 96
2.5.3	Reversible Enzyme Inhibition 98
2.5.3.1	General Rate Equation 98
2.5.3.2	Non-Competitive Inhibition and Graphic Representation of Inhibition Data 101
2.5.3.3	Competitive Inhibition 107
2.5.3.4	Uncompetitive Inhibition 111
2.5.3.5	Partially Non-competitive Inhibition 113

2.5.3.6	Partially Uncompetitive Inhibition	115
2.5.3.7	Partially Competitive Inhibition	117
2.5.3.8	Noncompetitive and Uncompetitive Product Inhibition	119
2.5.3.9	Substrate Inhibition	120
2.5.4	Enzyme Reactions with Two Competing Substrates	121
2.5.5	Different Enzymes Catalyzing the Same Reaction	123
2.6	Multi-substrate Reactions	124
2.6.1	Nomenclature	124
2.6.2	Random Mechanism	126
2.6.3	Ordered Mechanism	131
2.6.4	Ping-pong Mechanism	132
2.6.5	Product Inhibition in Multi-substrate Reactions	135
2.6.6	Haldane Relationships in Multi-substrate Reactions	135
2.6.7	Mechanisms with more than Two Substrates	136
2.6.8	Other Nomenclatures for Multi-substrate Reactions	138
2.7	Derivation of Rate Equations of Complex Enzyme Mechanisms	138
2.7.1	King-Altmann Method	138
2.7.2	Simplified Derivations Applying Graph Theory	144
2.7.3	Combination of Equilibrium and Steady State Approach	145
2.8	Kinetic Treatment of Allosteric Enzymes	147
2.8.1	Hysteretic Enzymes	148
2.8.2	Kinetic Cooperativity, the Slow Transition Model	149
2.9	pH and Temperature Dependence of Enzymes	151
2.9.1	pH Optimum and Determination of pK Values	151
2.9.2	pH Stability	153
2.9.3	Temperature Dependence	154
2.10	Isotope Exchange	158
2.10.1	Isotope Exchange Kinetics	159
2.10.2	Isotope Effects	163
2.10.2.1	Primary Kinetic Isotope Effect	163
2.10.2.2	Influence of the Kinetic Isotope Effect on V and K_m	164
2.10.2.3	Other Isotope Effects	165
2.11	Special Enzyme Mechanisms	166
2.11.1	Ribozymes	166
2.11.2	Polymer Substrates	167
2.11.3	Kinetics of Immobilized Enzymes	168
2.11.3.1	External Diffusion Limitation	169
2.11.3.2	Internal Diffusion Limitation	172
2.11.3.3	Inhibition of Immobilized Enzymes	173
2.11.3.4	pH and Temperature Behavior of Immobilized Enzymes	174
2.11.4	Transport Processes	175
2.11.5	Enzyme Reactions at Membrane Interfaces	178
2.12	Application of Statistical Methods in Enzyme Kinetics	185
2.12.1	General Remarks	185

2.12.2	Statistical Terms Used in Enzyme Kinetics	189
	References 190	
3	**Methods** 195	
3.1	Methods for Investigation of Multiple Equilibria	195
3.1.1	Equilibrium Dialysis and General Aspects of Binding Measurements 197	
3.1.1.1	Equilibrium Dialysis 197	
3.1.1.2	Control Experiments and Sources of Error	200
3.1.1.3	Continuous Equilibrium Dialysis 203	
3.1.2	Ultrafiltration 206	
3.1.3	Gel Filtration 207	
3.1.3.1	Batch Method 208	
3.1.3.2	The Method of Hummel and Dreyer 209	
3.1.3.3	Other Gel Filtration Methods 210	
3.1.4	Ultracentrifugation 211	
3.1.4.1	Fixed Angle Ultracentrifugation Methods 212	
3.1.4.2	Sucrose Gradient Centrifugation 214	
3.1.5	Surface Plasmon Resonance 218	
3.2	Electrochemical Methods 219	
3.2.1	The Oxygen Electrode 220	
3.2.2	The CO_2 Electrode 222	
3.2.3	Potentiometry, Redox Potentials 223	
3.2.4	The pH-stat 223	
3.2.5	Polarography 225	
3.3	Calorimetry 226	
3.4	Spectroscopic Methods 228	
3.4.1	Absorption Spectroscopy 230	
3.4.1.1	The Lambert-Beer Law 230	
3.4.1.2	Spectral Properties of Enzymes and Ligands 231	
3.4.1.3	Structure of Spectrophotometers 235	
3.4.1.4	Double Beam Spectrophotometer 237	
3.4.1.5	Difference Spectroscopy 238	
3.4.1.6	The Dual Wavelength Spectrophotometer 241	
3.4.1.7	Photochemical Action Spectra 242	
3.4.2	Bioluminescence 243	
3.4.3	Fluorescence 243	
3.4.3.1	Quantum Yield 243	
3.4.3.2	Structure of Spectrofluorimeters 244	
3.4.3.3	Perturbations of Fluorescence Measurements 246	
3.4.3.4	Fluorescent Compounds (Fluorophores) 247	
3.4.3.5	Radiationless Energy Transfer 252	
3.4.3.6	Fluorescence Polarization 254	
3.4.3.7	Pulse Fluorimetry 255	

3.4.4	Circular Dichroism and Optical Rotation Dispersion	*257*
3.4.5	Infrared and Raman Spectroscopy	*262*
3.4.5.1	IR Spectroscopy	*263*
3.4.5.2	Raman Spectroscopy	*263*
3.4.5.3	Applications	*264*
3.4.6	Electron Paramagnetic Resonance Spectroscopy	*264*
3.5	Measurement of Fast Reactions	*267*
3.5.1	Flow Methods	*268*
3.5.1.1	The Continuous Flow Method	*268*
3.5.1.2	The Stopped-flow Method	*271*
3.5.1.3	Measurement of Enzyme Reactions by Flow Methods	*274*
3.5.1.4	Determination of the Dead Time	*276*
3.5.2	Relaxation Methods	*277*
3.5.2.1	The Temperature Jump Method	*278*
3.5.2.2	The Pressure Jump Method	*281*
3.5.2.3	The Electric Field Method	*283*
3.5.3	Flash Photolysis, Pico- and Femto-second Spectroscopy	*283*
3.5.4	Evaluation of Rapid Kinetic Reactions (Transient Kinetics)	*285*
	References	*289*

Subject Index *293*

Symbols and Abbreviations

Units indicated in brackets.
Special, rarely used abbreviations are defined in the text.

A, B, C	ligands, substrates	K'	macroscopic equilibrium constant
[A], [B], [C]	concentration terms for ligands/substrates (similarly for enzymes, products, inhibitors etc.)	K_a	association constant
		K_{app}	apparent equilibrium constant
A	absorption measure	K_d	dissociation constant
AUC	area under the curve	K_g	equilibrium constant of a reaction
B	absolute bioavailability		
c	concentration	K_i	inhibition constant
CL	clearance	K_{ic}	competitive inhibition constant
D	diffusion coefficient		
e	Euler number ($e = 2.71828$)	K_{iu}	uncompetitive inhibition constant
E	enzyme, macromolecule		
E_a	activation energy	K_m	Michaelis constant
EC_{50}	effective concentration	K_{mA}	Michaelis constant for substrate A
F	relative intensity of fluorescence		
		$k_{1,2,3...}$	rate constant in forward direction
FRET	fluorescence resonance energy transfer	$k_{-1,-2,-3...}$	rate constant in reverse direction
$\Delta G°$	standard Gibbs energy		
G	electric conductance (S)	k_{cat}	catalytic constant
$\Delta H°$	standard reaction enthalpy	k_B	Boltzmann constant ($k_B = R/N = 1.3810^{-23}$ JK^{-1})
h	Planck constant (6.626×10^{-34} Js)		
		kat	Katal, enzyme unit according to the SI system (mol s^{-1})
h_s	transport coefficient of substrate		
		M	amount of the drug applied
I	inhibitor	M_r	relative molecular mass (dimensionless)
I	light intensity		
IC_{50}	inhibitory concentration	m	number of binding classes per macromolecule
J	flux (e.g. of a ligand from one compartment to another)		
		n	number of identical binding sites per macromolecule
IU	enzyme unit (international unit, µmol min^{-1}, 1 IU = 16.67 nkat, 1 nkat = 0.06 IU)		
		n_h	Hill coefficient
		N_A	Avogadro constant (6.022×10^{23} mol^{-1})
K	microscopic equilibrium constant	Or	ordinate intercept
		P, Q, R	products

Enzyme Kinetics. Principles and Methods. 2nd Ed. Hans Bisswanger
Copyright © 2008 WILEY-VCH Verlag GmbH & Co. KGaA, Weinheim
ISBN: 978-3-527-31957-2

Symbol	Definition	Symbol	Definition
P	polarization	$\bar{Y}$	fraction of ligands bound per binding site
R	gas constant (8.314 J K^{-1} mol^{-1})	a	normalized ligand concentration [A]/K$_d$
r	fraction of ligands bound per macromolecule	ε	molar absorption coefficient
ΔS	standard reaction entropy	Φ	optical rotation
Sl	slope	Φ_F	quantum yield
T	absolute temperature (K)	Φ_s	substrate resp. Thiele module
t	time (s)	η	viscosity
U	voltage (V)	η_e	efficiency factor
v	reaction velocity	η_{e1}	efficiency factor for first order reactions
v_0	initial velocity for $t=0$		
V	maximal velocity for substrate concentrations $\to \infty$	λ	wavelength (nm)
		Θ	ellipticity
V_d	distribution volume	ρ	density (kg m^{-3})
		τ	relaxation time

Introduction and Definitions

In the simplest form an enzyme reaction can be formulated:

$$E + A \underset{k_{-1}}{\overset{k_1}{\rightleftharpoons}} EA \overset{k_2}{\longrightarrow} E + P \, .$$

A substrate A reacts with an enzyme E to form an enzyme–substrate complex EA, which is converted to product in an irreversible step, releasing the free enzyme, which enters into a new reaction cycle. The single reaction steps are designated by rate constants k, marked with positive figures in the forward direction (k_1, k_2, k_3...) and with negative figures in the backward direction (k_{-1}, k_{-2}, k_{-3}...). A more detailed examination of even such an apparently simple reaction requires the consideration of more steps:

$$E + A \underset{k_{-1}}{\overset{k_1}{\rightleftharpoons}} EA \underset{k_{-2}}{\overset{k_2}{\rightleftharpoons}} E^*A \underset{k_{-3}}{\overset{k_3}{\rightleftharpoons}} E^*P \underset{k_{-4}}{\overset{k_4}{\rightleftharpoons}} EP \underset{k_{-5}}{\overset{k_5}{\rightleftharpoons}} E + P \, .$$

After initial binding of the substrate to the enzyme a loose association complex EA is formed in a rapid equilibrium. In the next step the bound substrate is adapted to a strong transition state by interaction with the residues of the catalytic center E*A and is prepared for conversion to the product E*P. Before the formation of free product P and enzyme an association complex EP is formed. The whole reaction sequence appears quite symmetrical and can be run through from both the forward and backward direction (P acting as substrate and A as product). Now five equilibria are considered and, for a complete description of the same reaction, ten rate constants must be known, instead of only the three assumed in the first scheme.

Actually the situation is even more complicated as, for most enzyme reactions, more components must be considered, such as two or more substrates (A, B, C...) and products (P, Q, R...), cofactors, inhibitors and activators (the latter two are also called *effectors*):

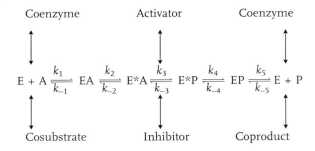

It is obvious that such complex mechanisms cannot be explored by only one or even a few experiments. Sophisticated theoretical treatments and extensive methodical approaches are required to unravel these mechanisms. To simplify the scheme, the various steps can be assigned to two different types, equilibrium and kinetic steps, and steps of each particular type can be treated similarly. Binding studies for the different ligands must be undertaken to investigate individual equilibria and in combination with kinetic studies and studies of conformational changes the central steps can be elucidated. Considering all the information the complete mechanism can be derived from the detailed results.

The investigation of binding equilibria requires the strict exclusion of any kinetic processes and is, in principle, different to kinetic studies, but it is an indispensable contribution to the understanding of enzyme mechanisms. Consequently, before treating enzyme kinetics in detail, equilibria will be discussed. Because of the strict absence of kinetic processes, this treatment is not restricted only to enzymes, but is applicable to any specific binding process and, since binding is understood usually to be the specific interaction of a low-molecular weight compound with a larger target, like a protein, receptor, DNA, etc., in this chapter the term *macromolecule* will be used synonymously with the term enzyme (both designated with E), while the following chapters on enzyme kinetics will be confined to enzymes. The low-molecular weight binding compound, which may be a substrate, an inhibitor, activator, hormone or any other specific binding metabolite, is called a *ligand*. It is assumed that there exists a distinct region for the ligand on the macromolecule, a specific *binding site*, in contrast to unspecific binding, due for example to ionic or hydrophobic interactions. The various mechanisms for equilibria between different ligands and macromolecules are gathered together under the term *multiple equilibria*.

The principal differences between equilibrium and kinetic investigations are summarized in Table 1. Equilibria are obviously time-independent and, therefore, the determination methods do not depend on time, measurements can be performed for long periods, while kinetic measurements are confined to the limited time when the reaction proceeds. In reality, however, it must be considered that biological substances, like enzymes, are not very stable, especially under experimental conditions and, therefore, equilibrium measurements should also be performed within a short time. A severe disadvantage of equilibrium measurements is that reversible binding causes no real change in the features

Table 1 Differences between equilibrium and kinetic studies.

Procedure	Equilibrium studies	Enzyme kinetic studies
Time dependence	Time independent rapid equilibrium	Time dependent directional progression
Constants	Thermodynamic constants: dissociation (association) constants	Kinetic constants: Michaelis constant, maximum velocity
Detection principal	Free and bound components chemically identical	Substrate and product chemically different
Detection sensitivity	Dependent on macromolecule concentration	Dependent on product formation
Macromolecule/enzyme amounts required	Macromolecule and ligand in comparable amounts $[E] \sim [A]$	Catalytic enzyme amounts $[E] \ll [A]$
Purity requirements	Macromolecule present in high purity	High enzyme activity, no requirement for high purity if no disturbing influences

of the components, while enzyme reactions proceed with chemical conversion of substrate to product, which can be used as a detection signal. Therefore, there is no need for high amounts of enzymes in enzyme kinetics, while with binding measurements the amount of bound ligand is directly proportional to the macromolecule concentration, which must be considerably high to enable accurate detection. The requirements for purity of the enzyme or the macromolecule are also different. For binding measurements high purity is usually needed, because the molar concentration must be known, while for enzyme kinetic determinations only disturbing influences, like side reactions, must be eliminated and inert contaminations are mostly unobjectionable. Finally, there are also major differences in the constants. From equilibrium treatments thermodynamic constants, like association or dissociation constants, are derived, while kinetic studies yield the more complex kinetic constants. On the other hand there are also similarities. Both types of constant are composed of rate constants, which are valid for both approaches, for example, inhibition constants, although determined kinetically, are really dissociation constants. Generally it can be stated that equilibrium studies are easier to treat theoretically but the experimental procedures are more difficult, while enzyme kinetics has a more complex theory but is easier to handle experimentally and, therefore, is more frequently applied.

Another main area is the treatment of fast reactions. This field can also be differentiated into kinetic methods which directly observe fast reactions, i.e. the continuous and stopped flow methods, while other techniques, like relaxation methods, deal with equilibria (although the deviation from equilibrium is a kinetic process). These techniques allow one to analyse complex mechanisms from a particular viewpoint and, besides the fact that fast processes become accessible, distinct rate constants can also be determined, rendering this approach as a valuable completion of equilibrium and kinetic studies with conventional methods.

Although different areas are treated in this book a uniform nomenclature is used throughout. For example, equilibrium treatments are connected to thermodynamics and deal mostly with association constants, while enzyme kinetics involve dissociation constants (the Michaelis constant is related to a dissociation constant), both types of constants describe principally the same equilibrium, only in a reversed sense. Since the main emphasis of this book is on enzyme kinetics, dissociation constants are used throughout. The terms A, B, C... will be used for any ligand, including substrates, binding specifically to a macromolecule or an enzyme. Only if it is necessary to discriminate between distinct types of ligands will different terms be used, e.g. I for the inhibitors, P, Q, R... for products. As far as possible the NC-IUB recommendations (Nomenclature Committee of the International Union of Biochemistry, 1982) and the IUPAC rules (International Union of Pure and Applied Chemistry, 1981) are regarded. Concentrations are indicated by square brackets ([A], etc.). The following reference list comprises standard textbooks relevant to the various fields treated in this book but is not a complete list of all such books.

References

Bisswanger, H. (2004) *Practical Enzymology*, Wiley-VCH, Weinheim.

Cornish-Bowden, A. (1976) *Principles of Enzyme Kinetics*, Butterworth, London, Boston.

Cornish-Bowden, A. (2004) *Fundamentals of Enzyme Kinetics*, 3rd edn, Portland Press Ltd. London.

Cornish-Bowden, A., Wharton, C. W. (1988) *Enzyme Kinetics*, IRL Press, Oxford.

Dixon, M., Webb, E. C. (1979) *Enzymes*, Academic Press, New York.

Edsall, J. T., Gutfreund, H. (1983) *Biothermodynamics*, J. Wiley & Sons, New York.

Eisenthal, R., Danson, J. M. (1992) *Enzyme Assays. A Practical Approach*, IRL Press, Oxford.

Engel, P. C. (1996) *Enzymology Labfax*, Academic Press, New York.

Fersht, A. (1977) *Enzyme Structure and Mechanism*, W. H. Freeman & Co., San Francisco.

Fromm, H. J. (1975) *Initial Rate Kinetics*, Springer, Berlin.

Johnson, K. A. (2003) *Kinetic Analysis of Macromolecules: A Practical Approach*, Oxford University Press, Oxford.

Klotz, I. M. (1986) *Introduction to Biomolecular Energetics Including Ligand–Receptor Interactions*, Academic Press, Orlando.

Kuby, S. A. (1991) *Enzyme Catalysis, Kinetics and Substrate Binding*, CRC Press, Boca Raton.

Laidler, K. J., Bunting, P. S. (1973) *The Chemical Kinetics of Enzyme Action*, 2nd edn, Clarendon Press, Oxford.

Leskovac, V. (2003) *Comprehensive Enzyme Kinetics*, Kluwer Academic, Dordrecht.

Marangoni, A. G. (2003) *Enzyme Kinetics. A Modern Approach*, Wiley-Interscience, Hoboken, New Jersey.

Page, M. (Ed.) (1984) *The Chemistry of Enzyme Action. New Comprehensive Biochemistry*, Vol. 6. Elsevier, Amsterdam.

Price, N. C., Stevens, L. (1989) *Fundamentals of Enzymology*, Oxford University Press, Oxford.

Purich, D. L. (Ed.) (1982) *Enzyme Kinetics and Mechanism, Methods in Enzymology*, Vol. 87, Academic Press, New York.

Purich, D. L. (1996) *Contemporary Enzyme Kinetics and Mechanism*, Academic Press, New York.

Purich, D. L. (1999) *Handbook of Biochemical Kinetics*, Academic Press, New York.

Roberts, D. V. (1977) *Enzyme Kinetics*, Cambridge University Press, Cambridge.

Segel, I. H. (1975) *Enzyme Kinetics*, J. Wiley & Sons, New York.

Taylor, K. B. (2002) *Enzyme Kinetics and Mechanisms*, Kluwer Academic Publishers, Dordrecht, NL, Boston, London.

Nomenclature rules

International Union of Pure and Applied Chemistry (1981) *Symbolism and terminology in chemical kinetics, Pure Appl. Chem.* 53, 753–771.

Nomenclature Committee of the International Union of Biochemistry (1982) *Symbolism and terminology in enzyme kinetics, Eur. J. Biochem.* 128, 281–291.

1
Multiple Equilibria

Chemical reactions are initiated by accidental collision of molecules, which have the potential (e.g. sufficient energy) to react with one another to be converted into products:

$$A + B \longrightarrow P + Q$$

In living matter it cannot be left to chance whether a reaction happens or not. At exactly the time required the respective compounds must be selected and converted to products with high precision, while at unfavorable times spontaneous reactions must be prevented. An important prerequisite for this selectivity of reactions is the highly specific recognition of the required compound. Therefore, any physiological reaction occurring in the organism is preceded by a specific recognition or binding step between the respective molecule and a distinct *receptor*. The exploration of binding processes is important for understanding biological processes. The receptors can be enzymes, but also non-enzymatic proteins like membrane transport systems, receptors for hormones or neurotransmitters, and nucleic acids. Generally, receptors are macromolecular in nature and thus considerably larger than the efficacious molecules, the *ligands*. For the binding process, however, they must be treated as equivalent partners (unlike for enzyme kinetics, where the enzyme as catalyst does not take part in the reaction).

As a precondition for binding studies specific binding must be established and unspecific association excluded. There exist many reasons for unspecific binding, like hydrophobic or electrostatic interactions (charged ligands can act as counterions for the surplus charges of proteins). A rough indicator for specific binding is the magnitude of the dissociation constant, which is mostly below 10^{-3} M (although there are exceptions, like the binding of H_2O_2 to catalase or glucose to glucose isomerase). Specific binding is characterized by a defined number of binding sites n, which is in stoichiometric relationship to the macromolecule. In contrast, unspecific binding has no defined number of binding sites, and thus the binding process is not saturable. Furthermore, the ligand can be replaced by structural analogs, while different or distantly related compounds are not accepted.

In the following the processes leading to a specific interaction between a ligand and a macromolecule will be described, i.e. how the ligand finds its bind-

Enzyme Kinetics. Principles and Methods. 2nd Ed. Hans Bisswanger
Copyright © 2008 WILEY-VCH Verlag GmbH & Co. KGaA, Weinheim
ISBN: 978-3-527-31957-2

1.1
Diffusion

A prerequisite for any reaction of a ligand with a macromolecule is the fact that the partners must find one another. In a free space a particle moves in a straight direction with a kinetic energy of $k_BT/2$, T being the absolute temperature and k_B the Boltzmann constant. According to Einstein's relationship a particle with mass m, moving in a distinct direction with velocity v possesses kinetic energy $mv^2/2$. Combining both relationships Eq. (1.1) follows:

$$v^2 = k_B T/m .\tag{1.1}$$

Accordingly, a macromolecule like the lactate dehydrogenase ($M_r=140\,000$) would move at a rate of 4 m s^{-1}, its substrate lactic acid ($M_r=90.1$) at 170 m s^{-1}, and a water molecule ($M_r=18$) at 370 m s^{-1}. Enzyme and substrate will fly past one another like rifle bullets. In the dense fluid of the cell, however, the moving particles are permanently hampered and deflected from linear movement by countless obstacles: water molecules, ions, metabolites, macromolecules and membranes and, actually, the molecule moves more like a staggering drunkard than in a straight progression. However, this tumbling increases the collision frequency and the probability of distinct molecules meeting one another.

The distance x covered by a molecule in solution within time t in one direction depends on the diffusion coefficient D according to the equation:

$$x^2 = 2Dt .\tag{1.2}$$

The diffusion coefficient is itself a function of the concentration of the diffusing compound, in dilute solutions it can be regarded as constant. It depends on the particle size, the consistency of the fluid and the temperature. For small molecules in water the coefficient is $D=10^{-5}$ cm^2 s^{-1}. A cell with the length 1 μm will be passed within 0.5 ms, 1 mm within 500 s, thus, for a thousandfold distance a millionfold time is required. This demonstrates that there exists no 'diffusion velocity', the movement of the molecules is not proportional to time, but to its square root. A diffusing molecule does not remember its previous position, it does not strive systematically for new spaces but searches new regions randomly in undirected movement. As an example, a 10 cm high saccharose gradient, used in ultracentrifugation for separation and molecular mass determination of macromolecules, has a life-span of about four months, taking $D=5\times 10^{-6}$ cm^2 s^{-1} for saccharose. The tendency of the gradient to equalize its concentration is considerably low.

Equation (1.2) describes the one-dimensional diffusion of a molecule. For diffusion in a three-dimensional space over a distance r the diffusion into the three space directions x, y and z is assumed to be independent of each other:

$$r^2 = x^2 + y^2 + z^2 = 6Dt \,. \tag{1.3}$$

Mere meeting of ligand and macromolecule is not sufficient to accomplish specific binding, rather the ligand must locate the binding site on the macromolecule. This is realized by translocation of the ligand volume $4\pi R^3/3$ by the relevant distance of its own radius R. After a time t_x the molecule has searched (according to Eq. (1.3) for $r=R$) a volume of:

$$\frac{6Dt_x}{R^2} \cdot \frac{4\pi R^3}{3} = 8\pi DRt_x \,. \tag{1.4}$$

The volume searched per time unit is $8\pi DR$, the probability of collision for a certain particle in solution is proportional to the diffusion coefficient and the particle radius.

At the start of a reaction $A + B \rightarrow P$ both participants are equally distributed in solution. Within a short time, molecules of one type, e.g. B, become depleted in the vicinity of the molecule of the other type (A) not yet converted, so that a concentration gradient will be formed. Consequently, a net flow Φ of B-molecules occurs in the direction of the A-molecules located at a distance r,

$$\Phi = \frac{dn}{dt} = DF\frac{dc}{dr} \,, \tag{1.5}$$

n is the net surplus of molecules passing through an area F within time t, c is the concentration of B-molecules located at a distance r from the A-molecules. This relationship in its general form is known as *Fick's First Law of Diffusion*. In our example of a reaction of two reactants, F has the dimension of a spherical surface with the radius r. Eq. (1.5) then changes into:

$$\left(\frac{dc}{dr}\right)_r = \frac{\Phi}{4\pi r^2 D'} \tag{1.6}$$

D' is the diffusion coefficient for the relative diffusion of the reactive molecules. Integration of Eq. (1.6) yields:

$$c_r = c_\infty - \frac{\Phi}{4\pi r D'} \tag{1.7}$$

where c_r is the concentration of B-molecules at the distance r and c_∞ the concentration at infinite distance from the A-molecules. The last corresponds approximately to the average concentration of B-molecules. The net flow Φ is proportional to the reaction rate and that is again proportional to the average concentration c of those B-molecules just in collision with the A-molecules, r_{A+B} being the sum of the radii of an A- and a B-molecule:

$$\Phi = kc_{r_{A+B}} \,. \tag{1.8}$$

where k is the rate constant of the reaction in the steady-state, where c_r becomes equal to $c_{r_{A+B}}$ and r equal to r_{A+B}. Inserted into Eq. (1.7), this becomes:

$$c_{r_{A+B}} = \frac{c_\infty}{1 + \frac{k}{4\pi r_{A+B} D'}} \cdot \qquad (1.9)$$

The net flow under steady-state conditions is:

$$\Phi = k_a c_\infty \qquad (1.10)$$

where k_a is the relevant association rate constant. Equations (1.8)–(1.10) may thus be rewritten:

$$\frac{1}{k_a} = \frac{1}{4\pi r_{A+B} D'} + \frac{1}{k} \cdot \qquad (1.11)$$

This relation becomes linear in a graph plotting $1/k_a$ against the viscosity η of the solution as, according to the *Einstein-Sutherland Equation*, the diffusion coefficient at infinite dilution D_0 is inversely proportional to the friction coefficient f and that again is directly proportional to the viscosity η:

$$D_0 = \frac{k_B T}{f} = \frac{k_B T}{6\pi\eta r} \cdot \qquad (1.12)$$

$1/k$ is the ordinate intercept. In the case of $k \gg 4\pi r_{A+B} D'$ the intercept is placed near the coordinate base, it becomes:

$$k_a = 4\pi r_{A+B} D' \cdot \qquad (1.13)$$

This borderline relationship is known as the *Smoluchowski limit* for translating diffusion, the reaction is *diffusion-controlled*. In contrast to this, in *reaction-controlled* reactions the step following diffusion, i.e. the substrate turnover, determines the rate. A depletion zone emerges around the enzyme molecule, as substrate molecules are not replaced fast enough. A *diffusion-limited dissociation* occurs, if the dissociation of the product limits the reaction. Viewing two equally reactive spheres with radii r_A and r_B and diffusion coefficients D_A and D_B, we obtain for Eq. (1.13):

$$k_a = 4\pi r_{A+B} D' = 4\pi (r_A + r_B)(D_A + D_B) \cdot \qquad (1.14)$$

By inserting Eq. (1.12) and with the approximation $r_A = r_B$ and with $D_0 = D_A = D_B$ we obtain:

$$k_a = \frac{8 k_B T}{3\eta} \cdot \qquad (1.15)$$

Thus the association rate constants for diffusion-controlled reactions are in the range 10^9–10^{10} M^{-1} s^{-1}.

Uniform values should be obtained if the rate constants are exclusively determined by diffusion. In reality, however, the values of the rate constants of diffusion-controlled reactions of macromolecules vary within a range of more than five orders of magnitude. The reason for this variation is that, for successful binding of the ligand, random collision with the macromolecule is not sufficient. Both molecules must be in a favorable position to each other. This causes a considerable retardation of the binding process. On the other hand, attracting forces could facilitate the interaction and direct the ligands towards their proper orientation. Under such conditions rate constants can even surpass the values of mere diffusion control. Quantitative recording of such influences is difficult as they depend on the specific structures of both the macromolecule and the ligand. Theories have been developed to establish general rules for ligand binding.

Ligand approach a macromolecule at a rate according to Eq. (1.13), but only those meeting the correct site in the right orientation will react. If the binding site is regarded as a circular area, forming an angle a with the center of the macromolecule (see Fig. 1.1), the association rate constant of Eq. (1.13) will be reduced by the sine of that angle:

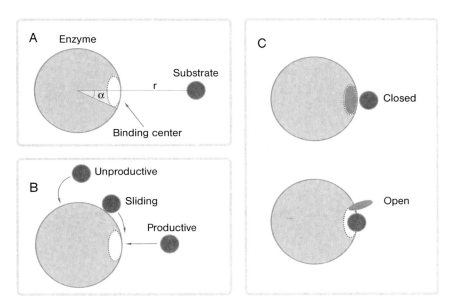

Fig. 1.1 Schematic illustration of the interaction of a substrate molecule with its binding site on the enzyme (A). B, productive and unproductive binding, sliding of the ligand along the surface; C, gating.

1 Multiple Equilibria

$$k_a = 4\pi r_{A+B} D' \sin \alpha \ . \tag{1.16}$$

The necessity of appropriate orientation between ligand and binding site should be considered by the introduction of a suitable factor, depending on the nature of the reactive groups involved. It is also suggested that the ligand may associate unspecifically to the surface of the macromolecule, where it dissociates in a two-dimensional diffusion to find the binding site (*sliding model*; Berg, 1985, Fig. 1.1 B). Such unspecific binding, however, is not able to distinguish between the specific ligand and other metabolites which may also bind and impede the two-dimensional diffusion. The *gating model* (Fig. 1.1 C) assumes the binding site to be opened and closed like a gate by changing the conformation of the protein, thus modulating the accessibility for the ligand (McCammon and Northrup 1981).

A basic limit for the association rate constant for the enzyme substrate is the quotient from the catalytic constant k_{cat} and the Michaelis constant K_m (cf. Section 2.2.1):

$$\frac{k_{cat}}{K_m} = \frac{k_{cat} k_1}{k_{-1} + k_2} \tag{1.17}$$

frequently around 10^8 M^{-1} s^{-1} for a diffusion-controlled reaction. For most enzyme reactions the reaction rate is determined more by the non-covalent steps during substrate binding and product dissociation rather than by the cleavage of bounds.

1.2
Interaction between Macromolecules and Ligands

1.2.1
Binding Constants

Binding of a ligand A to a macromolecule E

$$E + A \underset{k_{-1}}{\overset{k_1}{\rightleftharpoons}} EA \tag{1.18}$$

is described with the law of mass action, applying the association constant K_a:

$$K_a = \frac{k_1}{k_{-1}} = \frac{[EA]}{[A][E]} \tag{1.19a}$$

or its reciprocal value, the dissociation constant K_d:

$$K_d = \frac{k_{-1}}{k_1} = \frac{[A][E]}{[EA]} \tag{1.19b}$$

Both notations are used, the association constant more frequently for the treatment of equilibria, the dissociation constant for enzyme kinetics. Here the dissociation constant will be employed throughout. The association constant has the dimension of a reciprocal concentration (M^{-1}), the higher the numerical value, the higher the affinity. Conversely, dissociation constants possess the dimension of a concentration (M) and lower values indicate stronger binding. Eqs. (1.19 a, b) are not quite correct, in the place of concentrations c (e.g. [A]) activities $a = fc$ should be used. Since activity coefficients f approach one in very dilute solutions they can be disregarded for enzyme reactions.

If one reaction component is present in such a large excess that its concentration change during the reaction can be neglected, the absolute concentration can be included in the constant. This applies especially for water, if it takes part in the reaction, e.g. in hydrolytic processes:

$$A + H_2O \xrightleftharpoons{enzyme} P + Q \,.$$

As a solvent, with a concentration of 55.56 mol l^{-1}, water exceeds by far the nano- to millimolar amounts of the other components in an enzyme assay and any change in its concentration will hardly be detectable. Therefore, a binding constant for water cannot be determined and the reaction will be treated as if water is not involved:

$$K'_d = \frac{[A][H_2O]}{[P][Q]} = K_d[H_2O]\,; \quad K_d = \frac{[A]}{[P][Q]} \,.$$

Hydrogen ions, frequently involved in enzyme reactions, are treated in a similar manner. An apparent dissociation constant is defined:

$$K_{app} = K_d[H^+] \,.$$

Contrary to genuine equilibrium constants this constant is dependent on the pH value in the solution.

1.2.2
Macromolecules with One Binding Site

To determine the binding constants for a distinct system the mass action law (Eq. (1.19)) can be applied. However, the terms required for solution of the equation, the concentrations of the free macromolecule [E], the free substrate [A] and the enzyme-substrate complex [EA], are unknown. Only the total amounts of macromolecule $[E]_0$ and of ligand $[A]_0$ added to the reaction are known. They separate into free and bound components according to the mass conservation principle:

$$[E]_0 = [E] + [EA] \tag{1.20}$$

$$[A]_0 = [A] + [EA] \,. \tag{1.21}$$

Binding experiments yield the portion of the ligand bound to the macromolecule $[A]_{bound}$ (see Chapter 3). In the simple reaction with only one ligand binding to a macromolecule (Eq. (1.18)) [A]$_{bound}$ is equal to [EA]. Inserting Eq. (1.20) into Eq. (1.19b) eliminates the free macromolecule concentration [E]:

$$[A]_{bound} = \frac{[E]_0 [A]}{K_d + [A]} \quad . \tag{1.22}$$

This equation describes the binding of a ligand to a macromolecule with one binding site. It will be discussed in detail in the following section together with the analogous Eq. (1.23) for macromolecules with several identical binding sites.

1.3
Macromolecules with Identical Independent Binding Sites

1.3.1
General Binding Equation

Most proteins and enzymes found in living organisms are composed of more than one, mostly identical, subunit. For reasons of symmetry it can be taken that each of these subunits carries one identical binding site for the ligand, so that the number n of binding sites equals the number of subunits. This is a plausible assumption, but it must be stated that, in the strict sense, identity means equality of binding constants. If affinities of binding sites located on non-identical subunits are the same by chance, or if a single subunit possesses more than one binding site with similar binding constants (e.g. due to gene duplication), this will not be differentiated by binding analysis and requires additional experiments.

Binding of a ligand to identical sites on the same macromolecule can occur independently, otherwise the first bound ligand can influence the following binding steps. Such influences will be considered in Section 1.5, while here only independent binding is considered. Such binding processes are principally described by Eq. (1.22), since it should make no essential difference whether the binding occurs at a macromolecule with only one binding site, or whether n sites are gathered on the same macromolecule. If $[F]_0 = n[E]_0$ is assumed to be the total amount of binding sites, this can replace $[E]_0$ in Eq. (1.22):

$$[A]_{bound} = \frac{[F]_0 [A]}{K_d + [A]} = \frac{n [E]_0 [A]}{K_d + [A]} \tag{1.23}$$

The number of binding sites is indicated in the numerator, but as a further difference it must be considered, that [A]$_{bound}$ can no longer be equated with [EA], but comprises all partially saturated forms of the macromolecule:

$$[A]_{bound} = [EA] + 2[EA_2] + 3[EA_3] + \ldots n[EA_n] \tag{1.24}$$

1.3 Macromolecules with Identical Independent Binding Sites

In fact the macromolecule will be saturated stepwise:

$$E + A \rightleftharpoons EA \qquad K'_1 = \frac{[E][A]}{[EA]}$$

$$EA + A \rightleftharpoons EA_2 \qquad K'_2 = \frac{[EA][A]}{[EA_2]}$$

$$EA_2 + A \rightleftharpoons EA_3 \qquad K'_3 = \frac{[EA_2][A]}{[EA_3]}$$

$$\vdots \qquad \vdots$$

$$EA_{n-1} + A \rightleftharpoons EA_n \qquad K'_n = \frac{[EA_{n-1}][A]}{[EA_n]} \;.$$

Each step has its own dissociation constant. For independent binding all individual dissociation constants may be taken as equal and Eq. (1.23) will be obtained. Although these considerations lead to the correct binding equation, the derivation was simplified. The correct derivation, which is much more complicated, is given in Box 1.1.

Box 1.1 Derivation of the General Binding Equation

The dissociation constants of the individual binding steps are called *macroscopic* dissociation constants K', in contrast to *microscopic* (or *intrinsic*) binding constants K for binding to the individual sites of the macromolecule.

Macroscopic binding constants

$$E \underset{-A}{\overset{+A}{\rightleftharpoons}} \overset{K'_1}{} EA \underset{-A}{\overset{+A}{\rightleftharpoons}} \overset{K'_2}{} EA_2 \underset{-A}{\overset{+A}{\rightleftharpoons}} \overset{K'_3}{} EA_3$$

Microscopic binding constants

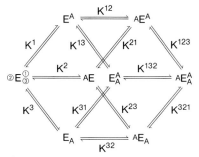

Scheme 1. Macroscopic and microscopic binding constants of a macromolecule with three identical binding sites. The E-form at the left in the lower scheme shows the relative orientation and the denomination of the binding sites. The constants are designated according to the sequence of occupation, the last figure indicating the actual occupation.

This is demonstrated in Scheme 1 for a macromolecule with three binding sites. The first binding step has one macroscopic dissociation constant K'_1, but three microscopic dissociation constants, designated as K^1, K^2, and K^3, according to the numbers of the binding sites $_2E^1_3$. Therefore, one ligand binding to the macromolecule can choose between three binding sites and, consequently, three different macromolecule species can be formed. For the second binding step three forms are also possible, but there are six ways to obtain these species, accordingly there exist six microscopic dissociation constants (K^{12} etc.). From these three forms three equilibria characterized by three microscopic binding constants (K^{123} etc.) lead to the one fully saturated macromolecule form. Obviously the complete binding process is described by three macroscopic and twelve microscopic dissociation constants (Scheme 1). The relationship between both types of constants can be established by applying the respective mass action laws. The macroscopic dissociation constant of the first binding step is defined as:

$$K'_1 = \frac{[E][A]}{[EA]} = \frac{[E][A]}{[E^A] + [_AE] + [E_A]}.$$

The microscopic binding constants are used to replace the individual macromolecule forms

$$K^1 = \frac{[E][A]}{[E^A]} \ ; \quad [E^A] = \frac{[E][A]}{K^1}$$

$$K^2 = \frac{[E][A]}{[_AE]} \ ; \quad [_AE] = \frac{[E][A]}{K^2}$$

$$K^3 = \frac{[E][A]}{[E_A]} \ ; \quad [E_A] = \frac{[E][A]}{K^3}$$

$$K'_1 = \frac{1}{\frac{1}{K^1} + \frac{1}{K^2} + \frac{1}{K^3}}.$$

If the three binding sites are identical, the microscopic constants can be equalized, $K^1 = K^2 = K^3 = K$, and both types of constants are related as $K' = K/3$.

Correspondingly, the second binding step is:

$$K'_2 = \frac{[EA][A]}{[EA_2]} = \frac{([E^A] + [_AE] + [E_A])[A]}{[_AE^A] + [E^A_A] + [_AE_A]}$$

$$K^{12} = \frac{[E^A][A]}{[_AE^A]} \ ; \quad [_AE^A] = \frac{[E^A][A]}{K^{12}} \quad \text{etc., hence}$$

$$K'_2 = \frac{K^{13}K^{21}K^{23} + K^{12}K^{13}K^{23} + K^{13}K^{21}K^{32}}{K^{13}K^{23} + K^{12}K^{23} + K^{13}K^{21}}.$$

For $K^{12} = K^{13} = \ldots = K$ results $K'_2 = K$.

The third binding step is:

$$K'_3 = \frac{[EA_2][A]}{[EA_3]} = \frac{([_AE^A] + [E^A_A] + [_AE_A])[A]}{[_AE^A_A]},$$

$$K^{123} = \frac{[_AE^A][A]}{[_AE^A_A]} \; ; \quad [_AE^A] = \frac{K^{123}[_AE^A_A]}{[A]} \text{ etc.}$$

For $K^{123} = K^{132} = \ldots = K$ results $K'_3 = 3K$.

Even if all microscopic dissociation constants are identical, they differ from the macroscopic ones and there are differences between each binding step. The general relationship between both types of dissociation constants for n binding sites is

$$K'_d = K_d \frac{i}{n - i + 1}, \quad (2)$$

i representing the respective binding step. Ligands occupying stepwise a macromolecule with identical sites have Ω possibilities of orientation, depending on the respective binding step i:

$$\Omega = \frac{n!}{(n-i)!i!} \quad (3)$$

For the derivation of the general binding equation a saturation function r is defined as the quotient from the portion of bound ligand to the total amount of the macromolecule:

$$r = \frac{[A]_{bound}}{[E]_0} = \frac{[EA] + 2[EA_2] + 3[EA_3] + \ldots n[EA_n]}{[E] + [EA] + [EA_2] + [EA_3] + \ldots [EA_n]}. \quad (4)$$

The concentrations of the individual macromolecule forms are not accessible experimentally and are replaced by the macroscopic dissociation constants:

$$K'_1 = \frac{[E][A]}{[EA]} \; ; \quad [EA] = \frac{[E][A]}{K'_1}$$

$$K'_2 = \frac{[EA][A]}{[EA_2]} \; ; \quad [EA_2] = \frac{[EA][A]}{K'_2} = \frac{[E][A]^2}{K'_1 K'_2}$$

$$K'_3 = \frac{[EA_2][A]}{[EA_3]} \; ; \quad [EA_3] = \frac{[EA_2][A]}{K'_3} = \frac{[E][A]^3}{K'_1 K'_2 K'_3}$$

$$\vdots \qquad \vdots \qquad \vdots$$

$$K'_n = \frac{[EA_{n-1}][A]}{[EA_n]} \; ; \quad [EA_n] = \frac{[EA_{n-1}][A]}{K'_n} = \frac{[E][A]^n}{K'_1 K'_2 K'_3 \cdots K'_n}.$$

Thus evolves:

$$r = \frac{\dfrac{[A]}{K'} + \dfrac{2[A]^2}{K'_1 K'_2} + \dfrac{3[A]^3}{K'_1 K'_2 K'_3} + \cdots \dfrac{n[A]^n}{K'_1 K'_2 K'_3 \cdots K'_n}}{1 + \dfrac{[A]}{K'_1} + \dfrac{[A]^2}{K'_1 K'_2} + \dfrac{[A]^3}{K'_1 K'_2 K'_3} + \cdots \dfrac{[A]^n}{K'_1 K'_2 K'_3 \cdots K'_n}}$$

$$= \frac{\displaystyle\sum_{i=1}^{n} \dfrac{i[A]^i}{\left(\prod_{j=1}^{i} K'_j\right)}}{1 + \displaystyle\sum_{i=1}^{n} \dfrac{[A]^i}{\prod_{j=1}^{i} K'_j}}. \tag{5}$$

In the case of independent identical binding sites the macroscopic binding constants of the individual binding steps according to Eq. (2) are replaced by a uniform microscopic constant K_d:

$$r = \frac{\displaystyle\sum_{i=1}^{n} i \left(\prod_{j=1}^{i} \dfrac{n-j+1}{j}\right)\left(\dfrac{[A]}{K_d}\right)^i}{1 + \displaystyle\sum_{i=1}^{n} \left(\prod_{j=1}^{i} \dfrac{n-j+1}{j}\right)\left(\dfrac{[A]}{K_d}\right)^i}. \tag{6}$$

The product terms of the numerator and denominator are binomial coefficients, which can be converted as follows:

$$\binom{n}{i} = \left(\frac{n!}{i!(n-i)!}\right),$$

so that Eq. (6) may be written in the form:

$$r = \frac{\displaystyle\sum_{i=1}^{n} i \binom{n}{i}\left(\dfrac{[A]}{K_d}\right)^i}{1 + \displaystyle\sum_{i=1}^{n} \binom{n}{i}\left(\dfrac{[A]}{K_d}\right)^i}.$$

Applying the binomial rule, the denominator can be converted as $(1+[A]/K_d)^n$. For the numerator the derived binomial rule applies:

$$r = \frac{n\left(\dfrac{[A]}{K_d}\right)\left(1 + \dfrac{[A]}{K_d}\right)^{n-1}}{\left(1 + \dfrac{[A]}{K_d}\right)^n}.$$

By reduction the already known form of the binding equation (Eq. (1.23)) will be achieved:

$$r = \frac{[A]_{bound}}{[E]_0} = \frac{n[A]}{K_d + [A]}. \tag{7}$$

Irvin Langmuir developed such an equation in 1916 for the adsorption of gases to solid surfaces and, therefore, the authorship of this equation is ascribed to him, although Adrian J. Brown and Victor Henri had already developed a similar equation in 1900, which was examined in detail by Leonor Michaelis and Maud Menten in 1913. This *Michaelis-Menten equation*, already exhibiting some modifications in comparison to Eq. (1.23), is of general importance for enzyme kinetics (see Section 2.2.1).

Equation (1.23) describes the relationship between the free and the bound ligand. By successive increase of the free ligand a saturation curve will be obtained (Fig. 1.2A), which follows mathematically the function of a right-angle hyperbola (this will be explained in Section 2.3.1, Box 2.1). At extremely high concentrations of A ($[A] \to \infty$) K_d in the denominator of Eq. (1.23) can be ignored and the curve approaches n, the number of binding sites. At the position where the free ligand concentration equals the value of the dissociation constant, $[A] = K_d$, $r = n/2$, according to Eq. (1.23). In this manner the K_d value can be determined from half saturation. Thus both the dissociation constant and the number of binding sites can be obtained from this saturation curve (Fig. 1.2A).

There exist three, principally equivalent modes of plotting binding data. The amount of bound ligand $[A]_{bound}$ obtained from the experiment can be plotted directly against the free ligand concentration $[A]$. Saturation will be reached at

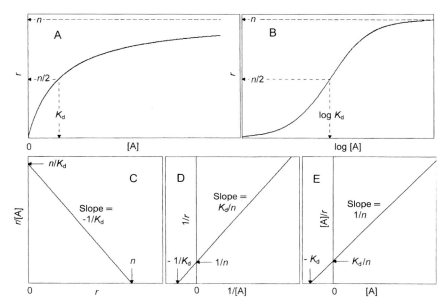

Fig. 1.2 Modes of representation of binding data. (A) Direct plot; (B) semi-logarithmic plot; (C) Scatchard plot; (D) double-reciprocal plot; (E) Hanes plot.

$n[E]_0$. It is more convenient to take the saturation function r by division of $[A]_{bound}$ by $[E]_0$, as discussed already. If r is further divided by n, the function $\overline{Y}$ results:

$$\overline{Y} = \frac{[A]_{bound}}{n[E]_0} = \frac{[A]}{K_d + [A]} \ . \tag{1.23a}$$

In this case the value of the saturation becomes 1. The function $\overline{Y}$ is used if different mechanisms are compared theoretically (without defining n) or, experimentally, where the portion of bound ligand is not directly known, as in spectroscopic titrations (see Section 1.3.2.2).

1.3.2
Graphic Representations of the Binding Equation

1.3.2.1 Direct and Linear Diagrams
Generally binding studies should yield 3 kinds of information:
- The affinity of the macromolecule for the ligand, represented by the value of the dissociation constant K_d.
- The number of binding sites n.
- The respective binding mechanism.

The goal of graphic representations is to obtain this information in a clear, unambiguous manner. There exist different kinds of graphic representations and it must be decided which will be the most appropriate for the respective experimental data. Usually the data will be represented in a variety of plots because special aspects will become more obvious in one type than in another, although, as a rule, missing information cannot be recalled by any representation.

The direct representation of binding data has already been discussed (Fig. 1.2 A). This is always recommended as a primary step, since the data suffer no distortion, especially with respect to the error distribution, due to recalculations, as in linear diagrams. A difficulty is the treatment of saturation, which is mostly underestimated, as saturation is actually reached at infinity. It must be considered that experimentally a continuous curve is not obtained, rather, a scattering set of data points. Thus determination of n, and also of K_d, depending on half saturation, may become difficult. Non-linear regression analysis improves the analysis.

Besides the problems of determination of the constants, the detection of possible alternative binding mechanisms (which will be discussed later) is more difficult with the direct, non-linear plot, because weak deviations from the normal function will easily be hidden behind the data scatter. Linear representations partly avoid such disadvantages but they reveal other limitations.

An alternative non-linear representation is the *semi-logarithmic plotting* of the saturation function r against log [A]. This diagram is recommended especially when larger concentration ranges are covered, which cannot be resolved completely in the direct plot. Sigmoidal curves are obtained in the semi-logarithmic representation (Fig. 1.2 B) and the logarithm of K_d is obtained from half saturation.

The similarity of the binding equation with the Michaelis-Menten equation holds also for the linearization methods, which will be discussed in detail in Section 2.3.1. Accordingly, there exist three simple linear transformations of the equations. One is the *double-reciprocal plot*, ascribed to Klotz (1946) (although he was not the original author; moreover, equivalent plots are designated differently for binding and kinetic treatments as will be discussed in Section 2.3.1.3). The reverse form of Eq. (1.23) is:

$$\frac{1}{r} = \frac{1}{n} + \frac{K_d}{n[A]} \, . \tag{1.25}$$

Plotting $1/r$ against $1/[A]$ should result in a straight line, intercepting the ordinate at $1/n$ and the abscissa at $-1/K_d$. Therefore, both constants can easily be obtained by extrapolation (Fig. 1.2 D). Alternative mechanisms show characteristic deviations from linearity. The double-reciprocal plot has the advantage of separation of the variables (in contrast to the other two linear diagrams), however, due to the reciprocal entry, strong distortions of the error limits result, being compressed to the high and expanded to the low ligand range. Linear regression is not applicable and especially the determination of n at the ordinate intercept often becomes dubious with scattering data.

Because n is an important value, the plot of *Scatchard* (1949) is preferred for the analysis of binding data. It is derived from Eq. (1.25) by multiplying by rn/K_d:

$$\frac{r}{[A]} = \frac{n}{K_d} - \frac{r}{K_d} \, . \tag{1.26}$$

Plotting $r/[A]$ versus r results in a straight line intersecting the abscissa at n and the ordinate at n/K_d (Fig. 1.2 C). In this diagram also the error limits do not remain constant, but increase towards high ligand concentrations, but the effect is lower than with the double-reciprocal diagram and linear regression is often applied. Although the variables are not separated, this is the most reliable linear diagram.

A third diagram is obtained by multiplying Eq. (1.25) by [A]:

$$\frac{[A]}{r} = \frac{[A]}{n} + \frac{K_d}{n} \, . \tag{1.27}$$

This diagram, known in enzyme kinetics as the *Hanes plot*, is seldom used for binding analysis. By plotting $[A]/r$ versus $[A]$, K_d/n follows from the ordinate and $-K_d$ from the abscissa intercept (Fig. 1.2 E). An advantage of this representation are the nearly constant error limits.

1.3.2.2 Analysis of Binding Data from Spectroscopic Titrations

Although methods for determination of binding are discussed later (see Chapter 3.4), theoretical aspects of the analysis will be discussed here. Spectroscopic titrations are convenient methods to study binding processes but the data need a special treatment as the diagrams discussed so far cannot be applied directly. The main difference to other binding methods is that the share of the free ligand [A] cannot be obtained directly by experiment and also the share of bound ligand results only as a relative spectral change, not as a molar concentration. The experimental procedure is usually the addition of increasing amounts of the ligand to a constant amount of the macromolecule in a photometric cuvette and the spectral change is recorded in dependence on the ligand concentration. Only the total amount of the added ligand $[A]_0$ is known, while for a plot as shown in Fig. 1.2A the free ligand concentration is required. In principal the same problem exists in enzyme kinetics where also the total substrate concentration is taken. However, because of the very low ('catalytic') enzyme concentrations the amounts of total and free ligand can be equated. This is not possible with binding measurements, where the macromolecule will be present in high concentrations to produce a detectable signal. Therefore, direct representation of the spectral change against $[A]_0$, as obtained from the experiment (*titration curve*), cannot be evaluated as discussed in Section 1.3.2.1 with K_d at half saturation. There exist different approaches to the evaluation of such titration curves, one of them, the Dixon plot, will be discussed in Section 2.3.1.1.

For direct evaluation of titration curves it can be assumed that in the low concentration range of the ligand for $[A]_0 < [E]_0$ nearly all ligand added will bind to the macromolecule, thus $[A]_0 \sim [A]_{bound}$ and no free ligand appears. Under these conditions a linear relation between the added ligand and $[A]_{bound}$ will occur, discernible by a linear increase in the spectral signal in the low ligand range. A tangent at this part of the titration curve represents the share of the bound ligand throughout (Fig. 1.3A). At higher ligand concentration only part of the ligand will bind and the remaining free ligand causes deviation of the saturation curve from the initial tangent. The spectral signal still increases upon further addition of ligand as long as free binding sites are available, but the increase will cease when all sites become occupied. Now the saturation curve tends to a saturation plateau, which can be indicated by an asymptotic line (also here it must be considered that saturation occurs actually at infinity). The optical signal at the position of the asymptotic line corresponds to the amount of ligand bound at saturation, and thus to all available binding sites $n[E]_0$. The concentration of $n[E]_0$ can be obtained directly from the abscissa coordinate of the intersection point of both the initial tangent and the asymptote (see Fig. 1.3A). The relative values of the optical signal at the ordinate can be converted to $\overline{Y}$ values, setting the saturation equal to $\overline{Y} = 1$. The total amount of ligand $[A]_0$ is the sum of free and bound ligand. Both shares can be obtained directly by a parallel line to the abscissa at any point of the curve. The distance from the ordinate axes to the tangent (abscissa coordinate) is $[A]_{bound}$ and that from there to the titration curve is [A]. In this manner all measured data can be converted into these two values, with the exception of the points in the low ligand range, which are used for

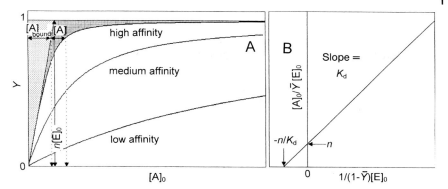

Fig. 1.3 Evaluation of spectroscopic titrations. (A) Direct plotting, (B) Stockell plot.

aligning the tangent. With the knowledge of [A]$_{bound}$ and [A], the conventional diagrams described in Section 1.3.2.1 and the respective evaluation of the constants can be performed. The severe disadvantage of this procedure is that it depends essentially on the alignment of the tangent. If there is a larger scatter, or if the assumption that at low ligand concentrations all ligand will be bound is not valid, the alignment will become incorrect. This is especially the case with low affinity binding, where there is a tendency to align the tangent too flat. Principally, the higher the affinity, the more the experimental curve approaches the two asymptotic lines, both these lines represent the case of infinite high affinity.

To circumvent the uncertainty of the initial tangent, the titration curve can be directly linearized according to a procedure suggested by *Stockell* (1959), where the free ligand concentration in Eq. (1.23) is replaced by [A]$_0$. The spectral signal is converted into values for $\overline{Y}$, saturation being defined as $\overline{Y}=1$. To derive a linear relationship $r = n\overline{Y} = n[EA]/[E]_0$ is inserted into Eq. (1.25), and [A]$_{bound} = n[EA]$:

$$\frac{1}{\overline{Y}} = 1 + \frac{K_d}{[A]_0 - n[EA]} = 1 + \frac{K_d}{[A]_0 - n\overline{Y}[E]_0}.$$

Transformation to

$$\frac{[A]_0}{\overline{Y}} - [A]_0 = n[E]_0(1 - \overline{Y}) + K_d$$

results in:

$$\frac{[A]_0}{[E]_0 \overline{Y}} = \frac{K_d}{[E]_0(1 - \overline{Y})} + n. \tag{1.28}$$

In this diagram (Fig. 1.3 B) a straight line should result and n and K_d can be obtained from the ordinate and abscissa intercepts, respectively. There still remains the uncertainty of the saturation asymptote, which is required for the definition of $\overline{Y}=1$. Therefore, the measurements must be extended far into the saturation range. This plot is very sensitive even for weak deviations from the theoretical function and a wrong saturation value may distort the whole curve. For this reason the Stockell plot is more difficult to interpret compared with the direct linearization methods of the binding equation in the case of alternative mechanisms or artificial influences.

For the evaluation procedure of *Job* (1928) the total concentrations of ligand and macromolecule are kept constant and only the molar proportions of both components are altered. X is the mol fraction of the macromolecule and Y that of the ligand, $X + Y = 1$. This is plotted against $[A]_{bound}$, determined, for example, by an optical signal or the enzyme activity. A curve results as shown in Fig. 1.4 and tangents are aligned at the positions $X=0$ and $Y=0$. Their common intercept has the value:

$$\frac{Y_i}{X_i} = \frac{K_d + nc_0}{K_d + c_0} \,. \tag{1.29}$$

X_i and Y_i are the mol fractions of macromolecule and ligand at the intercept, $c_0 = [E]_0 + [A]_0$ is the (constant) sum of the total concentrations of macromolecule and ligand. For $c_0 \gg K_d$ then $X_i/Y_i = n$. Here the stoichiometry of the binding can be taken from the ratios of the mol fractions at the tangent intercept. For $c_0 \ll K_d$ then $X_i/Y_i = 1$, the curve takes a symmetrical shape and the intercept always has the value 1, irrespective of the actual number of binding sites. This is a disadvantage of the Job plot. It can be circumvented as long as the sum of the macromolecule and ligand concentrations is higher than the value of the dissociation constant. If n is known, K_d can be calculated from Eq. (1.29), whereby the condition $c_0 \sim K_d$ should be regarded. K_d can also be obtained from the maximum of the curve in Fig. 1.4 according to

$$K_d = \frac{(an + a - n)^2 c_0}{4an} \,. \tag{1.30}$$

Here a represents the ratio of the actual measured value at the maximum, M_m, to the saturation value, M_∞.

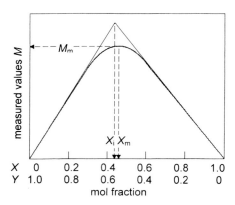

Fig. 1.4 Job plot for the evaluation of binding data.

1.3.3
Binding of Different Ligands, Competition

Due to the high binding specificity of proteins and especially of enzymes, usually only the physiological ligand or the enzyme substrate will be able to bind, while all other metabolites will be excluded. However, this selection cannot be absolute and compounds with high structural homology to the ligand may also be accepted. Knowing the configuration of the binding site or the active center such analogs can be designed and may, sometimes, bind even with higher affinity than the natural ligand. Such analogs may induce similar effects as the ligand, but mostly they are inactive and block the binding site for the native ligand, preventing its action and revealing an antagonistic effect. This *competition* for a distinct binding site of two or more compounds is a valuable tool to investigate specific binding, the action of drugs depends frequently on the antagonistic effect (e.g. β-receptor blocker). Competition is also a valuable tool in cases where binding of the ligand is difficult to detect, e.g. because of the lack of a measurable signal. In such cases a detectable second, e.g. fluorescent-labeled, ligand is applied. At first the binding characteristic and the dissociation constant of the labeled ligand is determined, thereafter the measurements are repeated in the presence of constant amounts of the unlabeled ligand and the dissociation constant for this ligand is obtained as described in the following.

The competition can be described by the scheme:

$$\begin{array}{c} E + A \xrightleftharpoons{K_A} EA \\ + \\ B \\ \updownarrow K_B \\ EB \end{array}$$

The binding affinities are expressed by the dissociation constants K_A and K_B for both compounds:

$$K_A = \frac{[E][A]}{[EA]} \quad \text{and} \quad K_B = \frac{[E][B]}{[EB]} \tag{1.31a}$$

The total amount of the macromolecule is

$$[E]_0 = [E] + [EA] + [EB] \,.$$

[E] and [EB] are replaced by K_A and K_B in Eq. (1.31a):

$$[E]_0 = \frac{K_A[EA]}{[A]}\left(1 + \frac{[B]}{K_B}\right) + [EA] \,.$$

By conversion the following expression for [EA] is obtained:

$$[EA] = \frac{[E]_0[A]}{[A] + K_A\left(1 + \dfrac{[B]}{K_B}\right)}.$$

For a macromolecule with n binding sites Eq. (1.32) results, as discussed already for Eq. (1.23):

$$r = \frac{n[A]}{[A] + K_A\left(1 + \dfrac{[B]}{K_B}\right)}. \tag{1.32}$$

The double-reciprocal relationship is:

$$\frac{1}{r} = \frac{1}{n} + \frac{K_A}{n[A]}\left(1 + \frac{[B]}{K_B}\right) \tag{1.33}$$

and the Scatchard equation:

$$\frac{r}{[A]} = \frac{n}{K_A\left(1 + \dfrac{[B]}{K_B}\right)} - \frac{r}{K_A\left(1 + \dfrac{[B]}{K_B}\right)}. \tag{1.34}$$

Compared with the general binding equation there are now two variable concentration terms, but as long as one of them (e.g. B) remains constant and only A is altered, the term within the brackets will also remain constant and the behavior corresponds essentially to the general binding equation with a hyperbolic curve (Fig. 1.5 A). The only difference is that the value of K_A is increased by the value of the term in brackets. If, in a second test series, another concentration of B is taken (but remains constant during the test series), the resulting curve will again be modified by a change in the apparent value of K_A. In this manner a series of hyperbolic curves are obtained. All can be linearized in the double-reciprocal plot (Fig. 1.5 B), the Scatchard plot (Fig. 1.5 C) and the Hanes plot (Fig. 1.5 D). The pattern of the lines is remarkable, with a common ordinate intercept in the double-reciprocal diagram, a joint abscissa intercept in the Scatchard plot and parallel lines in the Hanes plot. These patterns can be taken as indicative of a competition mechanism.

While the dissociation constant K_A for the first ligand can be obtained as already described in the absence of B, the constant for B, K_B, can be derived e.g. from the abscissa intercept $K_A(1+[A]K_B)$ in the double-reciprocal diagram from a knowledge of K_A.

Further procedures for the analysis of competition data are described in Section 2.5.3.3. It must, however, be considered, that, unlike with enzyme kinetic studies, competition is not always unequivocal and can easily be mixed up with the non-competitive mechanism, as is described in the following section.

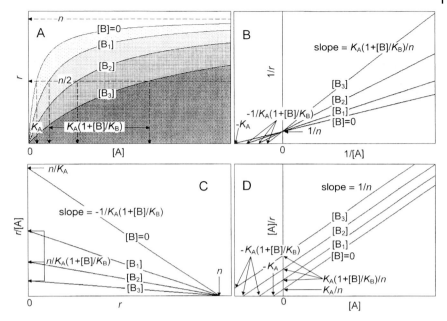

Fig. 1.5 Competition of a ligand for the same binding site. The concentration of ligand A is altered with ligand B at various, but constant amounts. (A) Direct plot, (B) double-reciprocal plot, (C) Scatchard plot, (D) Hanes plot.

1.3.4
Non-competitive Binding

A non-competitive binding mechanism exists if the second ligand induces the binding of the first one, but does not exclude its binding. While for competition it is assumed that both ligands bind to the same site, in the non-competitive mechanism both occupy different sites, which both influence one another, e.g. because of steric or electrostatic interactions.

$$
\begin{array}{ccc}
E + A & \underset{}{\overset{K_A}{\rightleftharpoons}} & EA \\
+ & & + \\
B & & B \\
{\Big\updownarrow} K_B & & {\Big\updownarrow} K'_B \\
EB + A & \underset{}{\overset{K'_A}{\rightleftharpoons}} & EAB
\end{array}
$$

Therefore the constants for binding to the free macromolecule, K_A and K_B, differ from those for the macromolecule occupied already with one ligand, K'_A and K'_B:

$$K'_A = \frac{[EB][A]}{[EAB]} \quad \text{and} \quad K'_B = \frac{[EA][B]}{[EAB]} \,, \tag{1.31 b}$$

and, considering also Eq. (1.31 a), they are linked

$$\frac{K_A}{K_B} = \frac{K'_A}{K'_B} \,. \tag{1.35}$$

The total amount of the macromolecule is:

$$[E]_0 = [E] + [EA] + [EB] + [EAB]$$

and the individual macromolecule forms can be eliminated by the constants defined in Eqs. (1.31 a, b):

$$[E]_0 = [E] + \frac{[E][A]}{K_A} + \frac{[E][B]}{K_B} + \frac{[E][A][B]}{K_A K'_B} \,,$$

$$[E] = \frac{[E]_0}{1 + \frac{[A]}{K_A} + \frac{[B]}{K_B} + \frac{[A][B]}{K_A K'_B}} \,.$$

The portion of $[A]_{bound}$ is:

$$[A]_{bound} = [EA] + [EAB] = \frac{[E][A]}{K_A} + \frac{[E][A][B]}{K_A K'_B} \,,$$

$$[A]_{bound} = \frac{\frac{[E]_0[A]}{K_A}\left(1 + \frac{[B]}{K'_B}\right)}{1 + \frac{[A]}{K_A} + \frac{[B]}{K_B} + \frac{[A][B]}{K_A K'_B}} \,.$$

The final equation for the non-competitive binding is obtained by replacing $[A]_{bound}$ by $r = [A]_{bound}/[E]_0$, assuming n binding sites and multiplying by K_A:

$$r = \frac{n[A]\left(1 + \frac{[B]}{K'_B}\right)}{K_A\left(1 + \frac{[B]}{K_B}\right) + [A]\left(1 + \frac{[B]}{K'_B}\right)} \,. \tag{1.36}$$

It is obvious that the equation will reduce to the normal binding equation if $K_B = K'_B$ (and, consequently $K_A = K'_A$), i.e. if there is no mutual interaction between both ligands. Transformation into the double-reciprocal form yields:

$$\frac{1}{r} = \frac{1}{n} + \frac{K_A\left(1 + \frac{[B]}{K_B}\right)}{n[A]\left(1 + \frac{[B]}{K'_B}\right)}. \qquad (1.37)$$

This will give a pattern of straight lines with a joint ordinate intercept, as shown in Fig. 1.5 B. Accordingly, the Scatchard plot

$$\frac{r}{[A]} = n\frac{\left(1 + \frac{[B]}{K'_B}\right)}{K_A\left(1 + \frac{[B]}{K_B}\right)} - r\frac{\left(1 + \frac{[B]}{K'_B}\right)}{K_A\left(1 + \frac{[B]}{K_B}\right)} \qquad (1.38)$$

will yield a pattern of straight lines as shown in Fig. 1.5 C (the same situation holds also for the Hanes plot, Fig. 1.5 D). Obviously both competitive and non-competitive binding are indistinguishable by graphic analysis, and this is a serious source of misinterpretation, the more so, as both corresponding mechanisms in enzyme kinetics are readily distinguishable by graphic analysis (see Section 2.5.3.2). The reason for this discrepancy may not be immediately clear. In enzyme kinetics there is a similar situation with the partially competitive inhibition, which yields just the same pattern in linearized diagrams as the competitive mechanism (Section 2.5.3.7) and, in fact, non-competitive binding must be regarded as analogous to this and not to the non-competitive inhibition. This discrepancy arises because in non-competitive inhibition only the enzyme substrate complex [EA] is enzymatic active, while the complex with both substrate and inhibitor bound [EAI] is inactive. In contrast, with partially competitive inhibition both complexes are assumed to be equally active. This is just the situation in binding experiments, where the share of ligand A actually bound to the macromolecule will be determined by experiment which will not differentiate between [EA] and [EAB], regarding both as equally active. To avoid this misinterpretation, there exists a simple control. Plotting the slopes of the straight lines of the double-reciprocal diagrams against the concentration of the second ligand, B, must yield a straight line (with $-K_B$ as abscissa intercept) for competitive binding, but for non-competitive binding there is a deviation from linearity. Such secondary diagrams can also be derived from the Scatchard and the Hanes representations and are discussed in more detail in Section 2.5.3.2.

1.4
Macromolecules with Non-identical, Independent Binding Sites

Various enzymes, membrane receptors and other macromolecules possess different binding sites for the same ligand. They may be located at the same subunit, but more often they are an indication of the presence of non-identical subunits. An example is the bacterial tryptophan synthase, consisting of two types of subunits (α, β), each binding indole as the intermediate of the enzyme reaction. Because the enzyme has the structure $\alpha_2\beta_2$, binding both to identical and non-identical sites occurs at the same time. Identical sites are called binding classes and one macromolecule can possess several (m) binding classes, each with several identical binding sites ($n_1, n_2, n_3\ldots$).

Obviously, a ligand binding to such a macromolecule will occupy the site with the highest affinity first, followed by occupation of the lower affinity sites, which require higher ligand concentrations. Assuming independent binding, each binding class will be saturated according to the general binding equation (Eq. (1.23)), so that the total binding process will be the sum of the individual saturation functions for each binding class:

$$r = \frac{n_1[A]}{K_{d1} + [A]} + \frac{n_2[A]}{K_{d2} + [A]} + \ldots \frac{n_m[A]}{K_{dm} + [A]}. \tag{1.39}$$

K_{d1}, K_{d2} etc. are the dissociation constants of the individual binding classes. Each binding process follows a normal hyperbolic binding curve and the resulting function is a superposition of different hyperbolae (Fig. 1.6 A). It shows a steep increase in the low concentration range of the ligand, where the high affinity site becomes occupied. At higher ligand concentrations, when this site becomes saturated, the low affinity sites will be occupied, resulting in a further, but smoother, rise of the curve. Although the curve does not have a pure hyperbolic shape, the deviation is difficult to recognize, especially with scattered data points; linearized plots are superior because they show characteristic deviations from linearity. Figures 1.6 B–D show the individual (linear) curves for a high and a low affinity site and the resulting composed function in the double-reciprocal, the Scatchard and the Hanes diagram, respectively.

It is easier to create a composed function from the partial functions than to resolve the individual functions for the separate binding sites from a composed function obtained by experimental results. There are several unknown values to be determined, such as the number of binding classes involved, the number of identical sites per binding class and the values of the dissociation constants and it is impossible to get all the information from one curve. As can be seen from Fig. 1.6, the individual functions are not merely the asymptotes to the extreme ranges of the resulting curve, although it may be assumed that at very low and very high ligand concentrations the high and low affinity sites, respectively, will be occupied preferentially. The Scatchard plot can be analysed by using the graphic method of Rosenthal (1967) (Fig. 1.7). The resulting curve may be considered to be composed of two straight lines, the slopes of which are initially taken from both end parts of the resulting curve and are moved in a parallel manner so that the sum of their ordinate intercepts corresponds to the ordinate intercept of the resulting curve. Lines drawn through the coordinate origin meet the resulting curve at a point P. Its coordinates are the sums of the coordinates of the respective intersection points of the individual curves, as described for Fig. 1.7. For an appropriate evaluation a computer analysis is strongly recommended (Weder et al. 1976).

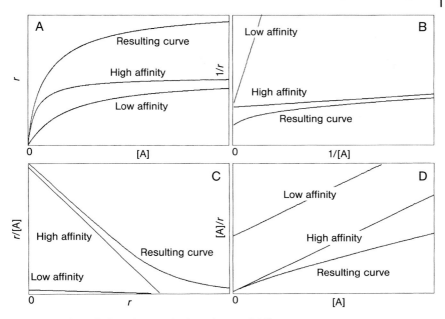

Fig. 1.6 Binding of a ligand to two binding classes of different affinity. The individual curves for the high and the low affinity site, and the resulting curve are shown. (A) Direct plotting, (B) double-reciprocal plot, (C) Scatchard plot, (D) Hanes plot.

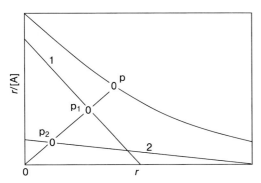

Fig. 1.7 Graphic analysis of a binding curve with two binding classes according to Rosenthal (1967). 1 and 2 are the lines of the separate binding classes. A straight line is drawn from the coordinate origin with the slope 1/[A], intersecting the individual lines at P_1 and P_2 and the resulting curve at P. The sum of the coordinates $[A]_{bound}/([A]_{bound}/[A])$ of the individual intersection points must yield the coordinates of the resulting curve, otherwise, the position of the individual lines must be changed.

Nevertheless, the analysis of such binding curves has only indicative character. On the one hand there is no essential difference in the resulting curves with two or with more binding classes and, on the other hand, there are also other binding mechanisms, showing similar curves, like negative cooperativity and half-of-the-sites-reactivity (see Section 1.5.6) or isoenzymes. Determination of the number and identity of the subunits of the macromolecule by other methods, like molecular mass determination, should be undertaken in parallel.

1.5
Macromolecules with Identical, Interacting Binding Sites, Cooperativity

1.5.1
The Hill Equation

About one hundred years ago it was observed the binding of oxygen to hemoglobin does not follow a hyperbolic saturation function, according to the binding equation, but has a characteristic S- or sigmoidal shape (Bohr 1904). Remarkably, the closely related myoglobin behaves quite normally (Fig. 1.8). Since this time this atypical behavior of hemoglobin has challenged a large number of scientists to derive theoretical approaches and to develop fundamental techniques, like X-ray crystallography of proteins and methods for the detection of fast reactions. No other biological compound has inspired the development of biochemistry so much as hemoglobin. This atypical saturation behavior acquired even more interest when similar curves were found with enzymes occupying key positions in the metabolism. It became obvious that an important regulatory principle of the cell is hidden behind this phenomenon.

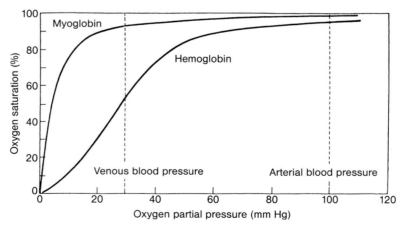

Fig. 1.8 Oxygen saturation curves for myoglobin and hemoglobin (according to M.F. Perutz, *Sci. Am.* 1978, *239*(6), 68–86).

Archibald Vivian Hill undertook, in 1910, a first attempt to explain this atypical behavior. He suggested that not only one but several (n) oxygen molecules bind simultaneously to the hemoglobin molecule:

$$E + nA \rightleftharpoons EA_n . \tag{1.40}$$

The dissociation constant according to the mass action law is defined as:

$$K_d = \frac{[E][A]^n}{[EA_n]} \tag{1.41}$$

and in analogy to Eq. (1.23) a binding equation can be derived for this mechanism, replacing [A] by $[A]^n$:

$$r = \frac{n[A]^n}{K_d + [A]^n} . \tag{1.42}$$

This *Hill equation* indeed yields sigmoidal saturation curves. It was the intention of Hill to determine the number of oxygen molecules, n, actually binding to hemoglobin. This can be achieved by linearization of Eq. (1.42), replacing r by $\overline{Y} = r/n$ (the number 1 in the expression $\overline{Y}/(1-\overline{Y})$ has the significance of the saturation value):

$$\frac{\overline{Y}}{1-\overline{Y}} = \frac{[A]^n}{K_d} .$$

In a logarithmic form the power n enters into the slope:

$$\log \frac{\overline{Y}}{1-\overline{Y}} = n \cdot \log[A] - \log K_d \tag{1.43}$$

if the left term is plotted against log [A] (Fig. 1.9). Presupposing the validity of Eq. (1.43) a straight line should be expected and the number, n, of oxygen molecules bound to hemoglobin should be derived directly from the slope. However, the function obtained from the experimental data looks quite different. Instead of a linear dependence a characteristic three-phase behavior is revealed, starting from a slope of exactly 1 at low ligand concentrations, increasing to a maximum slope for the hemoglobin saturation curve of $n=2.8$, and thereafter decreasing again to 1 near saturation of [A]. Obviously, the function obtained deviates in two essential respects from the prediction of Eq. (1.43), the missing linearity and a slope lower than the expected value for the 4 subunits. It must be emphasized that this three-phase shape is not a special feature of hemoglobin but is observed with all enzymes showing sigmoidal saturation behavior. As can be easily seen, Eq. (1.43) becomes the normal binding Eq. (1.23) for $n=1$, and for this the Hill plot will indeed yield a straight line with a slope of exactly one.

34 | 1 Multiple Equilibria

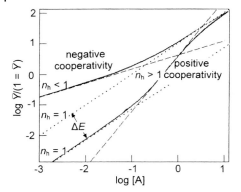

Fig. 1.9 Hill plot for positive and negative cooperativity. The dotted tangents to the curves in the lower and upper ligand ranges have a slope of 1, corresponding to normal hyperbolic binding. The Hill coefficient n_h is the slope of the dashed tangents to the maximum deviation.

Thus, the observed saturation behavior for sigmoidal curves appears to be a transition from two different, normal binding states at low and high ligand concentrations, respectively. This cannot be explained by the Hill equation and together with the wrong prediction of the number of binding sites n this equation may be regarded as useless. It is in fact not useful for describing sigmoidal saturation behavior, however, the diagram derived from this equation, still known as the *Hill plot*, proved to be a good graphic representation for any type of deviation from normal saturation behavior, as will be discussed later. It can also be used for the presentation of hyperbolic saturation curves, where both linearity and a slope of one is an indication of normal binding behavior. The abscissa intercept at half saturation, i.e. for $\log \overline{Y}/(1-\overline{Y}) = 0$, is K_d/n and becomes equal to K_d for $n=1$. Nevertheless, there is no real advantage over the other linearized diagrams to justify the circumstantial conversion of the experimental data.

1.5.2
The Adair Equation

Hill had no knowledge of the real structure of hemoglobin and did not realize that the number of binding sites was underestimated by applying his equation. It was 15 years later when G. S. Adair established that hemoglobin actually consists of four subunits and, thus, four oxygen molecules should bind. He derived an approach for the description of sigmoidal binding behavior, which, although some modifications have to be considered later, remains valid today in its fundamental aspects. He showed that the mechanism suggested by Hill is an oversimplification. If more than one ligand binds to a macromolecule, inevitably a consecutive binding process must be assumed. Even at high ligand concentrations binding will be initiated by occupation of one binding site by the first ligand, followed by binding of the second one, and so on, until all sites are occupied. This is formulated by the reaction sequence:

$$E + A \rightleftarrows EA$$
$$EA + A \rightleftarrows EA_2$$
$$EA_2 + A \rightleftarrows EA_3$$
$$\vdots \quad \vdots \quad \quad \vdots$$
$$EA_{n-1} + A \rightleftarrows EA_n$$

The sum of this reaction sequence:

$$E + nA \rightleftarrows EA_n$$

is identical with Scheme (1.40) from which Hill derived his equation. In fact he ignored intermediate binding steps and allowed only simultaneous binding of all ligands. Binding of single ligands is strictly forbidden by this mechanism and it remains to be explained how a macromolecule will manage to avoid binding of individual ligands and allow only occupation of all sites at the same time. Comparable processes can be imagined for crystallization and polymerization reactions. Each chemist knows from his own experience that crystallization, even from pure, oversaturated solutions, can require days or weeks or may not occur at all. However, addition of seed crystals or even scratching at the glass wall will immediately provoke the formation of crystals in the whole solution. Similar processes are observed by the formation of fibers, like actin and myosin, from their subunits, where also the first aggregation step is strongly disfavored. For a macromolecule with several binding sites, it must be assumed that their affinity for the ligand is negligible, but just at the moment when one ligand binds, all sites acquire a high affinity state and thus the binding of one ligand entails instantaneous binding of all others. Actually, such exclusive binding is not very realistic and macromolecules, like haemoglobin, possessing more binding sites cannot reject the binding of a single ligand molecule. However, this first binding can strongly favor the binding of the following ligands. Such mechanisms, which assume that one ligand supports binding of others is called *cooperativity*. The Hill equation describes an extremely strong, not very probable, cooperativity, while the approach of Adair describes this phenomenon on a more realistic basis.

The derivation of the Adair equation has already been anticipated in Box 1.1 by the derivation of the general binding equation. The saturation function is defined as r, the ratio of the bound ligand $[A]_{bound}$ to the total enzyme $[E]_0$, both expressed by the different enzyme forms:

$$r = \frac{[A]_{bound}}{[E]_0} = \frac{[EA] + 2[EA_2] + 3[EA_3] + \ldots n[EA_n]}{[E] + [EA] + [EA_2] + [EA_3] + \ldots [EA_n]} \; . \tag{1.44}$$

The intermediate enzyme forms are substituted by the macroscopic binding constants for the individual binding steps i.

$$K'_i = \frac{[EA_{i-1}][A]}{[EA_i]}.$$

In contrast to the general binding equation, the individual binding constants cannot be replaced by one single common constant and, therefore, the *Adair equation* reads:

$$r = \frac{\dfrac{[A]}{K'_1} + \dfrac{2[A]^2}{K'_1 K'_2} + \dfrac{3[A]^3}{K'_1 K'_2 K'_3} + \cdots \dfrac{n[A]^n}{K'_1 K'_2 K'_3 \ldots K'_n}}{1 + \dfrac{[A]}{K'_1} + \dfrac{[A]^2}{K'_1 K'_2} + \dfrac{[A]^3}{K'_1 K'_2 K'_3} + \cdots \dfrac{[A]^n}{K'_1 K'_2 K'_3 \ldots K'_n}}$$

$$= \frac{\sum_{i=1}^{n} \dfrac{i[A]^i}{\left(\prod_{j=1}^{i} K'_j\right)}}{1 + \sum_{i=1}^{n} \dfrac{[A]^i}{\prod_{j=1}^{i} K'_j}}. \tag{1.45}$$

Since now every binding step gets its individual binding constant, the change in the affinity from the first to the following binding steps can easily be demonstrated. Increasing affinity can be realized by decreasing values for the binding constants $K'_1 > K'_2 > K'_3 \ldots$ Under these conditions sigmoidal saturation curves are obtained and they show indeed the three-phase behavior in the Hill plot, as observed by applying real saturation data of hemoglobin with oxygen. Since the binding constants for each individual step cannot be obtained directly, they must be estimated and adapted until the theoretical curve fits the experimental data satisfactorily. The maximum steepness of the curve depends on the ratio of the individual constants, the more they differ, especially the higher the difference between the first and the last constant, the steeper the maximum slope. It can be seen that in any case the maximum slope ranges between 1 and n, the number of individual binding steps (usually identical with the number of binding sites, respectively, of identical subunits of the macromolecule). However, n cannot be surpassed by any combination of the constants. The maximum slope approaches n the higher the difference between the constants, while it approaches 1 the more the constants became equal to one another. From this consideration the value of 2.8 for oxygen binding to hemoglobin can be understood. The first oxygen raises the affinity for the following ones. If this rise is extremely strong, a value of 4 would be expected, corresponding to the four binding sites. In the case of only a moderate rise, a value between 1 and n will be obtained. So the maximum slope in the Hill plot is a measure of the cooperativity between the sites, a value near 1 meaning low cooperativity and a value near n high cooperativity. Different from the original assumption of Hill, the maximum slope indicates not the number of ligands bound or of binding sites on the macromolecule, but is a measure of cooperativity, the knowledge of binding sites being presupposed. To differentiate from n, the number of identical binding sites, the maximum slope in the Hill plot is designated as n_h (or h). Although n_h does not indicate the actual number of binding sites, it gives a hint for their minimum number, since $1 < n_h < n$. For example, the value of $n_h = 2.8$ found for hemoglobin shows that this macromolecule must be composed of at least 3 identical subunits (n can only be an integer).

A comment should be made about the significance of n: it stands for the number of *identical* binding sites, identical meaning of equal affinity, characterized by equal dissociation constants. If they are different, deviations as discussed in Section 1.4 will be obtained. However, no presupposition is made as to whether these binding sites are localized on one single subunit or protein chain (e.g. generated by gene duplication) or on separated subunits, nor there is any presupposition as to whether these separate subunits must be identical or can be different. Obviously identical subunits possess identical binding sites, while even apparently identical sites localized at the same polypeptide chain can differ in their binding constants, due to dissimilar constraints of the protein molecule. Therefore, identical binding sites are usually assumed to be located on identical subunits and n stands both for identical binding sites and identical subunits, although this must be taken with caution. Regarding hemoglobin, it consists of non-identical ($a_2\beta_2$) subunits, which is so far consistent with this consideration, as the binding constants can be taken as identical.

Although the Adair equation, in contrast to the Hill equation, is able to describe formally the experimental binding curves, it remains unsatisfactory as it is not based on a plausible binding mechanism. The Adair mechanism assumes that the binding steps, and not the binding sites of the macromolecule, differ in their affinity. In the absence of ligand all binding sites are regarded to be equal, and each binding step produces a defined change in the affinities of the still unoccupied binding sites. Consequently, the binding site of the macromolecule which becomes occupied last has to change its affinity n times, from K'_1 to K'_4 although it is not involved in the preceding binding steps. It is a theoretical mechanism, giving no explanation of how these affinity changes are achieved.

1.5.3
The Pauling Model

The first plausible description of cooperative phenomena was proposed in 1935 by Linus Pauling. He considered the macromolecule to consist of identical binding sites with an uniform binding constant K_d. He further assumed that the subunit occupied by a ligand confers a stabilizing effect on the unoccupied subunits enhancing their affinities, expressed by an interaction factor a. Considering the statistical factors described in Box 1.1, Eq. (2), the following constants can be ascribed to each individual binding step:

$$K'_{d1} = \frac{K_d}{4} \; ; \; K'_{d2} = \frac{2K_d}{3a} \; ; \; K'_{d3} = \frac{3K_d}{2a^2} \; ; \; K'_{d4} = \frac{4K_d}{a^3} \; .$$

Entering these constants into the Adair equation, the following binding function results:

$$r = \frac{\dfrac{4[A]}{K_d} + \dfrac{12a[A]^2}{K_d^2} + \dfrac{12a^3[A]^3}{K_d^3} + \dfrac{4a^6[A]^4}{K_d^4}}{1 + \dfrac{4[A]}{K_d} + \dfrac{6a[A]^2}{K_d^2} + \dfrac{4a^3[A]^3}{K_d^3} + \dfrac{a^6[A]^4}{K_d^4}} \; . \tag{1.46}$$

Although this relationship is simpler than the Adair equation it gives an intuitive description of the sigmoidal binding mechanism.

1.5.4
Allosteric Enzymes

The first attempts to explain the sigmoidal binding behavior concentrated on the immediate effect of oxygen on hemoglobin. Subsequently, it became obvious that this atypical binding behavior is not restricted to hemoglobin alone, but is a feature of numerous key enzymes and that it concerns not only one single ligand like oxygen or an enzyme substrate, but it can be influenced by other ligands, called *effectors*. For distinction the direct effects of the single ligand are denoted as *homotropic effects*, while influences from effectors are called *heterotopic effects*. These influences can either be positive or negative and the respective effector acts, correspondingly, as *activator* or *inhibitior*. The effectors act not by direct interaction with the first ligand, e.g. by displacement from its binding site (competition), rather they occupy a spatially separate binding site. This is called an *allosteric center* from the Greek words αλλος for different and στερεος for rigid. Accordingly, enzymes showing these features are called *allosteric enzymes*. The separate allosteric center permits the regulation of the enzyme by metabolites, which are completely different from the physiological ligands of the enzyme, like substrates, cofactors or coenzymes. An important regulatory principle is the *feedback inhibition*. Metabolic pathways are frequently controlled by their end products, which inhibit the first step of the pathway, so that intermediates will not accumulate. The final product of the pathway is quite different from the substrate or product of the enzyme catalyzing the initial step and will be not recognized by its catalytic site. Therefore it binds to an allosteric center, from which it influences, e.g. by conformational change, the catalytic efficiency of the enzyme. It had been observed that allosteric effectors confer a characteristic influence on the sigmoidal saturation function of the substrate. Inhibitors, although reducing the catalytic efficiency, increase the homotropic effect by intensifying the sigmoidal shape of the saturation curve, while activators raise the catalytic efficiency by weakening the homotropic effect, converting the sigmoidal shape of the saturation curve into a hyperbolic one. Theoretical approaches to explain cooperative effects with enzymes and related proteins must, therefore, also include heterotropic effects. It should be stressed, that *cooperativity*, i.e. increase of affinity of the same ligand upon consecutive binding and *allostery*, i.e. binding to spatially separated sites, are principally two independent phenomena, which may also occur separately in distinct enzymes. It is, however, an empirical observation that both features are usually combined in the same enzyme or protein system, since they both supplement one another and the regulatory power can be fully expressed only by combination of both phenomena. Therefore, it is justified to understand *allosteric enzymes* as a notation for enzymes revealing both cooperativity and allostery. Allostery is observed with various enzymes, but also with several non-enzyme proteins, like hemoglobin or the acetylcholine receptor and, therefore, in the following no differentiation is made between enzymes and proteins.

1.5.5
The Symmetry or Concerted Model

In 1965 Jacques Monod, Jeffries Wyman and Jean-Pierre Changeux presented the first comprehensive model for the description of allosteric enzymes in the publication *On the Nature of Allosteric Transition: A Plausible Model*. It became a guideline for the better understanding of regulatory mechanisms on enzymes. This *concerted or symmetry model* is based on certain presuppositions (see Fig. 1.10) which were deduced from observations made with hemoglobin and several allosteric enzymes:

1. An allosteric system is an oligomer composed of a defined number, n, of identical units (protomers). The protomer can either consist of a single subunit (polypeptide chain) or be composed of several non-identical subunits.
2. Protomers occupy equal positions in the macromolecule, there exists at least one symmetry axis.
3. The enzyme can accept at least two states of conformation termed T (*tense*) and R (*relaxed*), which differ in their energy potential. In the absence of ligand the transition from one into the other state occurs spontaneously, L being the equilibrium constant between both states:

$$L = \frac{[T]_0}{[R]_0} . \tag{1.47}$$

4. The molecular symmetry is preserved during the transition from one enzyme form to the other. At the same time all subunits of an enzyme molecule exist either in the T- or the R-state, intermediate forms with protomers in different conformations are excluded.
5. Both enzyme forms differ in their affinity for the ligand, T being the low affinity (or less active) and R the high affinity (or fully active) enzyme form, the ratio c of the dissociation constants for both forms is correspondingly:

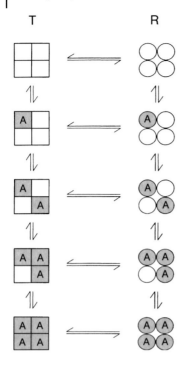

Fig. 1.10 Schematic representation of the conformational states and the fractional saturation of a tetramer macromolecule according to the symmetry model.

$$c = \frac{K_R}{K_T} < 1 . \tag{1.48}$$

6. In the absence of ligand the equilibrium L is in favor of the low affinity form T, i.e. L > 1.

The binding of the ligand to the two enzyme forms is described by the equilibria:

$$T \rightleftharpoons R$$

T + A $\rightleftharpoons$ TA	R + A $\rightleftharpoons$ RA
TA + A $\rightleftharpoons$ TA$_2$	RA + A $\rightleftharpoons$ RA$_2$
TA$_2$ + A $\rightleftharpoons$ TA$_3$	RA$_2$ + A $\rightleftharpoons$ RA$_3$
$\vdots$	$\vdots$
TA$_{n-1}$ + A $\rightleftharpoons$ TA$_n$	RA$_{n-1}$ + A $\rightleftharpoons$ RA$_n$.

The individual enzyme forms can be replaced by microscopic binding constants, which are assumed to be identical for all protomers in the same conformation:

1.5 Macromolecules with Identical, Interacting Binding Sites, Cooperativity

$$[TA] = [T]n\frac{[A]}{K_T} \qquad\qquad [RA] = [R]n\frac{[A]}{K_R}$$

$$[TA]_2 = [TA]\frac{(n-1)[A]}{2K_T} \qquad\qquad [RA]_2 = [RA]\frac{(n-1)[A]}{2K_R}$$

$$\vdots \qquad\qquad\qquad\qquad \vdots$$

$$[TA]_n = [TA_{n-1}]\frac{[A]}{nK_T} \qquad\qquad [RA]_n = [RA_{n-1}]\frac{[A]}{nK_R} \ .$$

From the fraction of the binding sites occupied by ligand

$$\overline{Y} = \frac{1}{n} \cdot \frac{([TA] + 2[TA_2] + \ldots n[TA_n]) + ([RA] + 2[RA_2] + \ldots n[RA_n])}{([T]_0 + [TA] + [TA_2] + \ldots [TA_n]) + ([R]_0 + [RA] + [RA_2] + \ldots [RA_n])} \tag{1.49}$$

the general saturation function for the symmetry model is obtained, replacing a for $[A]/K_R$, the ligand concentration, reduced by its dissociation constant:

$$\overline{Y} = \frac{Lca(1+ca)^{n-1} + a(1+a)^{n-1}}{L(1+ca)^n + (1+a)^n} \ . \tag{1.50}$$

Sigmoidal saturation curves are obtained when all three preconditions: $L>1$, $c<1$ and $n>1$ are fulfilled simultaneously. If only one fails, c or n becoming 1 or L approaching to low values, Eq. (1.50) reduces to the general binding equation:

$$\overline{Y} = \frac{a}{1+a} = \frac{[A]}{K_R + [A]} \ . \tag{1.23}$$

Conversely, the cooperativity, or the sigmoidicity, of the saturation curve becomes more intense, the more these preconditions are fulfilled, i.e. the larger L and n and the smaller c. In the direct non-linear plot (Fig. 1.11 A) such changes are less detectable, while the linear plots show characteristic deviations from a straight line. In the double-reciprocal plot (Fig. 1.11 B) the curve deviates towards the upper right, in the Hanes plot (Fig. 1.11 D) to the upper left, and in the Scatchard plot (Fig. 1.11 C) a maximum is passed. Further information about the cooperative systems can be obtained from the Hill plot (Fig. 1.9). As already mentioned, the curve progresses from a straight line with a slope of 1 at low ligand concentrations through a steeper section in the medium saturation range and returns to a straight line with a slope of 1 in the saturation range. Both sections with the slope of 1 represent simple binding characteristics, to the T-state in the very low and the R-state in the high saturation range. The distance between the two straight lines is an indication of the energy difference between the R- and T-states. The cooperative effect is greatest in the steepest area, where the system switches from the low affinity T-state to the high affinity R-state. The maximum slope is the Hill coefficient (n_h) indicating the strength of cooperativity (see below).

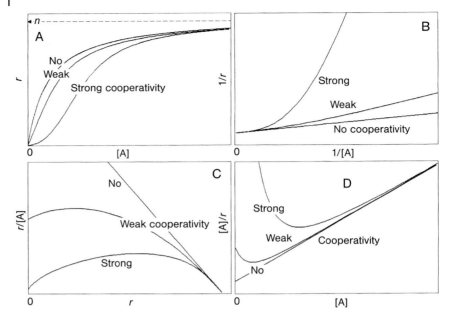

Fig. 1.11 Binding curves of cooperative systems according to the symmetry model in different representations. (A) Direct plot, (B) double-reciprocal plot, (C) Scatchard plot, (D) Hanes plot. No cooperativity: $L=c=1$; weak cooperativity: $L=5$, $c=0.1$; strong cooperativity: $L=100$, $c=0.01$.

The cooperative effect can be illustrated by considering that the first binding ligand will find only a few molecules in the high affinity R-state out of a surplus of non-binding molecules in the T-state. Binding will stabilize the R-state and withdraw it from the equilibrium. To restore the original equilibrium a molecule from the T-state will be converted into the R-state. Thus, for the following ligand, four additional binding sites (assuming $n=4$ for this example) are accessible. The number of accessible binding sites thus increases faster than the ligand concentration causing a disproportionate increase in binding or activity. This process will proceed until the pool of molecules in the T-state is depleted and the whole macromolecule population is shifted to the R-form. Then cooperative binding changes to normal binding and the slope in the Hill plot reduces to 1.

The relative size of the Hill coefficient between the limits $1 < n_h < n$ is determined by the values of L and c: the better the conditions $L \gg 1$ and $c \ll 1$ are fulfilled, the more n_h will approach the number of protomers n. In no case, however, can n be surpassed by n_h. Conversely, n_h cannot fall below 1 with any combination of L and c. The Hill coefficient thus proves to be a measure of the strength of cooperativity. The more it approaches the number of protomers, the more pronounced the cooperativity becomes. In the extreme case of $n_h = n$ the mechanism defined by the Hill equation (Eq. (1.42)) applies. In its strict definition the Hill coefficient indicates

Table 1.1 Relationship between the number of protomers n and the Hill coefficient n_h with heme proteins from different organisms (after Wyman 1967)

Protein	Source	n	n_h
Myoglobin	Mammalian	1	1
Myoglobin	Molluscs	2	1.5
Hemoglobin	Mammalian	4	2.8
Hemocyanin	Lobster	24	4
Chlorocruorin	*Spirographis*	~80	5
Erythrocruorin	*Arenicola*	>100	6

the reaction order with respect to the varied ligand. According to Eq. (1.40), n should only be an integer but since the mechanism depends on the strength of subunit–subunit interactions a fractional reaction order can also exist. The highest possible reaction order, i.e. maximal cooperativity, is achieved when all binding sites become simultaneously occupied. Therefore, the Hill coefficient is not a direct measure of the number of binding sites (or protomers), but ranges between 1 and n (as long as no other mechanism is responsible for the sigmoidal curve). There exists, however, no direct proportionality between n_h and n. An increase in n is not paralleled by a similar increase in n_h, even for identical values of L and n. In Table 1.1 oxygen-binding proteins from different organisms are compared with their number of protomers and the observed Hill coefficients. While n increases from 1 to 100 the Hill coefficient only rises to 6. This follows also from theoretical calculations.

Heterotopic effectors influence the equilibrium of R- and T-states by binding to allosteric centers. Activators act in the same manner as the cooperative ligand. They bind preferentially to the R-form and shift the equilibrium in this direction. L becomes diminished in the presence of the activator, the cooperativity will be attenuated and the Hill coefficient decreases. Consequently, in the presence of the activator the macromolecule will persist essentially in the R-state, so that the original cooperative ligand will find only the active R-state and bind to this in a quite normal, non-cooperative manner. Conversely, the inhibitor binds to and stabilises the T-form, increasing L and, subsequently, n_h, intensifying the cooperativity. Larger amounts of ligand are now required to shift the equilibrium towards the R-form, revealing an inhibitory effect.

The influence of effectors can be considered in Eq. (1.50) by modifying the equilibrium constant from L to L'. The meaning of L' is:

$$L' = L\left(\frac{1+d\beta}{1+\beta}\right)^n \cdot \left(\frac{1+e\gamma}{1+\gamma}\right)^n. \tag{1.51}$$

β and γ are the concentrations of inhibitor, or activator, reduced by their respective binding constants K_{Ri} and K_{Ra} to the R-form; $d = K_{Ri}/K_{Ti} > 1$ and $e = K_{Ra}/K_{Ta} < 1$ are the ratios of the binding constants for the R- and T-states of inhibitor and activator, respectively.

1.5.6
The Sequential Model and Negative Cooperativity

One year after the postulation of the concerted model D. E. Koshland, G. Nemethy and D. Filmer (1966) presented an alternative model for allosteric enzymes which describes the cooperative phenomena and heterotopic effects equally well. The general prerequisites are comparable, the macromolecule is assumed to be composed of several identical subunits and exists in at least two conformations, differing in their affinity. The low affinity or inactive T-state (for uniformity the terms from the concerted model are also applied here) prevails in the absence of ligand, the high affinity or fully active R-state in the presence of ligand. K_t is the equilibrium constant of both enzyme forms in the absence of ligand:

$$K_t = \frac{[T]}{[R]} \gg 1 . \tag{1.52}$$

There exist two substantial differences from the concerted model. Before postulating the sequential model Koshland developed the *induced-fit hypothesis*. It replaced the previous *lock-and-key model* of Emil Fischer (1894), which assumed that substrate specificity of enzymes is based on preformed rigid binding regions, into which only the proper substrate molecule can lock like a key. In comparison with this theory the induced-fit hypothesis predicted that the binding site would be created interactively by enzyme and substrate. Only the actual substrate is able to induce this adaptation. This hypothesis is a fundamental prerequisite for the sequential model. Unlike the concerted model, where the ligand is not actively involved in the shift from the T- to the R-state but only selects the form with higher affinity, the sequential model assumes that conformation transition is induced by the binding of the ligand. As a second difference from the concerted model a sequential transition is assumed, only subunits to which the ligand binds change into the R-form, all others remain in the T-state. The transition occurs stepwise in parallel with the saturation of the enzyme (Fig. 1.12).

Cooperativity originates from the interaction between the subunits. The intensity of the interaction depends on the conformational state of the neighboring subunits and is defined by interaction constants. They indicate the ratio of interacting (e.g. TT) to non-interacting subunits (T, T). As these are relative factors, the constant K_{TT} for the TT-interactions is defined as 1:

$$K_{TT} = \frac{[T][T][TT]}{[TT][T][T]} = 1 \tag{1.53}$$

$$K_{RT} = \frac{[T][R][TT]}{[RT][T][T]} = \frac{[R][TT]}{[RT][T]} \tag{1.54}$$

$$K_{RR} = \frac{[R][R][TT]}{[RR][T][T]} . \tag{1.55}$$

1.5 Macromolecules with Identical, Interacting Binding Sites, Cooperativity

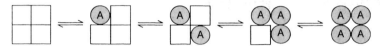

Fig. 1.12 Schematic representation of the conformation states and the fractional saturation of a tetrameric macromolecule according to the sequential model.

The interactions between the subunits can either be stabilizing (K_{RT} and $K_{RR} < 1$) or destabilizing (K_{RT} and $K_{RR} > 1$).

The saturation function for the sequential model is derived from the general form of the Adair equation (Eq. (1.45)):

$$\overline{Y} = \frac{1}{n} \cdot \frac{\frac{[A]}{\Theta_1} + \frac{2[A]^2}{\Theta_2} + \frac{3[A]^3}{\Theta_3} + \ldots \frac{n[A]^n}{\Theta_n}}{\Theta_0 + \frac{[A]}{\Theta_1} + \frac{[A]^2}{\Theta_2} + \frac{[A]^3}{\Theta_3} + \ldots \frac{[A]^n}{\Theta_n}} . \tag{1.56}$$

The terms Θ_0, Θ_1 etc. include all constants relevant for the respective binding step: the constant K_R for binding of the ligand to the R-state (binding to the lower affinity T-state is neglected), the constant K_t for the equilibrium between the two macromolecule forms, and the substrate concentration [A] considered in the equation with the power of the respective binding step i. The possible interactions between subunits determine the type of interaction constants to be considered for each binding step. This is demonstrated in Table 1.2 for the case of a macromolecule consisting of three identical subunits in a linear arrangement. Although such an arrangement is highly improbable, it is taken as a simple model to demonstrate the derivation of a rate equation in the sequential model. By inserting the Θ links into Eq. (1.56) the following equation results:

$$\overline{Y} = \frac{1}{3} \cdot \frac{\frac{[A]}{(K_{RT}^2 + 2K_{RT})K_R K_t} + \frac{2[A]^2}{(K_{RT}^2 + 2K_{RT}K_{RR})K_R^2 K_t^2} + \frac{3[A]^3}{2K_{RR}^2 K_R^3 K_t^3}}{1 + \frac{[A]}{(K_{RT}^2 + 2K_{RT})K_R K_t} + \frac{[A]^2}{(K_{RT}^2 + 2K_{RT}K_{RR})K_R^2 K_t^2} + \frac{[A]^3}{2K_{RR}^2 K_R^3 K_t^3}} . \tag{1.57}$$

Because this model rests on the respective types of interactions each aggregation state and each arrangement of subunits needs its own derivation and the above equation is valid only for the trimeric arrangement. In Fig. 1.13 some aggregation states and subunit arrangements are depicted. Obviously the number of possible arrangements increases with the number of subunits, e.g. for a tetramer there exist three symmetric orientations, linear, square and tetrahedral. This complicates the treatment of the model. Whereas for the concerted model one single equation can be applied for any oligomer, for the sequential model not only the number, but also the respective arrangement of subunits must be known. Furthermore, it must be considered that with higher aggregates different interactions can occur, even between identical subunits. For example a hexamer composed of two trimers will

1 Multiple Equilibria

Table 1.2 Conformation states and definitions of the Θ-values for a trimeric macromolecule with the subunits in a linear arrangement according to the sequential model

Enzyme conformation	Interaction constants	Θ values
Free enzyme		
TTT	$K_{TT}K_{TT}=1$	$\Theta_0=1$
1st Binding step		
TRT	$K_{RT}K_{RT} = K_{RT}^2$	$\Theta_1 = (K_{RT}^2 + 2K_{RT})K_R K_t$
TTR+RTT	$K_{TT}K_{RT} + K_{RT}K_{TT} = 2K_{RT}$	
2nd Binding step		
RTR	$K_{RT}K_{RT} = K_{RT}^2$	$\Theta_2 = (K_{RT}^2 + 2K_{RT}K_{RR})K_R^2 K_t^2$
RRT+TRR	$K_{RT}K_{RR}+K_{RR}K_{RT} = 2K_{RT}K_{RR}$	
3rd Binding step		
RRR	$K_{RR}K_{RR}=K_{RR}^2$	$\Theta_3 = K_{RR}^2 K_R^3 K_t^3$

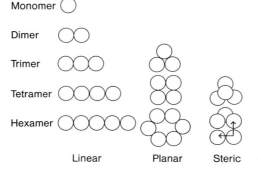

Fig. 1.13 Possible arrangements of subunits of differently aggregated macromolecules. Different horizontal and vertical subunit contacts for the hexamer are indicated by arrows at the bottom right.

possess other interactions within the trimer structure (Fig. 1.13, horizontal arrow), than those between the contact regions of the two trimers (Fig. 1.13, vertical arrow). For each type of contact site individual interaction constants must be defined.

Such complications render the application of the model more difficult and also the interaction constants are usually not accessible and must be estimated. However, the significance of both models rests not in their relative ease of treatment, but on their ability to gain a better understanding of regulatory mechanisms for which both models provide a clear conceptual basis. More information on the existence of one of these models for a distinct enzyme or protein requires detailed structural and conformational studies. One indication for the prevalence of one of the two models can be the relative position of the cooperative area (i.e. the maximum slope in the Hill plot) within the saturation function. In the sequential model it coincides exactly with the half-saturation range, in the concerted model this area shifts with rising n to the lower saturation range.

Heterotopic effects can be explained in the sequential model in a similar manner as in the concerted model. Allosteric activators reduce the cooperative effect by inducing the transition from the inactive to the active state like the cooperative ligand or substrate, while allosteric inhibitors strengthen the cooperative effect by stabilizing the T-state.

A special feature of the sequential model is the fact that interactions need not be stabilizing, but may also be destabilizing if K_{TR} and K_{RR} are larger than K_{TT}. The deviation from the normal hyperbolic saturation function is reversed to sigmoidal curves, rather it resembles that of non-identical independent binding centers (see Section 1.4), and in the linearized plots corresponding deviations result (see Fig. 1.6). In the Hill plot, instead of a maximum slope higher than 1 in the cooperative range, a minimum slope less than 1 is obtained. This anti-cooperative behavior, which is in contrast to normal cooperativity (also termed *positive cooperativity*) is defined as *negative cooperativity*. It is observed in several enzymes. The first example was the binding of NAD to glyceraldehyde-3-phosphate dehydrogenase and served as evidence for the validity of the sequential model (Convay and Koshland 1968).

In a strict sense many examples of negative cooperativity actually obey the mechanism of *half-of-the-sites reactivity*. The first ligand occupies one site of the macromolecule and interferes with the binding to the second site by steric or electrostatic interactions or by covalent reactions, like phosphorylation, so that occupation of the second, originally identical, site becomes aggravated. Thus only half of the original binding sites are saturated, the other half remains unsaturated or requires very high ligand concentrations for saturation. Although the binding behavior and the respective graphical representations are very similar it is not a genuine negative cooperativity mediated by interaction of subunits. Half-of-the-sites reactivity was observed e.g. with alcohol dehydrogenase, malate dehydrogenase and alkaline phosphatase (Levitzki and Koshland 1976).

Obviously, evaluation of binding curves which deviate from normal behavior in a sense like negative cooperativity is difficult because of several alternative explanations, such as half-of-the-sites reactivity, non-identical binding centers, different enzyme forms or isoenzymes. Additional information, especially from structural studies, is required to differentiate, with regard to negative cooperativity and half-of-the-sites reactivity, between identical subunits and non-identical subunits. A negative cooperative mechanism has been reported for many macromolecules like glyceraldehyde-3-phosphate dehydrogenase, CTP-synthetase, desoxythymidine kinase, receptors and binding of tRNA to ribosomes. The physiological advantage of negative cooperativity may be the greater insensitivity to fluctuations in the concentration of metabolites, like substrates or effectors. Due to the high affinity of the first binding step these systems are already very active and able to maintain a basic turnover at low substrate levels. A larger increase in the substrate concentrations causes only a small further activity increase, but the system is able to follow substrate variations over a wide range in a damped mode without reaching early saturation.

1.5.7
Analysis of Cooperativity

Observation of atypical, i.e. sigmoidal saturation behavior with a distinct system is a first indication for the prevalence of cooperativity. Since the main mechanisms are based on the change in binding constants (*K*-systems), for analysis binding measurements are recommended. Since changes in the substrate affinity influences the enzyme reaction via the K_m-value, measurements of enzyme activity, which are easier to perform, can also yield sigmoidal dependences. Alternatively, the two enzyme conformations can differ in their catalytic activity instead of in their affinity (*V*-systems). A combination of both binding and kinetic measurements will give valuable additional information. This holds also for the action of the effectors, which can also be studied by both techniques. For the analysis of sigmoidal curves, linearized plots are preferable to direct, non-linear representations, as deviations from the linear progression are easily detectable (Fig. 1.11). These curves can be linearized by entering $[A]^n$ instead of $[A]$, where the Hill coefficient n_h (indicating the reaction order for $[A]$) and not the number of binding sites n must be used. To establish a cooperative mechanism a larger number of measurements is required than for hyperbolic systems and a broader concentration range of the ligand has to be covered.

Deviations from normal behavior may, however, have other reasons. Sigmoidal saturation curves can be observed in multiple substrate reactions, but artificial effects can also cause such curves. Enzymes are often unstable in dilute solutions and when a test series is performed from high to low substrate concentrations, the rates for the latter experiments may slow down because of this effect. A further source of error is the underestimation of initial velocities in the lower substrate range, especially with high amounts of enzyme, as will be discussed in Section 2.3.2. With high enzyme concentrations the assumption $[A]_0 = [A]$ is also no longer valid, which can also lead to misinterpretation.

As already mentioned an estimation of the strength of cooperativity is the relationship between the Hill coefficient and the number of binding sites n, with positive cooperativity n_h ranging between 1 and n, and with negative cooperativity tending to be below 1. In the Hill plot sigmoidal saturation curves usually show a three phase course, from a straight line with $n_h = 1$ across a steeper region with a maximum Hill coefficient to again a straight line with $n_h = 1$. Both straight lines represent the two enzyme states and the distance between both asymptotes multiplied by $RT\sqrt{2}$ yields the difference between the free energies for the binding interaction of the first and the last ligand (Fig. 1.9). The respective dissociation constants for both states can be estimated from the ligand concentration at the position of half saturation: $\log\{\overline{Y}/(1-\overline{Y})\} = 0$ for $\overline{Y} = 0.5$. Since the cooperative range is the transition area between both states, and this cooperative range is usually at half saturation of the system, no defined dissociation constant can be obtained. Nevertheless, as for a given system half saturation is always at a distinct ligand concentration, a half saturation constant is defined also for cooperative systems, which is, in contrast to real dissociation constants, termed $S_{0.5}$.

Sometimes the R_s-value is taken as a measure of cooperativity. It is defined as the ratio of ligand concentration at 90% and 10% saturation, for a normal hyperbolic saturation curve the R_s-value is always 81. With positive cooperativity, the curve becomes steeper and the R_s-value decreases with the strength of cooperativity, while it increases with negative cooperativity (Table 1.3). The Hill coefficient and the R_s-value are not directly related. The Hill coefficient records cooperativity at a certain point, i.e. at maximum deviation, while the R_s-value covers a wider ligand range, but the connection to the number of protomers is lost. The estimation of the R_s-value is depicted in the semi-logarithmic diagram in Fig. 1.14, which is especially suited for the plotting of broad ligand concentration ranges applied with cooperative systems. Because in this plot normal binding curves also reveal a sigmoidal shape, a distinction can only be made by the steepness of the curve. The abscissa value of the turning point of the curve at half-saturation indicates the K_d-value in the case of normal binding behavior, or the $S_{0.5}$-value for cooperative systems.

Table 1.3 Comparison of the Hill coefficient n_h and the R_s-value (Taketa and Pogell 1965)

n_h	R_s
0.5	6570
1.0	81
2.0	9
4.0	3

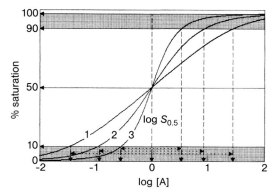

Fig. 1.14 Semi-logarithmic plot of saturation curves for the determination of the R_s-value from the ratio of ligand concentration at 90% and 10% saturation for 1) negative cooperative, 2) normal and 3) positive cooperative behavior. The ligand concentration at half-saturation (K_d- or $S_{0.5}$-value) is assumed to be 1.

1.5.8
Physiological Aspects of Cooperativity

Cooperativity is one of the most important regulatory principles in the metabolism and is found, besides hemoglobin, in many key enzymes of metabolic pathways, in membrane-bound enzymes where it is influenced by membrane fluidity, in transport systems and ATPases, in receptor–ligand binding (e.g. the estrogen receptor), in acetylcholine esterase involved in synaptic transfer, and in thrombin activity. The advantage of (positive) cooperative saturation behavior rests in the over-proportional reaction of the system upon ligand fluctuations and in the allosteric regulation frequently connected with the cooperative effect. Allosteric regulation may also occur with normal binding systems without any cooperativity, when an effector binds to a separated site, which influences the active site. However, a normal system is not able to react in such a sensitive manner as a cooperative system. Due to the steep increase in the sigmoidal saturation curve in the middle saturation range that usually correlates with the physiological range of ligand variation (Fig. 1.15), a slight concentration shift causes a large activity change. The action of effectors is not only confined to inhibition or activation, they can also render the system less sensitive to substrate variations. The activator elevates the system to full activity, the inhibitor brings it down to a minimal level.

The question may be raised, which of the two models is preferred in nature or do alternative mechanisms, not covered by these models, exist. Actually the essential predictions of these models have proved correct, e.g. identical subunits, distinct conformations differing in their affinity, and allosteric regulatory sites. In the following, thoroughly investigated examples of allosteric macromolecules will be presented and it will be shown that aspects of both models can be found, sometimes even in the same system. As shown in Fig. 1.16, both models occupy extreme positions among all conceivable combinations of conformation transitions. The concerted model permits only the uniform conformations bordered in the outer bands, the sequential model only the diagonal states of direct linkage of ligand binding and conformation transition. So both models

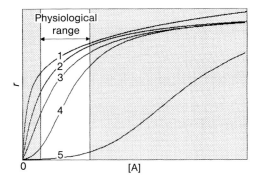

Fig. 1.15 Regulatory significance of allosteric enzymes. The physiological ligand range is highlighted 1) Negative cooperativity, 2) normal binding, 3) positive cooperativity with activator, 4) without effector and 5) with inhibitor.

1.5 Macromolecules with Identical, Interacting Binding Sites, Cooperativity

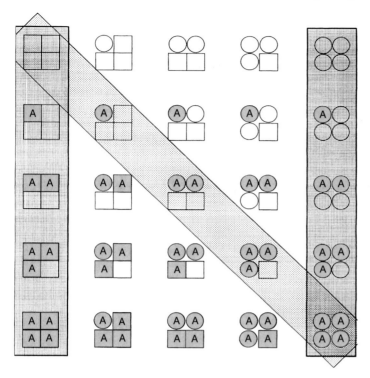

Fig. 1.16 Possible conformational and binding states of a tetrameric macromolecule. The lower affinity T-forms are presented as squares, the high affinity R-forms as circles. The vertical bars at the left and right enclose the states permitted in the concerted model, the diagonal bar the states assumed in the sequential model.

comprise already all plausible combinations. States not considered by them may also be included, but it is obvious that high cooperativity can only be obtained from the extreme positions. Therefore, an alternative model cannot be created without contributing additional aspects.

A test criterion for both models is their requirement of identical subunits. Cooperativity cannot be described with a monomeric macromolecule (with the exception that a macromolecule existing as a monomer in one state will aggregate to form the other state). Ribonuclease was the first example of an exclusively monomeric enzyme with sigmoidal saturation behavior. The cooperativity cannot be caused by interaction of subunits and thus is also not explained by the two models discussed so far. This phenomenon can, however, be explained by the ratio of the time dependences of the transition between the two states and the catalytic conversion of substrate to product, if the catalytic step is faster than the conformational transition. In contrast to both allosteric models, which are based on equilibria, here kinetic, non-equilibria states are considered. This is, therefore, termed *kinetic cooperativity* and can only be detected in the presence of the catalytic turnover, i.e. by observing product formation, while binding measurements will yield normal saturation behavior. A plausible model, the *slow transition model*, has been derived, which will be described in Section 2.8.2.

In the last decades more detailed information about allosteric systems has been gathered which refines the picture of this class of proteins and enzymes. The existence of separate centers for regulation and for action has been widely established, the average distance between the centers being 3.0–4.0 nm. A more unexpected feature is the fact that binding sites for substrates as well as regulatory sites can be located at subunit interfaces rather than at a distinct subunit. In muscle nicotine receptor the binding sites for acetylcholine are located at subunit boundaries.

It also turns out that the assumption of only two states is a simplification valid possibly only for distinct systems, while different sub-conformations are assumed in other cases, e.g. that distinct subunits in one (the T-) state can adopt conformations leading to the other (R-) state.

A further extension of the allosteric models is to membrane inserted structures, like membrane receptors and transmembrane ion channels, where the regulatory sites, to which (e.g.) the neurotransmitter binds, is at one side (synaptic side) of the membrane, with the active center on the opposite site of the membrane, so that the interaction between the two different sites is mediated by a transmembrane allosteric transition. Equilibrium (in the absence of the ligand) prevails between a silent resting state and an active (e.g. open channel) state, agonists stabilizing the active and antagonists the silent state. A consequence of transmembrane polarity given by the two opposite sites is the existence of only one symmetry axis, perpendicular to the membrane plane.

1.5.9
Examples of Allosteric Enzymes

1.5.9.1 Hemoglobin
Although not an enzyme, hemoglobin has given invaluable impetus for numerous theoretical and experimental approaches, like the cooperative models, or the advancement of fast kinetic techniques and X-ray structural analysis. The comparison of the sigmoidal characteristics for oxygen binding to tetrameric hemoglobin with the hyperbolic saturation behavior of the closely related monomer myoglobin demonstrates clearly the significance of the interaction of subunits for cooperativity. As an apparent contradiction to the postulates of the cooperative models, hemoglobin consists of two pairs of non-identical subunits $a_2\beta_2$ and should rather be regarded as a dimer consisting of two protomers. Accordingly, the Hill coefficient should not be greater than a value of 2, but a value of nearly 3 is actually found. The a- and β-subunits, however, possess not only a comparable structure, but also their affinities to oxygen are similar and thus they may be regarded as identical. X-ray crystallographic studies by Max Perutz

(1970, 1990) permit a detailed insight into the allosteric and cooperative machinery of hemoglobin. In the absence of oxygen (*desoxyhemoglobin*) hemoglobin is in a T-state of low affinity that is stabilized against the R-state of the oxygen-rich *oxyhemoglobin* by eight additional salt bridges between the subunits. Cleavage by carboxypeptidase of the C-terminal His and Tyr residues, which are involved in salt bridges, results in a non-cooperative form with high affinity for oxygen.

The divalent iron ion is complex-bound to the haem cofactor, coordinated by four porphyrine nitrogen atoms. In desoxyhemoglobin the iron exists in a high-spin state, emerging 0.06 nm out of the plane of the porphyrine ring, stabilized by a histidine residue on the fifth coordination site. The oxygen molecule binds to the sixth coordination site. This causes the iron to adopt the low-spin state and to move into the plane of the porphyrine ring, dragging along the histidine and inducing a conformational change in the R-state by cleaving the eight salt bridges between the subunits. The bound oxygen molecule stabilizes the R-state.

The significance of the sigmoidal saturation behavior for the regulation of the oxygen binding is demonstrated by its dependence on the concentration of protons (*Bohr effect*). The protons released from hydrogen carbonate in the blood capillaries bind to the terminal amino acids of hemoglobin and stabilize the T-state. The sigmoidicity of the saturation function becomes more pronounced, the binding capacity decreases and oxygen is released into the tissue. In contrast, higher oxygen binding caused by the elevated oxygen pressure in the lung releases protons from the hemoglobin, pH is lowered and the sigmoidicity decreases, induced by the stabilized R-state. The low pH in turn induces the release of CO_2 from hydrogen carbonate in the lung. 2,3-Bisphosphoglycerate stabilizes the T-state by connecting the β-subunits and decreases the oxygen binding capacity.

Recent investigations revealed aspects, which are not in accord with the mere symmetry model and require an extension. By encapsulation of hemoglobin in silica gel the T- and the R-states could be stabilized and it could be shown that subunits in the T-state can adopt R-like properties which is not merely consistent with the concerted model (Viappiani et al. 2004). Obviously heterotopic effects causing tertiary structural changes play a much greater role in determining the function of hemoglobin than do the homotopic T and R transitions in the quaternary structure (Yonetani et al. 2002).

1.5.9.2 Aspartate Transcarbamoylase

This enzyme from *Escherichia coli* clearly demonstrates the spatial separation of catalytic and regulatory centers on distinct polypeptide chains. The native enzyme molecule consists of six catalytic subunits (C, $M_r = 33\,000$), joined in two trimers, and six regulatory subunits (R, $M_r = 17\,000$) that form three dimers, resulting in a $(C_3)_2(R_2)_3$ structure. Catalytic and regulatory centers are 6 nm apart. The allosteric activator ATP and the inhibitor CTP both bind to the same region at the R-subunit. CTP stabilises the T-state and enhances the sigmoidal character of the substrate saturation function. ATP binds preferentially to the R-form and weakens the cooperativity of the substrate aspartate which also binds preferentially to the R-form. At the transition from the T- to the R-state the two catalytic trimers move apart by 1.1 nm and rotate by 12° in relation to each other, while the regulatory dimers rotate by 15° around the two-fold molecule axis. Because of this transition several amino acid residues important for the binding of aspartate move towards the active center and increase the affinity for the substrate. The removal of the regulatory subunits results in the loss of cooperativity and the regulation by ATP and CTP, while the catalytic activity is retained. With this enzyme it was also possible to demonstrate the concerted transition from the T-state to the R-state according to the symmetry model. The occupation of half of all binding sites by the transition state analog N-phosphoacetyl-L-aspartate (PALA) is sufficient to transfer the entire enzyme molecule into the R-state.

Aspartate transcarbamoylase is a good example of end-product inhibition. The enzyme catalyzes the initial reaction of the pyrimidine nucleotide biosynthesis pathway and is inhibited by CTP, the end-product of the pathway. The activator ATP is the end-product of the purine biosynthesis pathway. As for the nucleic acid biosynthesis, both nucleotides are required in an equal ratio, a surplus of purine nucleotides stimulates pyrimidine synthesis, which in turn is inhibited by a surplus of pyrimidine nucleotides (Kantrowitz and Lipscomp 1990).

1.5.9.3 Aspartokinase

Aspartokinase I: homoserine dehydrogenase I from *Escherichia coli* catalyzes the first and the third step of the threonine biosynthesis pathway. Methionine biosynthesis controlled by an aspartokinase II: homoserine dehydrogenase II, and lysine biosynthesis regulated by an aspartokinase III, both branch off from this pathway. Aspartokinase I: homoserine dehydrogenase I consists of four identical subunits ($M_r = 86\,000$). Each subunit has catalytic centers for both enzyme activities on two separate domains (*multifunctional enzyme*). The separate domains with their respective enzyme activities could be obtained by partial proteolysis or mutations. The aspartokinase domains retain their tetramer structure while homoserine dehydrogenase dissociates into dimers. In the native enzyme both activities are subject to end-product inhibition by threonine that shows a sigmoidal saturation pattern. This is more pronounced in the aspartokinase activity ($n_h \sim 4$) than in the homoserine dehydrogenase activity ($n_h \sim 3$). While the separate aspartokinase domain is still inhibited by threonine, the homoserine dehydrogenase domain becomes insensitive to this inhibition. Thus in the native enzyme both activities are regulated by one single regulatory binding site located on the aspartokinase domain. This was demonstrated by a one-step mutation where cooperativity for both activities was reduced by a comparable degree, i.e. to $n_h = 1.65$ for aspartokinase and $n_h = 1.45$ for homoserine dehydrogenase. It may be concluded from this that the native enzyme was formed by a fusion of the genes of two originally separate enzymes, an allosteric aspartokinase inhibited by threonine, and an originally unregulated homoserine dehydrogenase which was forced by fusion to adopt the allosteric properties.

1.5.9.4 Phosphofructokinase
Phosphofructokinase is the most important regulatory glycolyse enzyme. The corresponding reverse reaction step in gluconeogenesis is catalyzed by another enzyme, *fructose-1,6-bisphosphatase*. This necessitates a close regulatory linkage in order to avoid depletion of ATP by a *futile cycle* of the two counteracting reactions, the forward reaction consuming one ATP that cannot be regained in the reverse reaction. AMP is an activator of phosphofructokinase and an inhibitor of fructose-1,6-biphosphatase. Phosphofructokinase, a tetramer enzyme, is inhibited by phosphoenolpyruvate, which stabilizes the T-state. The substrate fructose-6-phosphate exhibits a cooperative effect. The transition from T- to R-state is effected by a counter-rotation of $7°$ of each two dimers, respectively. The binding of the inhibitor AMP to the tetramer fructose-1,6-biphosphatase causes a reorientation of two dimers by $19°$. In mammals both enzymes are additionally regulated by fructose-2,6-bisphosphate. The phosphofructokinase is allosterically activated and fructose-1,6-bisphosphatase is inhibited by negative cooperativity. Thus, both enzymes are subject to a reverse regulatory principle preventing a simultaneous parallel run of both reactions.

1.5.9.5 **Allosteric Regulation of the Glycogen Metabolism**
Biosynthesis and degradation of glycogen is also regulated by two allosteric enzymes. Here the allosteric control is additionally overlaid by a regulation by covalent modification, a phosphorylation governed by a cyclic cascade mechanism. *Glycogen synthase* is activated by glucose-6-phosphate and inhibited by AMP, while AMP activates and glucose-6-phosphate and ATP both inhibit the *glycogen phosphorylase*. The transition of glycogen phosphorylase from the T- to the R-state is accompanied by a relative rotation of the subunits against each other of $10°$. The quaternary structure of the enzyme is modified towards a more favorable folding, the catalytic center moving into the vicinity of the allosteric AMP binding center and the phosphorylation site. This enzymatic active R-state is stabilized on the one hand by AMP and on the other hand by phosphate residues covalently bound at the phosphorylation site.

1.5.9.6 **Membrane Bound Enzymes and Receptors**
The *nicotinic acetylcholine receptor* was first described in fish electric organ (Heidmann and Changeux 1978). Five subunits arrange in a pentameric ring-like assembly in the order $\alpha1$, γ, $\alpha1$, δ, and $\beta1$, i.e. 4 non-identical subunits, only $\alpha1$ contributing two copies. All four non-identical subunits, however, possess high sequence homology, obviously emanating from a fourfold gene duplication, so that the pentameric structure can be regarded as pseudo-symmetrical with a five-fold rotational axis. The five subunits consist of three domains: the hydrophilic, extracellular N-terminal domain carries the neurotransmitter binding site, four membrane-spanning segments forming together a transmembrane channel, and a hydrophilic domain to the cytoplasmic site, which is susceptible to phos-

phorylation and transmits the signal obtained from the neurotransmitter site into the cell.

There exist only two binding sites with both α1 subunits and a γ, respectively, a δ subunit for acetylcholine. These are located at the boundary between the subunits, the acetylcholine binding domain consists of three loops of the α subunit and three loops of the γ or δ subunit, respectively. This is in contrast to the symmetry model where the number of binding sites is assumed to be equal to the protomer number and subject to the same symmetry conditions.

The *G-protein-coupled receptors* (GPCRs) are generally viewed as monomeric allosteric proteins. They consist of seven transmembrane α helices. The ligand binding site is located between the transmembrane helices or the extracellular domain. The intracellular loop and the C-terminal segment interact with the G-protein. The active forms of GPCRs occur as transmembrane oligomers (dimers or higher oligomers), e.g. functional chimeras between muscarinic and adrenergic receptors. Upon ligand binding dimerization may occur.

The microbial *tl-lipase* binds with high affinity to the membrane. The catalytic triad Asp, His, and Ser of the catalytic center is accessible from the surface only through a pocket with a lid. This closes reversibly the access to the active site. Fluorescence studies using a tryptophan in the lid-helix revealed a two-state model where an inactive closed-lid state ($K_d = 350$ µM) and an active open-lid state ($K_d = 53$ µM) could be discerned (Berg and Jain 2002).

1.6
Non-identical, Interacting Binding Sites

The description of the binding of ligands to identical, to non-identical independent, and to identical interacting binding sites should consequently be completed by the treatment of ligands binding to non-identical, interacting binding sites. However, such cases have not yet been convincingly identified. Hemoglobin may be such an example due to its α and β subunits, but because of their similar binding constants they behave like identical subunits. Different independent binding sites cause – as shown in Section 1.4 – a deviation from the normal binding pattern (Fig. 1.6) which is just opposite to positive cooperativity with identical interacting binding sites (Fig. 1.9), as can easily be seen from the double-reciprocal plots comparing Fig. 1.6B and Fig. 1.11B. With positive cooperativity the curve deviates to the upper right, with differing binding sites to the lower right. At comparable intensity both effects will compensate each other, resulting in a straight line as in normal binding patterns. Even at different intensities of both effects they will partially compensate one another and only the predominant mechanism can manifest itself in a weakened form. The same applies for the simultaneous existence of positive and negative cooperativity (e.g. if the initial binding step increases and the final one decreases the affinity), as the latter shows a similar curvature as binding to non-identical sites. The significance of such superpositions in the sense of a counter-regulation or fine tuning may be discussed, incomplete compensations of counteracting effects may be responsible for inhomogeneities sometimes observed in saturation curves. In the absence of convincing examples, however, it remains open how far counteracting mechanisms within the same system actually exist.

On the other hand superposition of congeneric effects like negative cooperativity and binding to non-identical subunits will result in an amplification, but there are also no convincing examples.

References

Diffusion

Berg, H.C. (1983) *Random Walks in Biology*, Princetown University Press, Princetown, New Jersey.

Berg, O.G. (1985) Orientation constraints in diffusion-limited macromolecular association, *Biophys. J. 47*, 1–14.

McCammon, J.A., Northrup, S.H. (1981) Gated binding of ligands to protein, *Nature 293*, 316–317.

Noyes, R.M. (1961) Effects of diffusion rates in chemical kinetics, *Prog. React. Kinet. 1*, 129–160.

Binding Equilibria

Adair, G.S. (1925) The hemoglobin system. The oxygen dissociation curve of haemoglobin, *J. Biol. Chem. 63*, 529–545.

Klotz, I.M. (1985) Ligand-receptor interactions: facts and fantasies, *Quart. Rev. Biophys. 18*, 227–259.

Langmuir, I. (1916) The constitution and fundamental properties of solids and liquids, *J. Am. Chem. Soc. 38*, 2221–2295.

Competition

Thomä, N., Goody, R.S. (2003) What to do if there is no signal: using competition experiments to determine binding parameters, in: Johnson, K.A. *Kinetic Analysis of Macromolecules: A Practical Approach*, Oxford University Press, Oxford.

Graphic Methods

Huang, C.Y. (1982) Determination of binding stoichiometry by the continuous variation method: The Job plot, *Methods Enzymol. 87*, 509–525.

Job, P. (1928) Recherches sur la Formation de Complexes Minéraux en Solution, et sur leur Stabilité, *Ann. Chim. (Paris) 9*, 113–203.

Klotz, I.M. (1946) The application of the law of mass action to binding by proteins. Interaction with calcium, *Arch. Biochem. 9*, 109–117.

Rosenthal, H.R. (1967) A graphic method for determination and presentation of binding parameters in a complex system, *Anal. Biochem. 20*, 515–532.

Scatchard, G. (1949) Attractions of proteins for small molecules and ions, *Ann. N.Y. Acad. Sci. 51*, 660–672.

Stockell, A. (1959) The binding of diphosphopyridine nucleotide by yeast glyceraldehyde-3-phosphate dehydrogenase, *J. Biol. Chem. 234*, 1286–1292.

Weder, H.G., Schildknecht, J., Lutz, L.A., Kesselring, P. (1974) Determination of binding parameters from Scatchard plots. Theoretical and practical considerations, *Eur. J. Biochem. 42*, 475–481.

Allosteric Enzymes, Cooperativity

Berg, O.G., Jain, M.K. (2002) *Interfacial Enzyme Kinetics*, J. Wiley & Sons, Chichester.

Bohr, C. (1904) Die Sauerstoffaufnahme des genuinen Blutfarbstoffes und des aus dem Blute dargestellten Hämoglobins, *Zentralblatt Physiol. 23*, 688–690.

Changeux, J.-P., Edelstein, S.J. (2005) Allosteric mechanisms of signal transduction, *Science 308*, 1424–1428.

Convay, A., Koshland, D. E. (1968) Negative cooperativity in enzyme action, *Biochemistry* 7, 4011–4023.

Heidmann, T., Changeux, J.-P. (1978) Structural and functional properties of the acetylcholine receptor protein in its purified and membrane-bound states, *Ann. Rev. Biochem.* 47, 317–357.

Hill, A. V. (1910) The possible effects of the aggregation of molecules of hemoglobin on its dissociation curves, *J. Physiol.* 40, iv–vii.

Janin, J. (1973) The study of allosteric proteins, *Prog. Biophys. Mol. Biol.* 27, 77–120.

Kantrowitz, E. R., Lipscomp, W. N. (1990) Aspartate transcarbamoylase: The molecular basis for a concerted allosteric transition, *Trends Biochem. Sci.* 15, 53–59.

Koshland, D. E., Nemethy, G., Filmer, D. (1966) Comparison of experimental binding data and theoretical models in proteins containing subunits, *Biochemistry* 5, 365–385.

Levitzki, A., Koshland, D. E. (1976) The role of negative cooperativity and half-of-the-sites reactivity in enzyme regulation, *Curr. Top. Cellular Reg.* 10, 1–40.

Lipscomp, W. N. (1991) Structure and function of allosteric enzymes, *Chemtracts – Biochem. Mol. Biol.* 2, 1–15.

Monod, J., Wyman, J., Changeux, J.-P. (1965) On the nature of allosteric transition: A plausible model, *J. Mol. Biol.* 12, 88–118.

Neet, K. E. (1980) Cooperativity in enzyme function: Equilibrium and kinetic aspects, *Methods Enzymol.* 64, 139–192.

Pauling, L. (1935) The oxygen equilibrium of hemoglobin and its structural interpretation, *Proc. Natl. Acad. Sci. USA* 21, 186–191.

Perutz, M. (1970) Stereochemistry of cooperative effects in haemoglobin, *Nature* 228, 726–739.

Perutz, M. (1990) *Mechanisms of Cooperativity and Allosteric Regulation in Proteins*, Cambridge University Press, Cambridge.

Perutz, M., Wilkinson, A. J., Paoli, G., Dodson, G. (1998) *Annu. Rev. Biophys. Biomol. Struct.* 27, 1–34.

Taketa, K., Pogell, B. N. (1965) Allosteric inhibition of rat liver fructose 1,6-diphosphatase by adenosine 5'-monophosphate, *J. Biol. Chem.* 240, 651–662.

Viappiani, C., Bettati, S., Bruno, S., Ronda, L., Abbruzzetti, S., Mozzarelli, A., Eaton, W. A. (2004) New insight into allosteric mechanisms from trapping unstable protein conformations in silica gels, *Proc. Natl. Acad Sci. USA* 101, 14414–14419.

Wyman, J. (1967) Allosteric linkage, *J. Am. Chem. Soc.* 89, 2202–2218.

Wyman, J., Gill, S. J. (1990) *Binding and Linkage*, University Science Books, Mill Valley, CA.

Yonetani, T., Park, S., Tsuneshige, A., Imai, K., Kanaori, K. (2002) Global Allostery Model of Hemoglobin, *J. Biol. Chem.* 277, 34508–34520.

2
Enzyme Kinetics

In contrast to the time-independent *multiple equilibria*, enzyme kinetics (from the Greek word χινησισ – motion) deal with time-dependent enzyme reactions outside the equilibrium – the reaction strives to attain the equilibrium state. The study of enzyme reactions is a valuable and often relatively simple approach to elucidate mechanisms of enzyme catalysis and regulation. Both fields complement one another, the investigation of equilibria covers some areas of enzyme kinetics, like the initial binding process preceding the catalytic step. Enzyme kinetic studies also reveal information about binding processes. Cooperative phenomena are investigated both by binding and enzyme kinetic studies and the examination of enzyme reactions with more than one substrate can be greatly assisted by binding studies. While the laws of multiple equilibria are applicable to any binding processes in the cell, enzyme kinetics is (with few exceptions, like transport systems) restricted to enzymes. Enzyme kinetics studies usually start with the investigation of the behavior of the enzyme substrate and its conversion into product. Next steps are the examination of the role of cofactors, inhibitors or activators. A prerequisite for any enzyme kinetic study is knowledge of the chemical reaction order. Therefore this topic will be treated before discussing the intrinsic laws of enzyme kinetics.

2.1
Reaction Order

The order of a chemical reaction with respect to the individual components is defined as the power of the component concentration included in the rate equation. The overall reaction order is the sum of all component orders. A reaction:

$$2A + B \rightleftharpoons P$$

is third order in the forward direction, second order for the reactant A and first order for B and also for the reverse reaction.

Enzyme Kinetics. Principles and Methods. 2nd Ed. Hans Bisswanger
Copyright © 2008 WILEY-VCH Verlag GmbH & Co. KGaA, Weinheim
ISBN: 978-3-527-31957-2

2.1.1
First Order Reactions

The simplest chemical reaction is the spontaneous conversion of an educt A into a product P, as in the case of the radioactive decay:

$$A \xrightarrow{k_1} P.$$

The reaction rate v can be determined either from the time-dependent decrease in A or from the increase in P and is directly proportional to the amount of A:

$$v = -\frac{d[A]}{dt} = \frac{d[P]}{dt} = k_1[A]; \qquad (2.1)$$

k_1, the first order rate constant has the dimension s^{-1}, it is independent of the concentration. Integration of Eq. (2.1) from 0 to time t yields:

$$-\int_{A_0}^{A_1} \frac{d[A]}{[A]} = \int_{t_0}^{t} k_1 dt .$$

$$\ln[A] = \ln[A]_0 - k_1 t , \qquad (2.2)$$

$$[A] = [A_0]e^{-k_1 t} . \qquad (2.3)$$

Decrease in substrate or increase in product proceeds exponentially with time (Fig. 2.1 A). The curves can be linearized according to Eq. (2.2) by semi-logarith-

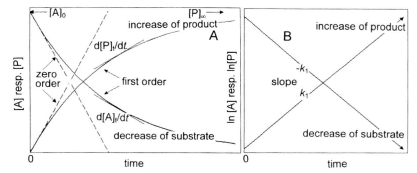

Fig. 2.1 Progress curves of zero and first order reactions following the formation of product or the degradation of substrate in a direct (A) and a semi-logarithmic (B) diagram. The evaluation of turnover rates by tangents is shown in (A).

mic plotting of substrate or product concentrations against time t. The first-order rate constant can be derived from the slope (Fig. 2.1 B) or from the half-life time $t_{1/2}$, the time required for the conversion of half of the initial amount of substrate. [A] becomes $[A]_0/2$ and $\ln([A]_0/[A]) = \ln 2$:

$$k_1 = \frac{\ln 2}{t_{1/2}} = \frac{0.69}{t_{1/2}} . \tag{2.4}$$

2.1.2
Second Order Reactions

Two substrates react with one another:

$$A + B \xrightarrow{k_1} P .$$

The turnover rate is proportional to the decrease in both A and B and the increase in P (the reaction is only discussed in the forward direction and for this it does not matter whether one, two or more products are formed):

$$v = -\frac{d[A]}{dt} = -\frac{d[B]}{dt} = \frac{d[P]}{dt} = k_1[A][B] . \tag{2.5}$$

The dimension of the second order rate constant k_1 (s^{-1} M^{-1}) also includes a concentration term. Integration of Eq. (2.5) gives:

$$k_1 t = \frac{1}{[A]_0 - [B]_0} \ln \frac{[B]_0 [A]}{[A]_0 [B]} . \tag{2.6}$$

The turnover rate depends now on two variables, A and B, and straight lines will no longer be obtained in the semi-logarithmic diagram. To solve Eq. (2.6) one of the variables must be regarded as constant. This can be done if one reactant, e.g. $[B]_0$, is present in large excess, so that its concentration change during the reaction course can be regarded as insignificant. The reaction then is of *pseudo-first order*:

$$v = k_1[A][B]_0 = k_1'[A] . \tag{2.7}$$

The constant $[B]_0$ is included in the pseudo-first order rate constant $k_1' = k_1[B]_0$. Under these conditions the reaction can be treated as first order reaction. Straight lines in the semi-logarithmic plot (Fig. 2.1 B) confirm the predominance of these conditions. Their slope, divided by the concentration of the constant reaction participant, gives the second order rate constant. This plot serves also to differentiate between pseudo- and true first order by changing the initial concentration of the constant reactant. While first order reactions are independent of the concentration and should not show any shift of the slope, with pseudo-first order reaction the slope will be changed by the same factor as the concentration. In this manner it can be distinguished for example whether a conformation change of an enzyme occurs spontaneously or is induced by ligand binding.

If pseudo-first order conditions cannot be established, i.e. if it is not possible to add one reactant in very high concentration, both reactants can be used in exactly the same amounts, $[A]_0 = [B]_0$. In this case Eq. (2.5) simplifies to:

$$v = -\frac{d[A]}{dt} = k_1 [A]^2 \tag{2.8}$$

and integration to t

$$-\int_{A_0}^{A_1} \frac{d[A]}{[A]^2} = \int_{t_0}^{t} k_1 dt$$

gives

$$\frac{1}{[A]} = \frac{1}{[A]_0} + k_1 t . \tag{2.9}$$

In a diagram of $1/[A]$ vs. t a linear function with slope k_1 will be obtained. The second order rate constant k_1 can also be obtained from the half-life time:

$$k_1 = \frac{1}{t_{1/2}[A]_0} . \tag{2.10}$$

2.1.3
Zero Order Reactions

It has been mentioned that the simplest chemical reactions are of first order and so it should not be possible to have a simpler reaction. However, reactions exist which are even independent of the reactant concentrations and are consequently designated as *zero order*. This can be achieved by the action of a catalyst present in very limited amounts. In this case the reaction rate is dictated solely by the amount of catalyst which remains unchanged during the reaction, irrespective of the actual reactant concentration. Although being intrinsically second order, the reaction can be regarded as independent of the reactant concentration:

$$v = -\frac{d[A]}{dt} = \frac{d[P]}{dt} = k . \tag{2.11}$$

Integration with respect to time gives a linear relationship:

$$[A] = [A]_0 - kt . \tag{2.12}$$

Therefore zero order reactions can be identified by the linear progression of the substrate decay or product formation (Fig. 2.1 A). The slope yields the zero order rate constant.

2.2
Steady-State Kinetics and the Michaelis-Menten Equation

2.2.1
Derivation of the Michaelis-Menten Equation

The Michaelis-Menten equation is the fundamental equation of enzyme kinetics, although is was originally derived for the simplest case of an irreversible enzyme reaction, converting a single substrate into a product. Examples of this reaction type can be cleavage reactions (peptidases, proteases, nucleases, etc.) or isomerizations (considering only the forward reaction). Since an irreversible reaction course is assumed it does not matter whether only one product or more products are formed:

$$A + E \underset{k_{-1}}{\overset{k_1}{\rightleftarrows}} EA \xrightarrow{k_2} E + P .$$

The time-dependent variations of the individual reactants are expressed by the differential equations:

$$\frac{d[A]}{dt} = -k_1[A][E] + k_{-1}[EA] \qquad (2.13)$$

$$\frac{d[E]}{dt} = -k_1[A][E] + (k_{-1} + k_2)[EA] , \qquad (2.14)$$

$$\frac{d[EA]}{dt} = k_1[A][E] - (k_{-1} + k_2)[EA] , \qquad (2.15)$$

$$\frac{d[P]}{dt} = k_2[EA] = v . \qquad (2.16)$$

The *turnover rate* v is defined as the product formation. This depends on, and is therefore directly proportional to, the amount of the enzyme–substrate complex EA, according to Eq. (2.16). [EA] depends on the concentrations of the reactants. To solve Eqs. (2.13)–(2.16) the time-dependent concentration changes of the reactants must be known, which is difficult in practice, especially for [E] and [EA]. Figure 2.2 shows a simulation of the changes in all reactants assuming appropriate constants. Three phases can be differentiated:
1. A short initial phase (pre-steady-state or burst), where the [EA] complex is formed and free enzyme decreases. The turnover rate is low in this region.
2. A medium phase, the turnover rate attains its highest value, the concentration of the [EA] complex remains nearly constant.

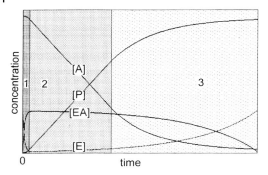

Fig. 2.2 Time-related changes of the reactants of an enzyme-catalyzed reaction. 1) Pre-steady-state phase, 2) steady-state phase, 3) substrate depletion.

3. A depletion phase, where the substrate becomes exhausted, the [EA] complex decays and turnover rate decreases, finally to zero.

Variation of the rate constants alters the relative ranges of the three phases. If all three constants are of comparable size, the medium phase becomes short and the concentration of [EA] will not be constant at any point. However, it is plausible to assume the preceding equilibrium to be fast compared with the enzyme catalysis, i.e. $k_1 \approx k_{-1} > k_2$. Under these conditions the medium phase becomes considerably longer as Fig. 2.2 shows. Both decay of substrate and formation of product proceed in a linear (zero order) manner, while the concentration of the [EA] complex remains nearly constant, because formation and decay of the [EA] complex keep just in balance. This can be regarded as a quasi-equilibrium state, which is maintained for a limited period. In contrast to a true *equilibrium* this phase is designated as *steady-state*. Since the reaction rate depends directly on the concentration of the [EA] complex, the linear, zero order conversion of substrate to product reflects just the steady-state range, during which the time-dependent changes of [EA], and consequently also of [E], can be taken to be zero, $d[EA]/dt = d[E]/dt = 0$. Then Eq. (2.14) and (2.15) simplify to:

$$k_1[A][E] = (k_{-1} + k_2)[EA] .$$

Substitution of [E] according to the principle of mass conservation, $[E]_0 = [E] + [EA]$, yields the term

$$[EA] = \frac{k_1[A][E]_0}{k_1[A] + k_{-1} + k_2}$$

by entering this into Eq. (2.16) a relationship between turnover rate and substrate amount is obtained:

$$v = \frac{d[P]}{dt} = k_2[EA] = \frac{k_2[E]_0[A]}{\frac{k_{-1}+k_2}{k_1}+[A]}, \qquad (2.17)$$

which represents the final rate equation in the form of the rate constants. Because the individual rate constants are not directly accessible they are converted into *kinetic constants*. The expression $(k_1+k_2)/k_1$ contains three rate constants which are combined into a common constant K_m, the Michaelis constant, while $k_2[E]_0$ is replaced by V, the maximum velocity. It is assumed that the total enzyme amount $[E]_0$ remains constant during the reaction and if all available enzyme molecules participate in the reaction, the highest possible turnover is achieved. Eq. (2.17) in the form of the kinetic constants thus reads:

$$v = \frac{V[A]}{K_m + [A]}. \qquad (2.18)$$

This equation, based on the steady-state theory, was originally derived in 1925 by George Edward Briggs and John Burton Sanderson Haldane. Up to now it is regarded as valid and thus of fundamental importance for enzyme kinetics. Because it is derived only for the simple case of an irreversible single-substrate reaction, its significance for more complex reactions, including reversibility, more substrates, cofactors, inhibitors etc. remains to be proven and this will be discussed in the following sections. A similar relationship was presented in 1902 by Adrian J. Brown in Birmingham and by Victor Henri in Paris for the description of the invertase reaction. Leonor Michaelis studied together with his Canadian coworker Maud Leonora Menten the validity of the Henri equation with the enzyme invertase in 1913 in Berlin. Their particular merit was the perception that enzyme reactions require strictly standardized conditions with respect to temperature, pH value and ionic strength. However, in contrast to the steady-state theory, these early studies were based on the equilibrium assumption. The balance of equilibrium between free components and the enzyme–substrate complex (*Michaelis complex*) was assumed to occur very quickly, compared with the catalytic turnover, so that k_2 can be neglected and the concentration of the enzyme–substrate complex depends only on k_1 and k_{-1}. In this case v is a direct measure of [EA] or $[A]_{bound}$ in a fast binding equilibrium. Thus the derivation of the rate equation is analogous to that of the binding equation (Eq. (1.23)) and would result in:

$$v = \frac{V[A]}{K_d + [A]} = \frac{V[A]}{\frac{k_{-1}}{k_1}+[A]}. \qquad (2.19)$$

In this expression K_m is replaced by the dissociation constant K_d. The derivation of the rate equation on the basis of the steady-state theory is nearer to reality as the catalytic constant k_2 can often not be neglected, and [EA] is determined by both the attainment of equilibrium and the catalytic rate. Correspondingly, the constant K_m, determined by kinetic measurements, deviates frequently from the dissociation constant K_d obtained from binding measurements. Nevertheless, appreciating the merits of Michaelis and Menten the term *Michaelis-Menten equation* has been kept also for the equation developed by Briggs and Haldane, as well as the term *Michaelis constant* for K_m.

Due to the principal analogy between the Michaelis-Menten equation and the general binding equation (1.23), the same evaluation procedures are applicable. The *saturation function* $[A]_{bound}$ (resp. $r=[A]_{bound}/[E]_0$) corresponds to the reaction rate v, saturation is reached at $n[E]_0$ (resp. n) there, and here at V. K_d is replaced by the Michaelis constant K_m. A principal difference, however, exist with respect to the experimental procedure. For binding measurements the concentration of the free ligand [A], which is a variable in the general binding equation, must be known. This is principally the same with the Michaelis-Menten equation. However, for kinetic measurements only catalytic amounts of the enzyme are needed, so that $[E]_0 \ll [A]$ and the portion of [EA] can be neglected. Instead of [A], which is difficult to determine, the total amount of substrate added to the enzyme assay $[A]_0=[EA]+[A] \sim [A]$ can be taken.

The Michaelis-Menten equation is characterized by two constants:
- *Michaelis constant* K_m, dimension M, related to the dissociation constant and gives an indication of the affinity of the substrate, low K_m values indicating high affinities.
- *Catalytic constant* k_{cat}, dimension s^{-1}, is a measure of the turnover rate of the enzyme.

From the Michaelis-Menten equation K_m and the maximum velocity V can be obtained directly by experiment, while for $k_{cat}=V/[E]_0$ the molar amount of the applied enzyme must be known. As it is assumed that $[E]_0$ remains constant during the reaction, V is also considered as a constant. However V values from different experiments are often difficult to compare, because the actual amount of the enzyme and its specific activity are not always determined exactly.

The ratio $k_{cat}/K_m = k_{cat}\, k_1/(k_{-1}+k_2)$ is defined as the *catalytic efficiency* or *specifity constant*, large values indicating high specificity. It has the dimension $M^{-1}\,s^{-1}$ of a second order rate constant.

2.3
Analysis of Enzyme Kinetic Data

2.3.1
Graphical Representations of the Michaelis-Menten Equation

2.3.1.1 Direct and Semi-logarithmic Representations
According to the Michaelis-Menten equation the dependence of the reaction velocity v on the substrate concentration [A] yields a hyberbolic saturation curve similar to that obtained with the general binding equation (1.23) in the *direct diagram* (Fig. 2.3A). A steep increase at low substrate amounts slows at higher concentrations and reaches a constant saturation value, the maximum velocity

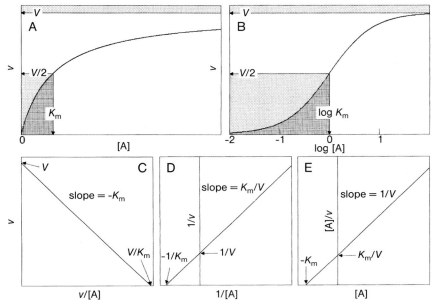

Fig. 2.3 Non-linear and linear representations of the Michaelis-Menten equation. (A) Direct diagram, (B) semi-logarithmic diagram, (C) Eadie-Hofstee diagram, (D) double-reciprocal (Lineweaver-Burk) diagram, (E) Hanes plot. The definitions of K_m and V are indicated.

V. V will actually be attained only for $[A] \to \infty$, in this case K_m in the denominator can be neglected and $v = V$. The substrate concentration at the position $V/2$ has just the same value as K_m and so in a direct diagram V can be obtained from the saturation and K_m from the half saturation (Fig. 2.3A). This implies, however, some uncertainty, because it is often overlooked that V is really defined for infinite substrate concentrations and the decline in the saturation curve entices one to underestimate V, and hence also K_m. Certainly, infinite substrate amounts cannot be realised by experiment, and even high amounts are frequently problematic due to inhibitory influences or limited solubility. Therefore, often a multiple of K_m, e.g. five- or tenfold, is taken as saturation. In Table 2.1 the actual velocities obtained at different degrees of saturation are shown and it becomes evident, that the deviation can be considerable, e.g. tenfold K_m as the substrate concentration yields only about 90% of the maximum velocity.

Practically the situation is even more complicated, because experimental data are always prone to error scattering and a given set of data can be matched by various curves (Fig. 2.4). Therefore, experiments must be done repeatedly and it is essential, that the experiment covers the whole range of the curve – as a rule – within a power of ten lower and higher than the K_m value. Figure 2.4 shows the cases of data covering either only the lower or the higher concentration range and in both cases the uncertainty of the resulting curve becomes noteworthy.

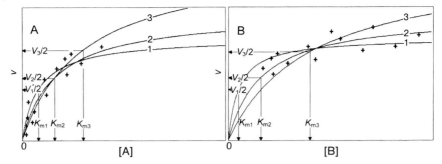

Fig. 2.4 Uncertainties in the determination of kinetic constants of the Michaelis-Menten equation with strong scattering data and inappropriate ranges for the substrate concentration. In (A) only the lower and in (B) only the higher substrate concentration range is considered. Three curves are shown all of which can be adapted to the given points.

Table 2.1 Enzyme turnover rates expressed as % of the real maximum velocity V at an x-fold surplus of substrate compared with K_m (taken as 1). The right column shows the apparent K_m values determined at $V/2$, when instead of the real V the rates obtained by the respective substrate surplus (second column) are taken

Substrate surplus (x-fold K_m)	Turnover rate (% V)	K_m from turnover rate
2	66.7	0.50
5	83.3	0.71
10	90.9	0.83
20	95.2	0.91
30	96.8	0.94
40	97.6	0.95
50	98.0	0.96
100	99.0	0.98
∞	100	1

2.3 Analysis of Enzyme Kinetic Data

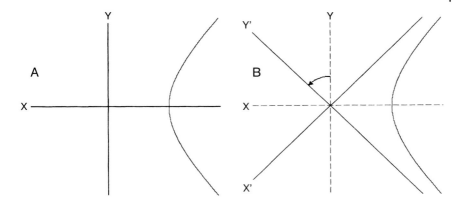

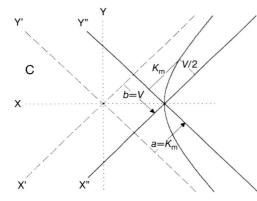

Fig. 2.5 Transformation of a right-angle hyperbola into the shape of a Michaelis-Menten curve. (A) Right-angle hyperbola, (B) rotation of the axes by 45°, (C) shift of the axes by the increments a and b. The correlations to the kinetic constants are indicated.

Although the shape of the curve obeying the Michaelis-Menten equation and, similarly, the general binding equation is usually called *hyperbolic*, the relation to the mathematical hyperbola function is not very obvious. Box 2.1 and Fig. 2.5 illustrate this relationship.

Box 2.1 Relationship between Michaelis-Menten and Hyperbola Equation

The saturation curve resulting from the Michaelis-Menten equation is asserted to have the shape of a right-angle hyperbola, although its equation and shape seems quite different. The general hyperbola equation is:

$$\frac{X^2}{A^2} - \frac{Y^2}{B^2} = 1 . \tag{1}$$

For a right-angle hyperbola $A = B$ (Fig. 2.5 A)

$$X^2 - Y^2 = A^2 . \tag{2}$$

Rotation of the coordinates by 45°: $X = X' \cos a - Y' \sin a$; $Y = X' \sin a + Y' \cos a$. Considering $\sin 45° = \cos 45° = 0.7071$ yields (Fig. 2.5 B):

$$(X' - Y')^2 - (X' + Y')^2 = 2A^2,$$

$$-2X'Y' = A^2.$$

Shifting of the axes by the increments a and b:
$X' = X'' + a$; $Y' = Y'' - b$ (Fig. 2.5 C):

$$-2(X'' + a)(Y'' - b) = A^2$$

$$Y'' = \frac{-A^2/2 + ab + bX''}{X'' + a}. \tag{3}$$

If $ab = A^2/2$ is chosen, the equation simplifies to:

$$Y'' = \frac{bX''}{a + X''}. \tag{4}$$

Now the analogy becomes obvious and the Michaelis-Menten equation will be obtained for $Y'' \approx v$, $X'' \approx [A]$, $a \approx K_m$, and $b \approx V$.

An alternative non-linear representation is the *semi-logarithmic plot* of v against log [A], which yields S-shaped (sigmoidal) curves even for normal saturation curves (Fig. 2.3 B). Such a diagram is recommended if data spread over a wide concentration range of the substrate are be plotted.

Non-linear representations have the disadvantage that deviations from the normal curve, caused by alternative mechanisms, artificial influences or systematic errors, are difficult to recognize and to evaluate. Negative cooperativity or non-identical binding centers yield curves very similar to the hyperbolic saturation function. Also other deviations, e.g. sigmoidal shapes, may easily be overlooked if weakly pronounced or in the case of large error scattering, especially with the semi-logarithmic plot.

In spite of the disadvantages of non-linear representations of the data, they have their benefits. In fact, an appropriate enzyme kinetic analysis will not be confined to only one diagram, and the direct representation of the data without any transformation, as required for the linearization methods, is always recommended. The data with their respective error distribution can be examined without any distortions. Non-linear regression methods like that described by Cornish-Bowden (1984) for the Michaelis-Menten equation are frequently more reliable in the determination of the constants than linear regressions in linearized diagrams. Irrespective of the regression method actually applied they must be critically judged, especially for possible deviations from the normal curve progression.

A method proposed by Malcom Dixon in 1965 for the determination of the Michaelis constant considers a fact frequently overlooked in enzyme kinetics. As substrate concentration in the diagrams the actual amount $[A]_0$ present in the test assay is usually taken, although according to the Michaelis-Menten equation the free concentration $[A]$ is required. Both values differ by the amount of the enzyme substrate complex $[EA]$: $[A]_0 = [A] + [EA]$. Since for enzyme tests only catalytic enzyme amounts are necessary, and thus $[E]_0 \ll [A]_0$, $[EA]$ can usually be neglected and $[A]_0$ be taken instead of $[A]$. If, however, this condition is no longer valid, e.g. with enzymes with low activity or with binding measurements, the free substrate concentration $[A]$ will deviate considerably from $[A]_0$. This is addressed with the Dixon method, which is applicable also for the analysis of binding measurements by spectroscopic titration (Section 1.3.2.2). A disadvantage is that the saturation value (V) must be determined by an asymptotic line and cannot be extrapolated as in linearized representations. As shown in Fig. 2.6A a tangent is drawn to the saturation curve at the coordinate origin and a straight line connects the coordinate origin and the point in the curve at $v = V/2$. The intercepts of both straight lines with the saturation asymptote V correlate with the substrate concentrations $[A]_0'$ and $[A]_0''$ at the abscissa. The difference between both values is K_m (or K_d). Subtracting this difference from the asymptote leftward, the remaining distance from this point to the ordinate correlates with the total enzyme amount $[E]_0$. With the aid of a connecting line between this point on the asymptote and the coordinate origin, the distribution of components in the solution can be determined at each point of the saturation curve. Imagining a horizontal line through any point of the saturation curve, the distance on this line between the ordinate and the connecting line correlates with the amount of the EA complex at this particular saturation. The distance from there to the saturation curve equals the free substrate $[A]$, while the distance from the same point to the perpendicular on $[E]_0$ indicates the free enzyme $[E]$.

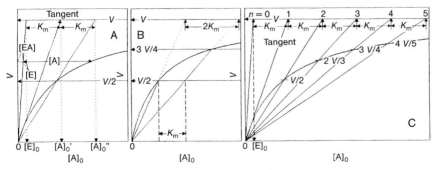

Fig. 2.6 Determination of kinetic constants. (A) Method of Dixon (1965), (B) plot of Kilroy-Smith, (C) diagram of Dixon (1972).

To avoid uncertainties in the drawing of tangents, *Kilroy-Smith* (1966) modified this method (Fig. 2.6 B). Connecting lines are drawn between the ordinate origin and the points V/2 or 3V/4 of the saturation curve. K_m results from the distance of the intercepts of a horizontal line for V/2 with these two connecting lines. By extending the connecting lines to the asymptote V, this distance equals $2K_m$. *Dixon* (1972) modified this method further. Straight lines are drawn from the origin through the points $v = (n-1)V/n$ (for $n = 0, 1, 2, 3$, etc.) i.e. V/2, 2V/3, 3V/4, 4V/5, etc., intersecting the asymptote V at distances of each K_m (Fig. 2.6 C). The line for $n = 1$ is the tangent of the original method to the origin of the saturation curve (Fig. 2.6 A). If the section for K_m is again subtracted towards the left, the connecting line for $n = 0$ is obtained, the distance to the ordinate indicating, as mentioned above, $[E]_0$. This method is also applicable for the determination of inhibition constants (see Section 2.5.3.2).

For the calculation of this method the concentrations of free substrate or ligand [A] and enzyme [E] are substituted by the total amounts minus the share bound as enzyme complex. For the dissociation constant K_d the term:

$$K_d = \frac{[A][E]}{[EA]} = \frac{([A]_0 - [EA])([E]_0 - [EA])}{[EA]}$$

$$K_d = \left(\frac{[A]_0}{[EA]} - 1\right)([E]_0 - [EA])$$

(2.20)

is obtained.

For the equation of the curve tangent in the coordinate origin it may be assumed that $[A]_0$ is small and [EA] may be neglected for $[E]_0$:

$$K_d = \frac{[A]_0}{[EA]}[E]_0 - [E]_0 \;.$$

The equation of the tangent for enzyme kinetic conditions under consideration of $v = k_{cat}[EA]$, and $V = k_{cat}[E]_0$, assuming a rapid equilibrium and setting $K_m \sim K_d$ is:

$$\frac{[A]_0}{v} = \frac{K_m}{V} + \frac{1}{k_{cat}} \;.$$

(2.21)

The tangent intersects the saturation asymptote at $v = V$, substrate concentration at this point is $[A]'_0 = K_m + V/k_{cat}$. At half saturation $v = V/2$ is $[EA] = [E]_0/2$ or $[E]_0 - [EA] = [E]_0/2$. Following Eq. (2.20) the substrate concentration $[A]'_0$ at this point is:

$$K_m = [A]''_0 - \frac{V}{2k_{cat}} = [A]'_0 - \frac{V}{k_{cat}}$$

(2.22)

$$\frac{V}{k_{cat}} = 2([A]'_0 - [A]''_0)$$

$$K_m = [A]'_0 - 2([A]'_0 - [A]''_0) = 2[A]''_0 - [A]'_0 \;.$$

According to Eq. (2.22) this gives:

$$K_m[A]'_0 - [E]_0$$

$$[E]_0 = [A]'_0 - K_m .$$

From this equation $k_{cat} = V/[E]_0$ or the absorption coefficient for the EA complex for spectroscopic titrations can be obtained.

For the conversion after Kilroy-Smith (1966) the concentration $[EA]^{\#} = v/k = 3V/4k$ is inserted at $v = 3V/4$:

$$[E]_0 - [EA]^{\#} = \frac{V}{k} - \frac{3V}{4k} = \frac{V}{4k} . \tag{2.23}$$

Inserting this term in Eq. (2.20), considering substrate concentration $[A]_0^{\#}$ at $v = 3V/4$, gives:

$$K_m = \left([A]_0^{\#} \frac{4k}{3V} - 1\right) \frac{V}{4k} = \frac{[A]_0^{\#}}{3} - \frac{V}{4k} . \tag{2.24}$$

Finally, by inserting Eq. (2.22)

$$K_m = \frac{2[A]_0^{\#}}{3} - [A]_0'' \tag{2.25}$$

is obtained.

2.3.1.2 Direct Linear Plots

The *direct linear plot* of Eisenthal and Cornish-Bowden (1974) represents a completely different plotting method. The Michaelis-Menten equation is modified into a linear equation with the (imaginary) variables V as ordinate and K_m as abscissa (v and [A] being regarded as constants):

$$V = K_m \frac{v}{[A]} + v . \tag{2.26}$$

The corresponding straight line intersects the ordinate at v and the abscissa at –[A].

If, conversely, the values for [A] are plotted on the abscissa (with negative signs), the corresponding values for v on the ordinate and each pair of values is connected by a straight line, a bundle of straight lines is obtained with a common intercept to the right of the ordinate with the X and Y coordinates K_m and V, respectively (Fig. 2.7 A). Because of error scattering actually no common intercept point will be obtained, rather a cloud of intercepts of the various lines. The constants are then derived from the average value (*median*) of all X and Y coordinates of the individual intercepts (Fig. 2.7 B). Some of the intercepts can fall

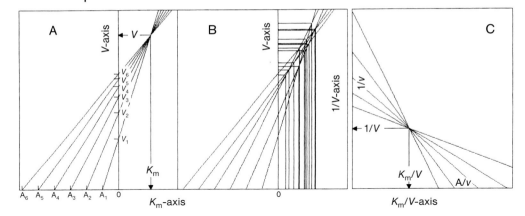

Fig. 2.7 Direct linear plots. (A) Substrate concentrations are plotted on the negative branch of the abscissa, the v values on the ordinate, both are connected by straight lines. (B) The same plot as (A) with error scattering. (C) Reciprocal plot. The mode of determination of the constants is indicated.

far outside the plot, so that it must be drawn in various dimensions. An improvement offers the reciprocal form of Eq. (2.26) (Cornish-Bowden and Eisenthal 1978):

$$\frac{1}{V} = \frac{1}{v} - \frac{K_m}{V[A]} \; . \tag{2.27}$$

$1/v$ is plotted on the ordinate and $[A]/v$ on the abscissa (Fig. 2.7C). The joint section of the straight lines, which is now located inside the plot, has the X- and Y-coordinates K_m/V and $1/V$, respectively.

A third transformation gives the equation:

$$\frac{1}{K_m} = \frac{V}{K_m v} - \frac{1}{[A]} \; . \tag{2.28}$$

By entering −1[A] against $v/[A]$, a common intercept results with the coordinates $1/K_m$ and V/K_m. As in the previous plot the reciprocal entry causes a scale distortion.

Direct linear plots do not require regression methods but deviations from normal behavior are difficult to detect because they are hidden behind the error scattering. Actually a more or less characteristic distortion of the intercept cloud is indicative of non-hyperbolic behavior. The same problem holds also for the analysis of enzyme inhibitions and multiple substrate reactions, which cause, theoretically, characteristic shifts of the common intercepts (see Section 2.5.3.2).

2.3.1.3 Linearization Methods

Several disadvantages mentioned for non-linear plots can be eliminated by applying linearization methods. The kinetic constants can be derived easily from axis intercepts or from the slopes of the straight lines. Deviations from the Michaelis-Menten law result in characteristic deviations from the linear progression and the type of the deviation is an indication of the respective mechanisms (e.g. cooperativity) or of artificial influences. A further important advantage of linearization methods is the analysis of enzyme kinetic methods when two or more ligands are varied, as in enzyme inhibitions or multiple substrate reactions. The respective mechanisms can be identified from the resulting straight line pattern.

It must, however, be borne in mind that no analysis method is able to provide information which is not included in the data. This is demonstrated by the example of the enzyme catalase. It is difficult to determine the K_m value of the substrate H_2O_2 with usual methods, because high concentrations of H_2O_2 inactivate the enzyme, so that only the lower substrate range can be tested and saturation cannot be attained. Consequently, half saturation for K_m determination is not accessible. However, since the lower part of the saturation curve is available, at least this region can be linearized, e.g. in the double-reciprocal plot (which will be discussed later), and by extrapolation of the obtained line, K_m should be easily read from the abscissa intercept, without requirement of data from the inaccessible saturation area. However, applying this procedure, it will be seen that the available data of the lower substrate region gather far to the upper right of the plot, and a straight line through these points will practically extrapolate to the coordinate origin and exact ordinate and abscissa intercepts cannot be given.

There exist three simple linear transformations of the Michaelis-Menten equation. All three were first proposed by Woolf and mentioned in Haldane's and Stern's book *'Enzyme Chemistry'* (1932) without finding particular resonance. Later, other authors described individual linearization methods in extensive publications. Against the generally accepted rule that a method should be designated according to the first author, these methods are mostly named after the later ones. Although not quite just, it is practical, as otherwise all three diagrams must be called 'Woolf plot' and may not be distinguished. To complicate the situation the same plots were described by several authors enabling numerous combinations of names, some considering the authorship of Woolf (e.g. Scatchard, Eadie, Eadie-Scatchard, Eadie-Hofstee, Woolf-Hofstee plots). As the linear transformation of the Michaelis-Menten equation can hardly be regarded as a profound mathematical discovery, the frequently applied names will be preferred for a clear distinction of the diagrams and Woolf's merit may be honored by this remark.

As expressed in its name the *double-reciprocal diagram* (also called *Lineweaver-Burk plot*) is based on the reciprocal form of the Michaelis-Menten equation:

$$\frac{1}{v} = \frac{1}{V} + \frac{K_m}{V} \cdot \frac{1}{[A]} \; . \tag{2.29}$$

Plotting the reciprocal velocity $1/v$ against the reciprocal substrate concentration $1/[A]$ yields a straight line intersecting the ordinate at $1/V$, and the abscissa at

$1/K_m$ (Fig. 2.3 D). This is the most frequently used linearization method for the Michaelis-Menten equation, but also the least appropriate one (Markus et al. 1976). The essential disadvantage is the uneven distribution of data. Due to the reciprocal transformation equidistant substrate concentrations will be compressed towards the coordinates and stretched in the opposite direction. If substrate concentrations are chosen to yield equal distances in this plot to compensate for this shortcoming, they do not cover the saturation curve satisfactorily. The dependent variable v suffers the same uneven distortion. Assuming an absolute, constant error throughout, the error limits for v are compressed towards the ordinate, and extended strongly towards the upper right (Fig. 2.8 B). Linear regressions, which treat all deviations equally result in severe deviations and large failures in the determination of the constants may occur. Already faint errors at low substrate concentrations become grossly extended and will deflect the regression line. Therefore regression analysis must consider appropriate weighting factors (Wilkinson 1961). It is, at first glance, a contradiction, that just this distortion is the reason for frequent application of this plot, but the compression of errors at high substrate concentrations gives the impression of minor error scattering and the data appear more reliable. A real advantage of this plot compared with other linearization methods is the individual display of the variables v and [A], separated by the coordinates. Deviations from straight lines due to alternative mechanisms to the Michaelis-Menten equation or artifi-

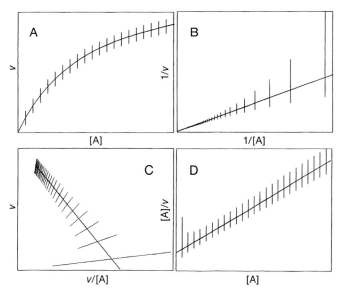

Fig. 2.8 Error limits in different diagrams of the Michaelis-Menten equation assuming an absolute constant error.
(A) Direct diagram, (B) double reciprocal plot, (C) Eadie-Hofstee plot, (D) Hanes plot.

cial influences, as well as inhibition and multiple substrate mechanisms result in characteristic patterns and can easily be identified in this plot.

Multiplication of the reciprocal Michaelis-Menten equation (2.29) by [A] yields the expression for the *Hanes plot*:

$$\frac{[A]}{v} = \frac{[A]}{V} + \frac{K_m}{V}, \tag{2.30}$$

Straight lines with the slope $1/V$, the abscissa intersection $-K_m$ and the ordinate intersection K_m/V are obtained, if $[A]/v$ is plotted against [A] (Figs. 2.3 E and 2.8 D). The error limits are only slightly distorted to low substrate concentrations, so that simple linear regressions can be applied. However, the variables [A] and v are not separate, substrate concentrations are included on both axes.

Multiplication of Eq. (2.29) by vV and conversion gives the equation for the *Eadie-Hofstee plot*:

$$v = V - K_m \cdot \frac{v}{[A]}. \tag{2.31}$$

Plotting v against $v/[A]$ gives V from the ordinate intercept and $-K_m$ from the slope (Figs. 2.3 C and 2.8 C). This diagram corresponds to the Scatchard plot frequently used for binding studies, only the axes are exchanged (see Section 1.3.2.1). The variables are not separated and there is also a distortion of error limits, extending to higher substrate concentrations, which is, however, not as drastic as in the double-reciprocal plot.

2.3.2
Analysis of Progress Curves

The graphical representations of the Michaelis-Menten equation described so far are based on the assumption that turnover rates v are measured in separate assays at various substrate concentrations and the values for the initial velocities and initial substrate concentrations $[A]_0$ are transferred onto the respective plots. However, if one observes one single enzyme reaction, the initial substrate $[A]_0$ will finally disappear: $[A]_\infty = 0$ (for irreversible reactions), and during the reaction all substrate concentrations between these two limits will be passed. The ordinate values of *progress curves*, e.g. registered as absorption in a photometric enzyme test, indicate the actual substrate (or product) concentration at the respective time. The slopes of tangents at any point of the curve represents the corresponding reaction rates $v = d[P]/dt$ (Fig. 2.1). Obviously, a single progress curve contains all information for Michaelis-Menten kinetics and $[A]_0$ must just be chosen large enough so that the reaction will run through the whole saturation range from high to low substrate amounts and will provide all [A] and v values for complete evaluation of the Michaelis-Menten equation. However, apart from the inaccuracy in applying tangents on distinct regions of an experimental curve with its error deviations and artificial influences the most severe drawback of this procedure results from two inevitable features of enzyme reactions, product inhibition and reversibility (the latter may be neglected for quasi-irreversible reactions). These complications can be considered in the treatment of the progress curve (Balcolm and Fitch 1945). The progress curves are divided into equal time segments (e.g. 12 s), at which the respective substrate concentrations are determined (e.g. via absorption). The corresponding rate is obtained from the slope of the line connecting two neighboring points.

2.3.2.1 Integrated Michaelis-Menten Equation

To improve the direct evaluation of progress curves the integrated Michaelis-Menten equation is used, which describes the progress of enzyme reactions obeying the Michaelis-Menten equation and can be evaluated similarly to the analysis of hyperbolic saturation curves, e.g. by linearization. A progress curve obtained from an enzyme test, e.g. from a photometric measurement, can be transferred point by point to a corresponding plot as described in the following, but the real advantage of this method is the online recording and analyzing of the progress curve in a computer, immediately after each measurement.

The Michaelis-Menten equation

$$v = -\frac{d[A]}{dt} = \frac{V[A]}{K_m + [A]} \tag{2.18}$$

is modified

$$-\frac{K_m + [A]}{[A]} d[A] = -\frac{K_m}{[A]} d[A] - d[A] = V dt$$

and integrated from $[A]_0$ at time $t=0$ until $[A]$ at time t:

$$-K_m \int_{[A]_0}^{[A]} \frac{d[A]}{[A]} - \int_{[A]_0}^{[A]} d[A] = V \int_0^t dt . \tag{2.32}$$

The resulting integrated form of the Michaelis-Menten equation reads:

$$K_m \ln \frac{[A]_0}{[A]} + [A]_0 - [A] = Vt . \tag{2.33}$$

The Michaelis-Menten equation was first derived in this form by Victor Henri in 1902. Linear relationships are obtained by further conversions (Walker and Schmidt 1944; Jenning and Niemann 1953):

$$\frac{[A]_0 - [A]}{t} = V - K_m \frac{\ln \frac{[A]_0}{[A]}}{t} , \tag{2.34 a}$$

2.3 Analysis of Enzyme Kinetic Data

or in logarithms of ten:

$$\frac{[A]_0 - [A]}{t} = V - K_m \frac{2.3 \cdot \log \frac{[A]_0}{[A]}}{t}. \qquad (2.34\,b)$$

If formation of product is measured instead of substrate consumption one may set: $[P] = [A]_0 - [A]$, assuming an irreversible reaction. $[P]_\infty$ is the product concentration at the end of the reaction:

$$\frac{[P]}{t} = V - K_m \frac{\ln \frac{[P]_\infty}{[P]_\infty - [P]}}{t}, \qquad (2.35\,a)$$

$$\frac{[P]}{t} = V - K_m \frac{2.3 \cdot \log \frac{[P]_\infty}{[P]_\infty - [P]}}{t}. \qquad (2.35\,b)$$

$([A]_0 - [A])/t$ against $\ln([A]_0/[A])/t$ or $[P]/t$ against $\ln([P]_\infty/([P]_\infty - [P]))/t$ (plotted either as ln or as log) yields a straight line with a slope of $-K_m$ and the ordinate intercept V (Fig. 2.9 A). This is the most frequent representation of the integrated Michaelis-Menten equation. In a second linearization method $-1/K_m$ is

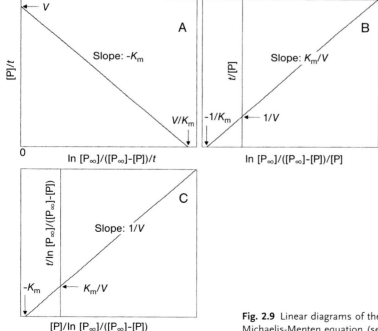

Fig. 2.9 Linear diagrams of the integrated Michaelis-Menten equation (see text).

taken from the abscissa section and $1/V$ from the ordinate intercept, according to the equation:

$$\frac{t}{[P]} = \frac{2.3 \cdot \log \frac{[P]_\infty}{[P]_\infty - [P]}}{[P]} \cdot \frac{K_m}{V} + \frac{1}{V} \qquad (2.36)$$

(Fig. 2.9 B). A third linearization method is based on the formula:

$$\frac{t}{2.3 \cdot \log \frac{[P]_\infty}{[P]_\infty - [P]}} = \frac{[P]}{V \cdot 2.3 \cdot \log \frac{[P]_\infty}{[P]_\infty - [P]}} + \frac{K_m}{V} \ . \qquad (2.37)$$

$1/V$ as slope, K_m/V as ordinate intercept and $-K_m$ as abscissa intercept are obtained (Fig. 2.9 C).

Similar to the direct analysis methods for the Michaelis-Menten equation, the representations of its integrated form can also identify more complex mechanisms, like enzyme inhibition or multiple substrate reactions, detectable from the line patterns or from deviations from the straight line.

2.3.2.2 Determination of Reaction Rates

To apply the previously discussed evaluation methods reaction rates must be determined and because this is of fundamental importance for enzyme kinetics, some principal aspects should be discussed. As shown in Fig. 2.2, the time progression of enzyme-catalyzed reactions can be divided into three phases, a pre-steady-state phase, the steady-state phase and the region of substrate depletion. Since enzyme kinetic relationships are based on the validity of the steady-state equation, for rate determination only the middle linear area is of importance, excepting the previously discussed direct analysis of progress curves.

The recording of the steady-state phase requires the continuous registration of progress curves, e.g. by optical measurements (*continuous tests*). With a large number of enzyme reactions, however, substrate or product alterations cannot be followed by a directly detectable signal during the reaction. In such cases the turnover has to be determined stepwise by stopping the enzyme reaction after distinct times and subsequent analysis of the reaction products, e.g. by color reaction, separation by HPLC, or radioactive labelling (*stopped test*). The progress of the reaction can be followed by performing several parallel tests, stopping each after a different time and completing the various time values for a progress curve. To detect the initial linear phase a series of time values in this region must be taken. Because this is a laborious procedure, testing is often simplified, taking only a single time value at a distinct time, e.g. 5 min, according to a standard protocol of the respective enzyme assay. It is assumed that the time value lies within the linear range of the progress curve and the rate is simply derived from the slope of the line between the blank at $t=0$ and the measured value (*single point measurement*). For enzyme kinetic studies such a method is completely inadequate, not only because of the large errors of single measurements but also because the duration of the linear steady-state range is limited and depends on several factors, like the enzyme and substrate concentration. It can easily be shorter than the chosen test time.

Coupled assays are frequently applied to make accessible enzyme reactions that are difficult to detect. The test reaction is connected to continually detectable reactions, like NADH-dependent dehydrogenases, whereby the product of one reaction serves as the substrate for the other. This is also a problematic principle for enzyme kinetic studies. It must be ensured that under each condition the test reaction and not the coupled detection reaction remains rate limiting. This is usually established by applying a surplus of the detection enzyme relative to the test enzyme but, by investigating a broad variety of parameters necessary for enzyme kinetic studies, it cannot be excluded that conditions may change, favoring the activity of the test enzyme.

Since the Michaelis-Menten equation was derived for the condition $d[EA]/dt = 0$, it is only valid within the linear steady-state region. The better the condition $[E]_0 \ll [A]_0$ is fulfilled, the more pronounced the linear region becomes. While with saturating substrate amounts linear progress curves are easy to detect, the steady-state phase becomes increasingly shorter and more difficult to detect at lower substrate concentrations, especially below the K_m value (Fig. 2.10 A) and sometimes, linearity disappears completely. As the steady-state phase should proceed immediately after the start of the reaction, a tangent is usually aligned to the curve at $t = 0$ to determine an 'initial rate', even if no linear part becomes visible (*tangent meth-*

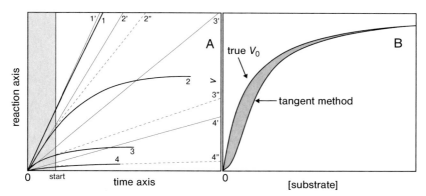

Fig. 2.10 Determination of the initial rate by the tangent method. (A) The curved, thick lines represent real progress curves (1, 2, 3...). The straight thin lines are the true initial rates of the progress curves at $t = 0$ (1', 2', 3' ...). The shaded area is the period between initiation of the reaction at $t = 0$ and start of the registration (start). The dotted lines are the tangents to the progress curves at the start of registration (1", 2", 3"...). (B) Plot of the true initial rates v_0 and of tangents to the curves at the start of registration versus substrate concentration.

od). The advantage of this procedure is that influences of the arising product, like product inhibition and reverse reaction, can be disregarded. However, apart from the uncertainty as to whether the steady-state equation is valid even for the non-linear initial region, the reaction starts actually at the moment when enzyme is added to the test assay, however it will take some time (at least several seconds) for mixing, starting the (e.g. photometric) measurement and the registration, and then the rate measured is no longer an initial reaction rate. This causes an underestimation of the reaction rate, and the failure becomes more severe, the lower the initial substrate concentrations. The analysis of such data may result in sigmoidal instead of hyperbolic curves (Fig. 2.10 B).

In such cases, the enzyme amount should be reduced. This improves steady-state conditions and time progression and thus the linear region is extended. On the other hand, the reaction rate will become slower upon reducing the enzyme concentration, and to compensate for this the sensitivity of the measuring method must be increased. If, however, the velocity falls below the detection level, the initial reaction rate can be estimated by graphical methods. Before discussing this topic, an apparent contradiction should be mentioned. The Michaelis-Menten equation was derived assuming the steady-state condition $d[EA]/dt=0$. The derivation was independent of substrate concentration and should be valid as long as the condition $[E]_0 \ll [A]_0$ holds. This condition often leads to the misinterpretation that the enzyme must be saturated to guarantee the substrate surplus and hence linearity of progress curves may only be expected at substrate saturation. This assumption overlooks the fact that surplus of substrate does not automatically mean enzyme saturation, rather, saturation depends on the equilibrium constant. For example, for an enzyme concentration of $[E]_0 = 10^{-9}$ M a substrate concentration of 10^{-6} M means a 1000-fold surplus. If $K_m = 10^{-5}$ M, the amount of the enzyme–substrate complex according to $K_d = [A][E]/[EA]$ will be $[EA] = 10^{-10}$, i.e. the enzyme is only saturated by 10%.

2.3.2.3 Graphic Methods for Rate Determination

An improvement to the tangent method was described by Lee and Wilson (1971). Two random points on the progress curve (e.g. 0 and 30% reaction turnover) are connected by a straight line. Its slope correlates with the reaction rate at the substrate concentration of the average value of both points (here 15%). A further improvement is the *secant method* by Waley (1981). Here also a connecting line is drawn between two points of a progress curve $[A]_1$, t_1, and $[A]_2$, t_2 (Fig. 2.11 A). Its slope $([A]_1-[A]_2)/(t_1-t_2)$ corresponds to the slope of a tangent to the curve and thus to the rate for a third substrate concentration $[A]_3$ with the value

$$[A]_3 = \frac{[A]_1 - [A]_2}{\ln \frac{[A]_1}{[A]_2}} \qquad (2.38\,a)$$

respectively

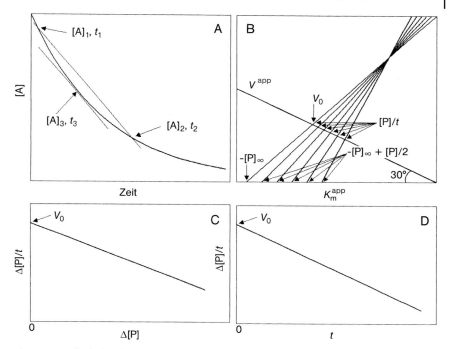

Fig. 2.11 Methods for determination of the actual and the true initial reaction rate. (A) Secant method (Waley 1981), (B) direct linear plot (Cornish-Bowden 1975), (C) method of Boeker (1982), (D) method of Alberty and Koerber (1957).

$$[P]_3 = [A]_0 - \frac{[P]_2 - [P]_1}{\ln\frac{[A]_0 - [P]_1}{[A]_0 - [P]_2}}, \qquad (2.38\,\text{b})$$

if the formation of product is recorded. The concentration of [A]₃ can be obtained from the absorption values A at corresponding times with photometric measurements, A_0 and A_∞ being the absorption values at the start and the end of the reaction:

$$[A]_3 = \frac{[A]_0}{A_0 - A_\infty} \cdot \frac{A_1 - A_2}{\ln\frac{A_1 - A_\infty}{A_2 - A_\infty}}. \qquad (2.38\,\text{c})$$

v can be determined with the secant method even with single point measuring, the slope of the connecting line between the start and the measuring point of the reaction being determined. The method requires the absence of product inhibition, but even in its presence v can be obtained by a simple modification, as long as this inhibition is competitive, which mostly applies. In this case two secants are drawn, one e.g. from 0–20%, the other from 0–40%. Both slopes correlate to two rates v_1 and v_2 with two corresponding substrate concentrations $[A]_1$ and $[A]_2$. The actual initial rate is obtained from the equation:

$$v = \frac{[A]_0([A]_1 - [A]_2)}{[A]_1[A]_2\left(\dfrac{1}{v_2} - \dfrac{1}{v_2}\right) + [A]_0\left(\dfrac{[A]_1}{v_1} - \dfrac{[A]_2}{v_2}\right)} \; . \tag{2.39}$$

Conversely the existence of product inhibition can be detected by comparing the slopes of the two secants. The ratio v/v' lies between 1 and 1.145 for an uninhibited reaction, in the presence of product inhibition it is higher, up to a maximum of 2.2.

2.3.2.4 Graphic Determination of True Initial Rates

While the methods discussed so far serve to show the determination of turnover rates with the aid of tangents, the method of Cornish-Bowden (1975), based on the integrated Michaelis-Menten equation, permits the determination of the true initial rate v_0 at $t=0$. The constants K_m and V in the integrated Michaelis-Menten equation (Eq. (2.33)) are only identical with those of the original Michaelis-Menten equation (Eq. (2.18)) if this equation is valid for the complete progress curve. This holds, however, only for irreversible single-substrate reactions. With more complex mechanisms, like product inhibition, reversibility of reactions or two or more substrates K_m and V do not remain constant and are termed apparent constants (K_m^{app}, V^{app}). Equation (2.33) then becomes

$$V^{app} = \frac{[P]}{t} + \frac{K_m^{app}}{t}\ln\frac{[P]_\infty}{[P]_\infty - [P]} \; . \tag{2.40}$$

[P] being the product concentration at time t and $[P]_\infty$ at the end of the reaction, at equilibrium. In a plot of V^{app} against K_m^{app}, straight lines with ordinate intercepts $[P]/t$ and abscissa intercepts $-[P]/\ln[P]_\infty/([P]_\infty-[P])$ are obtained. Actually the procedure is reversed. The axis intercept values are obtained from the progress curve and entered onto the corresponding axes. Without severe loss of accuracy the abscissa value can be simplified to $-[P]_\infty + 1/2[P]$. The corresponding axis sections are connected by straight lines in a direct linear plot and meet in a point to the right of the ordinate (Fig. 2.11 B). The connecting straight line between this point and the abscissa value $[P]_\infty$ intersects the ordinate at V^{app} $[P]_\infty/(K_m^{app}+[P]_\infty)$, the true initial rate v_0 of the reaction. As the slopes of the straight lines taken from a distinct progress curve must not differ significantly, the joint intercept is frequently difficult to detect. It becomes more visible if the ordinate is inclined towards the abscissa at an angle of 20° to 30°, which does not affect the result. For the determination of v_0 five straight lines are sufficient. Under experimental conditions frequently an intercept cloud results instead of the expected joint intercept. The connecting straight line may then be drawn through the cloud center and still gives sufficient accuracy. For complete accuracy, connecting lines must be drawn from $[P]_\infty$ to each individual intercept and the median of all ordinate intercepts should be calculated.

The method of Boeker (1982) is also based on the integrated Michaelis-Menten equation. Here $\Delta[P]/t$ is plotted against $\Delta[P]$. $\Delta[P]=[P]-[P]_0$ is the product formed during reaction ($[P]_0$ not necessarily being zero). The plot gives a linear relationship, v_0 being obtained by extrapolation to $\Delta[P]_0$ (Fig. 2.11 C). $[P]_\infty$ need not be known. In the related method of Alberty and Koerber (1957) $\Delta[P]/t$ is plotted against t and v_0 is obtained by extrapolation to $t=0$ (Fig. 2.11 D).

The general problem of the analysis of progress curves with the integrated Michaelis-Menten equation in the presence of product inhibition will be discussed in Section 2.4.3.

2.4
Reversible Enzyme Reactions

2.4.1
Rate Equation for Reversible Enzyme Reactions

The Michaelis-Menten equation was derived for an irreversible enzyme reaction, where substrates are transformed into products, but products cannot be converted to substrates. Although in terms of thermodynamics every reaction must be regarded as reversible, for a great number of enzyme reactions, like phosphatases or peptidases, and also the invertase reaction, which served as model for the Michaelis-Menten equation, the back reaction is grossly disfavored by the energetic barrier and it is justified, for simplicity, to take such a reaction as quasi-irreversible. However, the majority of all enzyme reactions, like isomerases, dehydrogenases or transaminases, are fully reversible. The substrates will not be completely converted to product, rather the system strives for an equilibrium state, which can similarly be achieved whether the reaction starts with substrates or with products. Therefore, the original Michaelis-Menten equation will be of limited value if it is restricted to irreversible reactions.

To derive a rate equation for a reversible enzyme reaction according to the scheme:

$$A + E \underset{k_{-1}}{\overset{k_1}{\rightleftarrows}} EA \underset{k_{-2}}{\overset{k_2}{\rightleftarrows}} E + P \tag{2.41}$$

the reverse reaction must be considered in the differential equations for E and EA, and in the expression for the velocity v:

$$\frac{d[E]}{dt} = (k_{-1} + k_2)[EA] - (k_1[A] + k_{-2}[P])[E] , \tag{2.42}$$

$$\frac{d[EA]}{dt} = -(k_{-1} + k_2)[EA] + (k_1[A] + k_{-2}[P])[E] , \tag{2.43}$$

$$\frac{d[P]}{dt} = k_2[EA] - k_{-2}[E][P] = v \,. \tag{2.44}$$

From Eqs. (2.42) and (2.43) [E] and [EA] can be calculated applying the relationship for the total enzyme amount $[E]_0 = [E] + [EA]$:

$$[E] = \frac{(k_{-1} + k_2)[E]_0}{k_1[A] + k_{-2}[P] + (k_{-1} + k_2)} \,, \tag{2.45}$$

$$[EA] = \frac{(k_1[A] + k_{-2}[P])[E]_0}{k_1[A] + k_{-2}[P] + (k_{-1} + k_2)} \,. \tag{2.46}$$

Insertion into Eq. (2.44) gives the rate equation in the form of the rate constant:

$$v = \frac{(k_1 k_2 [A] + k_{-1} k_{-2} [P])[E]_0}{k_1[A] + k_{-2}[P] + (k_{-1} + k_2)} \,. \tag{2.47}$$

Since rate constants are difficult to determine, Eq. (2.47) is converted into the form of the kinetic constant by multiplying numerator and denominator by $(k_{-1}+k_2)/k_1 k_{-2}$. K_{mA} and V_1 are the Michaelis constant and the maximum velocity for the forward reaction, K_{mP} and V_2 the respective constants for the reverse reaction:

$$K_{mA} = \frac{k_{-1} + k_2}{k_1} \,; \quad K_{mP} = \frac{k_{-1} + k_2}{k_{-2}} \,;$$

$$V_1 = k_2[E]_0 \,; \quad V_2 = k_{-1}[E]_0 \,;$$

$$v = v_1 - v_2 = \frac{K_{mP} V_1 [A] - K_{mA} V_2 [P]}{K_{mA} K_{mP} + K_{mP}[A] + K_{mA}[P]} \,. \tag{2.48}$$

Equation (2.48) includes the Michaelis-Menten equations for both the forward and the reverse reaction, which will become obvious if [P] or [A] is taken as zero:

$$v_1 = \frac{V_1[A]}{K_{mA} + [A]} \,; \quad v_2 = \frac{V_2[P]}{K_{mP} + [P]} \,. \tag{2.18}$$

This shows that the Michaelis-Menten equation is valid even in the presence of a reverse reaction, as long as the product is negligible. However, during the reaction the product concentration will steadily rise, and the commencing reverse reaction will increasingly affect the reaction rate. Thus the initial slope of the progress curve will be the same as for an irreversible reaction, as long as [P] can be considered as zero. However, in comparison to the original Michaelis-Menten equation the progress curve will deviate earlier from linearity (zero order conditions), when [P] accumulates in a high enough amount to support the reverse reaction. This demonstrates, on the one hand, the universality of the Michaelis-Menten equation, being applicable even for reverse reactions, on the other hand its validity range is restricted to a narrower time period for [P] ≈ 0. Therefore enzyme kinetic studies must concentrate strictly on initial rates of the progress curve at time t_0.

The same consideration is valid for the reverse reaction when starting with product, which now acts as substrate and, in this case, the substrate concentration must be zero at t_0. In this manner a reverse reaction obeying Eq. (2.48) can be analysed, by applying the simple Michaelis-Menten equations (Eq. (2.18)) from both the substrate and the product side. By this means all four kinetic constants, K_{mA}, K_{mP}, V_1, and V_2, and, with their knowledge, also the four rate constants can be determined:

$$k_1 = \frac{V_1 + V_2}{K_{mA}[E]_0}; \quad k_{-1} = \frac{V_2}{[E]_0}; \quad k_2 = \frac{V_1}{[E]_0}; \quad k_{-2} = \frac{V_1 + V_2}{K_{mP}[E]_0}.$$

Obviously from the initial rates of the forward and reverse reaction much information can be obtained. However, the increase in the reverse reaction alters the course of the progress curves and the linearization methods based on the integrated Michaelis-Menten equation (Eqs. (2.34)–(2.36)) no longer yield straight lines. For the analysis of reverse reactions Eq. (2.48) must be integrated, giving the relationship:

$$\frac{[A]_0 - [A]}{t} = \frac{K_{mP} V_1 + K_{mA} V_2}{K_{mP} - K_{mA}} - \frac{K_m K_{mP}}{K_{mP} - K_{mA}}$$
$$\times \left(1 + \frac{(V_1 + V_2)[A]_0}{K_{mP} V_1 + K_{mA} V_2}\right) \frac{\ln\left(1 - \frac{[A]_0 - [A]}{[P]_e}\right)}{t}. \quad (2.49)$$

Straight lines are now also obtained for reversible reactions, if $([A]_0-[A])/t$ is plotted against $\ln\{1-([A]_0-[A])/[P]_e\}/t$. This treatment, however, requires knowledge of $[P]_e$, the product concentration at the equilibrium of the reaction. If $[P]_e$ or V_2 are unknown, K_{mA} and V_1 can be obtained from the starting region ($[P]=0$) by plotting $([A]_0-[A])/t$ against $\ln([A]_0/[A])/t$ according to Foster and Niemann (Section 2.4.3) at various $[A]_0$ values, and K_{mP} and V_2 are obtained similarly from the reverse reaction. This approach, however, resembles the common method of determination of initial rates at different substrate concentrations, without applying the integrated Michaelis-Menten equation.

2.4.2
The Haldane Relationship

Unlike irreversible reactions, in reversible ones the substrate will only be converted until equilibrium is reached, where the turnover rates in both directions compensate each other and the overall reaction rate becomes $v=0$. In this case the denominators in Eqs. (2.47) and (2.48) are cancelled:

$$k_1 k_2 [A]_e = k_{-1} k_{-2} [P]_e \quad (2.50)$$

$$K_{mP} V_1 [A]_e = K_{mA} V_2 [P]_e \ . \tag{2.51}$$

$[A]_e$ and $[P]_e$ are substrate and product concentrations at equilibrium. Applying Eqs. (2.50) and (2.51) the equilibrium constant K_e for the whole reaction can be written as:

$$K_e = \frac{[E]_e [P]_e}{[E]_e [A]_e} = \frac{[P]_e}{[A]_e} = \frac{k_1 k_2}{k_{-1} k_{-2}} = \frac{K_P}{K_A} = \frac{K_{mP} V_1}{K_{mA} V_2} \ . \tag{2.52}$$

K_e is the ratio of the equilibrium concentrations of the reactants, as well as the ratio of the dissociation constants for the enzyme–product and the enzyme–substrate complexes $K_P = k_2/k_{-2}$ and $K_A = k_{-1}/k_1$, as well as the ratio of rate constants for the forward and the back reaction. J. B. S. Haldane showed that there also exist relationships between the kinetic constants and the thermodynamic equilibrium constants and that Michaelis constants and maximum velocities of forward and reverse reactions depend on each other. To guide a reaction into the direction of product ($K_e \gg 1$), conditions must be: $V_1 \gg V_2$, respectively $K_{mA} \ll K_{mP}$, while in the reverse case the formation of substrate will be preferred. If, for example by site-directed mutation, the specificity of an enzyme for its substrate and, thus, the K_{mA} value is altered, then inevitably another constant must change, e.g. K_{mP}, to keep K_e constant.

2.4.3
Product Inhibition

A further inevitable consequence of reversibility is the inhibition of the enzyme by its own product. The original derivation of the Michaelis-Menten equation for an irreversible reaction assumes only a binding equilibrium for the substrate. However, the product formed is also bound to the enzyme, even if the reverse reaction is negligible, and it will block the binding of a new substrate molecule as long as it remains bound to the enzyme. Thus product inhibition must be expected for each enzyme, only in case of weak product binding will it not be very pronounced. During the course of the reaction the inhibitory effect is not constant but increases with product formation. To analyze this inhibition, product is added in a certain concentration [P] when the reaction is started by addition of substrate. Assuming the reverse reaction to be negligible, in Eq. (2.48), $V_2 = k_{-1}[E]_0 = 0$, and, for the same reason, the Michaelis constant of product K_{mP} reduces to the binding constant $K_P = k_2/k_{-2}$ which then has the significance of an inhibition constant:

$$v = \frac{K_P V_1 [A]}{K_{mA} K_P + [A] K_P + [P] K_{mA}} = \frac{V_1 [A]}{K_{mA}\left(1 + \frac{[P]}{K_P}\right) + [A]} \ . \tag{2.53}$$

In its double-reciprocal form the equation is:

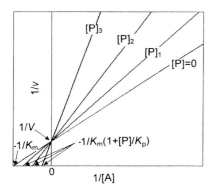

Fig. 2.12 Competitive product inhibition in a double-reciprocal plot. The mode of determination of the kinetic constants is indicated.

$$\frac{1}{v} = \frac{1}{V_1} + \frac{K_{mA}\left(1 + \frac{[P]}{K_P}\right)}{V_1[A]}.$$ (2.54)

In the double reciprocal diagram (Fig. 2.12), a group of straight lines is obtained for different product concentrations, with a common ordinate intercept at $1/V$ and the abscissa intercepts $-1/K_{mA}(1+[P]/K_P)$. With K_{mA} known (determined in the absence of product), K_P can be calculated. The common ordinate intercept indicates that V will not be changed by this inhibition, i.e. at infinite concentration the substrate will overcome the inhibitory effect of the product, because product will be completely displaced by the great surplus of substrate molecules. This is a *competitive product inhibition*, i.e. product competes for the substrate binding site at the catalytic center of the enzyme.

To derive Eq. (2.53) for product inhibition it has been assumed that the product concentration will remain constant, at least during the short time of determination of the initial rate. However, unlike other inhibitors, which remain really constant, the product increases continuously during the course of the reaction. This becomes particularly evident if, instead of initial rates, complete progress curves are evaluated, applying the integrated Michaelis-Menten equation, as discussed in Section 2.3.1.5. Since product inhibition is expected with any enzyme reaction, these derivations can only be used with the assumption of negligible product inhibition. However, for a more general treatment product inhibition should be included (Foster and Niemann 1953):

$$\frac{[A]_0 - [A]}{t} = \frac{VK_P}{K_P - K_{mA}} - \frac{K_{mA}(K_P + [A]_0)}{K_P - K_{mA}} \cdot \frac{\ln\frac{[A]_0}{[A]}}{t}.$$ (2.55)

Different initial concentrations of $[A]_0$ yield a group of straight lines with a common ordinate intercept at $\pm V/(1-K_{mA}/K_P)$, when $([A]_0-[A])/t$ is plotted against $\ln([A]_0/[A])/t$. The lines show a negative slope for $K_{mA} < K_P$ (Fig. 2.13), and a positive slope for $K_{mA} > K_P$. For $K_{mA} = K_P$ the straight lines form a right

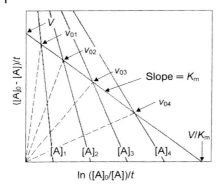

Fig. 2.13 Evaluation of a progress curve in the presence of product inhibition with the integrated Michaelis-Menten equation according to the method of Foster and Niemann (1953).

angle with the abscissa. Source lines with the slope of the respective initial substrate concentration $[A]_0$ cut the straight progress lines at the y-coordinate v_0, the real initial rate at $t=0$. The intercepts of the source lines with the progress lines lie on a straight line, as shown in Fig. 2.13, representing an ideal progress curve in the absence of product inhibition. Its slope is K_{mA}, the ordinate intercept V and the abscissa intercept V/K_{mA}. Knowing these constants, K_P can be calculated. The value of the slopes of the individual progress curves is $K_{mA} (K_P+[A]_0)/(K_P-K_{mA})$ according to Eq. (2.55). By replotting the slopes against $[A]_0$ the abscissa intercept is $-K_P$.

Competitive product inhibition is the most common type of product inhibition, but product can also affect enzyme reactions in other ways. In reactions with more than one substrate, the product will displace only the substrate from which it directly emerged, but not the cosubstrate. Therefore other inhibition types, *non-competitive* and *uncompetitive product inhibition*, are also observed. They are described in Section 2.5.1.

Product inhibition should not be confused with *feedback or end product inhibition*. This is a general regulation principle in metabolic pathways where the final product of the chain inhibits the activity of the first enzyme and switches off the whole pathway to prevent accumulation of intermediates. A prominent example is aspartate transcarbamoylase, the first enzyme in the pyrimidine biosynthetic pathway, which is inhibited by the end product CTP. Due to the multiple steps of the chain, the end product is structurally completely different from the substrate or product of the first enzyme, and will no longer be recognized by the catalytic center. Therefore the enzyme possesses a separate *allosteric* regulatory site, specific for the feed back inhibitor. Binding of the inhibitor to this site influences indirectly (e.g. by conformational change) the catalytic center (Section 1.5.4).

2.5
Enzyme Inhibition

2.5.1
Unspecific Enzyme Inhibition

When, upon adding a substance to an enzyme assay, a decrease in the velocity of the reaction is observed, the observation will be ascribed as enzyme inhibition and the respective substrate as an inhibitor, although this need not be the case. Enzyme inhibition, in its strict sense, is defined as a reduction in enzyme activity caused by specific binding of a ligand (inhibitor) to a defined binding site at the enzyme, like a catalytic or regulatory center. Reduction of the reaction velocity, however, can also be caused by other factors. Decrease in temperature, change in the pH value, the ion strength or the polarity of the solvent will affect the enzyme activity. Usually, such effects can be avoided under controlled assay conditions, however, they may be an unrecognized side effect of the added substance. If this substance is applied in an acid or alkaline form, if it is precooled, or if it contains e.g. stabilizing additives, a decrease in reaction velocity will be observed upon addition, and this decrease will apparently depend on the concentration of the substance, higher amounts producing larger effects, although being completely independent of the respective substance.

Even if the substance really reacts with the enzyme, the interaction need not be specific, rather it may be due to unspecific surface effects, which may disturb the native enzyme structure (Fig. 2.14). Charged compounds, like metal ions, but also anions, will displace counter-ions of charged groups at the enzyme surface. Hydrophobic compounds, and especially detergents like dodecyl sulphate, will react with hydrophobic protein regions and chaotropic substances e.g. urea or ammonium sulfate will disturb the hydrate shell. All these effects cannot be

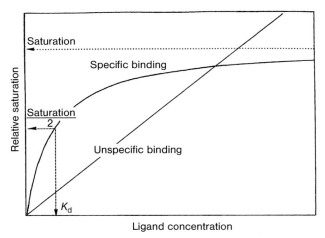

Fig. 2.14 Comparison of specific and unspecific ligand binding.

regarded as specific inhibition. In experiments they can be recognized and differentiated from specific inhibition by the following features:
- no defined saturation level
- no direct competition with analogous substances
- high concentrations (<1 mM) required.

2.5.2
Irreversible Enzyme Inhibition

2.5.2.1 General Features of Irreversible Enzyme Inhibition
Usually enzyme inhibition is assumed to be reversible: a substance, e.g. a structural analog of the substrate, binds to a distinct site, like the active center in a reversible manner and dissociates upon dilution or becomes displaced by high amounts of the genuine substrate. However, many substances bind to the enzyme in an irreversible manner. Principally, irreversible binding can be unspecific or to a specific site and especially for the latter case two types of irreversible binding must be differentiated:
- covalent binding
- extremely tight, non-covalent binding ($K_d < 10^{-8}$ M).

It is not in any way obvious, whether a particular substance reacts covalently or in a non-covalent manner, although, from the chemical view point, this should be easy to differentiate. For example, halogen derivatives are frequently applied as analogs for ligands, like substrates, because modifications with chloride or fluoride produce no severe change in the overall structure, so that the analog will still be accepted by the enzyme, but often loses its function as a substrate. Moreover, halogen derivatives are reactive compounds, which may react with functional groups, like amino groups. This reaction can occur either at the specific binding site, when the substance acts as a ligand analog and will inhibit the respective enzyme function or it can react with any accessible reactive group on the protein surface. In the latter case the inhibitory effect depends on the extent of structural alterations and may be regarded as unspecific.

Because irreversible and reversible inhibition must be treated differently (see below), before undertaking detailed inhibition studies, the respective type of inhibition should be examined. Any method of separating the reversible binding ligand, such as dialysis, ultrafiltration or gel filtration, can be applied and the original enzyme activity should be recovered, in contrast to irreversible binding where the enzyme remains inactive. It must, however, be considered that the loss in activity can be caused by the separation method and it will not always be clear whether such losses must be ascribed to the influence of the method or the inhibitor.

A rapid test for irreversible binding is dilution of the enzyme. Usually the enzyme activity will decrease according to the dilution factor, and this will be the case also if an irreversible inhibitor is bound to part of the enzyme molecules. A reversible inhibitor, however, will dissociate and the enzyme will show a higher activity than expected by mere dilution.

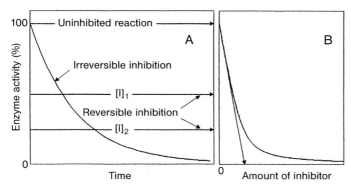

Fig. 2.15 Time course of an irreversible and a reversible inhibition (A) and determination of the share of irreversibly bound inhibitor (B).

A convenient and reliable test is the determination of the time dependence of the inhibition. Due to the rapid binding equilibrium a reversible inhibitor will bind to and inhibit the enzyme immediately and the degree of inhibition will remain constant if tested during a distinct time period, e.g. 1 h (provided that the enzyme remains generally stable during this time). In contrast to this behavior, the enzyme activity will decrease in an exponential (pseudo-first order) manner with the irreversible inhibitor (Fig. 2.15 A). This is because the reaction is intrinsically second order and, therefore, dependent on the concentration of the reactants, enzyme and inhibitor. To visualize this effect the inhibitor concentration must be very low. For an efficient irreversible inhibitor one molecule is enough to react with and inactivate one enzyme molecule, so the stoichiometric amount of the enzyme concentration would be sufficient to inactivate the enzyme completely. This test can also be used to determine the stoichiometry of binding of the inhibitor to the enzyme. In the case of strong binding the enzyme activity declines nearly linearly in the lower and middle concentration range of the inhibitor. The share of inhibitor bound per added enzyme is obtained by extrapolation to the abscissa (Fig. 2.15 B).

There exist two essential types of irreversible inhibition:
- suicide substrates
- transition state analogs.

Both are valuable tools for the study of enzyme mechanisms and some prominent examples for both types are presented.

2.5.2.2 Suicide Substrates

Suicide substrates are substrate analogs which are able to take part in the catalytic mechanism but, instead of being converted to product and released, a covalent linkage is formed between the substrate and the catalytic site which irreversibly

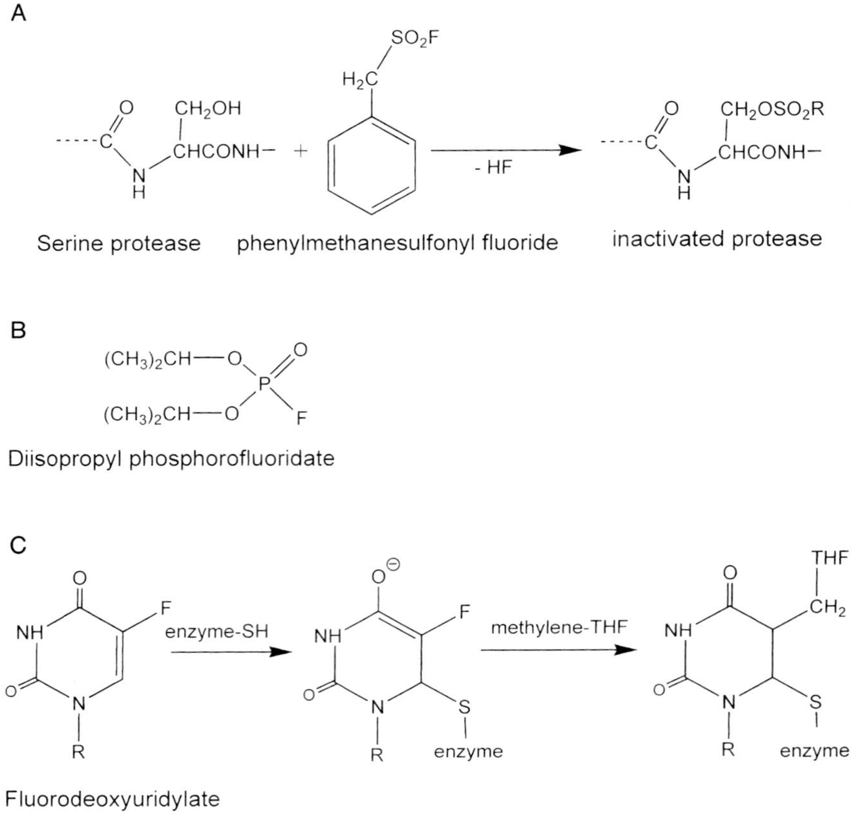

Fig. 2.16 Suicide inhibitors. (A) Reaction of a serine protease with phenylmethanesulfonyl fluoride, R=phenyl–CH$_2$–, (B) diisopropyl phosphorofluoridate, (C) Reaction of fluorodeoxyuridylate with thymidylate synthase and methylene tetrahydrofolate, R=deoxyribosylmonophosphate.

blocks the enzyme, the enzyme kills itself by accepting these inhibitors. Because the covalent binding must be activated and closed by the enzyme, these inhibitors will help to elucidate the catalytic mechanism. Striking examples are inhibitors of serine proteases, which are used to inactivate them in order to avoid degradation of enzymes, e.g. by isolation procedures. Figure 2.16 A shows the reaction scheme for the frequently used inhibitor phenylmethanesulfonyl fluoride (PMSF). It reacts with a half life of 1 h in neutral aqueous solution and must, therefore, be applied freshly. A more efficient, but very toxic reagent is diisopropyl phosphorofluoridate (DFP), which also inhibits the acetylcholine esterase and has been used as a warfare agent (Fig. 2.16 B).

Another prominent suicide inhibitor is fluorouracil, which is converted in the cell to 5-fluorodeoxyuridylate and reacts covalently with both the thymidylate synthase and the cofactor methylene tetrahydrofolate (Fig. 2.16 C). Because it inhibits DNA synthesis fluorouracil is applied as a drug in cancer therapy.

2.5.2.3
Transition State Analogs

In the course of the catalytic mechanism the substrate passes a transition state, where it is bound very tightly and its configuration becomes influenced by interaction with groups of the catalytic center in order to reduce the activation energy and to force the molecule into the direction of the product. Under normal reaction conditions the duration of the transition state is very short, within the range of picoseconds. Analogs, which mimic the transition state, bind very strongly and, even if the binding is not covalent, remain tightly bound and cannot be easily removed by separation methods like dialysis or gel filtration. As an example of a transition state the decarboxylation reaction of the pyruvate dehydrogenase complex is shown (Fig. 2.17 A). The substrate pyruvate binds to the C2-position of the thiazolium ring of the cofactor thiamin diphosphate, forming an ylide structure. Carbon dioxide will be released from the unstable intermedi-

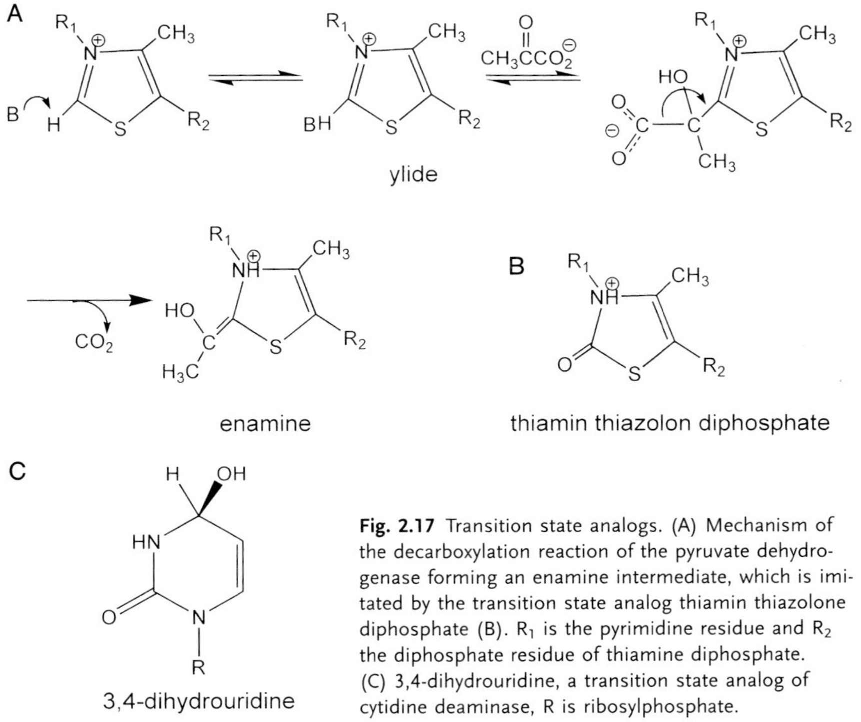

Fig. 2.17 Transition state analogs. (A) Mechanism of the decarboxylation reaction of the pyruvate dehydrogenase forming an enamine intermediate, which is imitated by the transition state analog thiamin thiazolone diphosphate (B). R_1 is the pyrimidine residue and R_2 the diphosphate residue of thiamine diphosphate. (C) 3,4-dihydrouridine, a transition state analog of cytidine deaminase, R is ribosylphosphate.

ate, while an enamine residue remains bound to the cofactor. Thiamine thiazolone diphosphate is a derivative of the cofactor with an oxygen bound to the C2-position (Fig. 2.17 B). It has a binding constant four orders of magnitude lower than the physiological cofactor and will not be removed from the enzyme even after 40 h dialysis. As a further example, 3,4-dihydrouridine, a transition state analog of cytidine deaminase, binds 2×10^8 times more tightly than the substrate cytidine (Fig. 2.17 C).

2.5.2.4 Analysis of Irreversible Inhibitions

If the inhibition has been identified to be of the irreversible type, the usual analysis for reversible inhibition mechanisms cannot be applied, because binding occurs in at least two steps, a fast initial step to a reversible association complex EI, followed by a slower step of irreversible reaction of the inhibitors with the enzyme to form an inactive EI_i complex:

$$E + I \underset{k_{-1}}{\overset{k_1}{\rightleftharpoons}} EI \overset{k_2}{\longrightarrow} EI_i \ .$$

This scheme resembles that of the irreversible enzyme reaction applied for the derivation of the Michaelis-Menten equation (Eq. (2.18)) and, therefore, also the derivation is similar for both cases. The total enzyme amount $[E]_0$ will be distributed between the three enzyme forms:

$$[E]_0 = [E] + [EI] + [EI]_i = [E]_a + [EI]_i \ . \tag{2.56}$$

Both [E] and [EI] can be regarded as active enzyme forms $[E]_a$, because the surplus of substrate applied in the enzyme test will displace the inhibitor from [EI]. In analogy to the Michaelis-Menten equation it is assumed that $[I] \gg [E]_0$, and under this condition the time-dependent formation of the inactive enzyme form [EI] is directly proportional to the concentration of the reversible [EI] complex:

$$\frac{d[EI]_i}{dt} = \frac{d([E]_0 - [E]_a)}{dt} = k_2[EI] \ .$$

The total enzyme amount $[E]_0$ remains constant, $d[E]_0/dt = 0$:

$$-\frac{d[E]_a}{dt} = k_2[EI] \ . \tag{2.57}$$

[EI] can be replaced from Eq. (2.56), considering the inhibition constant for the reversible binding of inhibitor, $K_i = [E][I]/[EI]$:

$$[E]_a = [EI] + [E] = [EI] + \frac{[EI]K_i}{[I]}$$

thus [EI] is eliminated from Eq. (2.57):

2.5 Enzyme Inhibition

$$\frac{d[E]_a}{dt} = \frac{k_2[E]_a}{1 + \frac{K_i}{[I]}}$$

Integration from $[E]_0$ at time $t=0$ to $[E]_a$ at time t:

$$-\int_{[E]_0}^{[E]_a} \frac{d[E]_a}{[E]_a} = \int_{t=0}^{t} \frac{k_2 dt}{1 + \frac{K_i}{[I]}}$$

results in:

$$\ln \frac{[E]_a}{[E]_0} = -\frac{k_2 t}{1 + \frac{K_i}{[I]}} \quad (2.58\,a)$$

or in logarithm to the base ten:

$$\log \frac{[E]_a}{[E]_0} = -\frac{k_2 t}{2.3\left(1 + \frac{K_i}{[I]}\right)} \quad (2.58\,b)$$

The time dependence of the enzyme activities in the presence of different inhibitor concentrations is plotted as log $([E]_a/[E]_0)$ against the incubation time t and yields straight lines as shown in Fig. 2.18 A. The expression for the slopes Sl

$$Sl = \frac{k_2}{1 + \frac{K_i}{[I]}} \quad (2.59)$$

plotted reciprocally against the reverse inhibitor concentrations yield a linear function:

$$\frac{1}{Sl} = \frac{1}{k_2} + \frac{K_i}{k_2[I]} \quad (2.60)$$

with $1/k_2$ and $-1/K_i$ as ordinate and abscissa intercepts, respectively (Fig. 2.18 B).

If a second ligand A (e.g. the substrate) competes for the inhibitor binding site, Eq. (2.58) has to be extended by the term $1+[A]/K_A$ in the denominator:

$$\ln \frac{[E]_a}{[E]_0} = \frac{k_2 t}{\left(1 + \frac{K_i}{[I]}\right)\left(1 + \frac{[A]}{K_A}\right)} \quad (2.61)$$

The equation for the slope has then to be extended correspondingly:

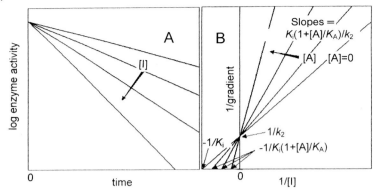

Fig. 2.18 Time dependence of irreversible inhibition. (A) Semi-logarithmic diagram for different inhibitor concentrations [I]. (B) Secondary plot of the slopes from plot (A) for several substrate concentrations [A].

$$\frac{1}{Sl} = \frac{1}{k_2} + \frac{K_i\left(1 + \frac{[A]}{K_A}\right)}{k_2[I]} \ . \tag{2.62}$$

Different straight lines are obtained for varying concentrations of A, as shown in Fig. 2.18 B. K_A can be calculated either from their intercept with the abscissa or be taken from a secondary plot of the slopes of this diagram plotted against [A].

2.5.3
Reversible Enzyme Inhibition

2.5.3.1 General Rate Equation

If there is no indication to the contrary, enzyme inhibition is assumed to be caused by reversible, specific interaction of a distinct substance, the inhibitor, with the enzyme molecule, reducing its activity. There are different possibilities of interaction and, thus, different inhibition mechanisms exist. For each of these mechanisms a special equation, describing the dependence of the enzyme velocity on substrate and inhibitor can be derived, according to the general rules for steady-state equations, as already shown for the simple and the reversible Michaelis-Menten equation (Eqs. (2.18) and (2.48)). Here a differing approach is used. A general inhibition mechanism is defined, including all essential inhibition types, and a general rate equation is derived, comprising all rate equations of the individual inhibition mechanisms. The rate equations of the single mechanisms can be easily obtained by simplifying the general scheme. It may be argued that an arbitrary number of inhibition mechanisms may be defined and a general scheme must include a lot of possibilities. However, only plausible assumptions must be considered. They are described by the following statements:

- The inhibitor binds to the enzyme forming an EI complex,
- The inhibitor can also bind to the enzyme–substrate complex, forming an EAI complex,
- The EAI complex may be inactive or active.

It can be seen that these apparently simple rules comprise all plausible mechanisms, since, for any mechanism, binding must inevitably be the first step, but nothing is stated about the binding site, which may or may not be at the catalytic center, or about the influence on the enzyme velocity, which may be due to direct interaction or mediated via conformational changes. This general mechanism can be formulated by the following reaction scheme:

$$
\begin{array}{ccccc}
E + A & \underset{k_{-1}}{\overset{k_1}{\rightleftharpoons}} & EA & \xrightarrow{k_2} & E + P \\
+ & & + & & \\
I & & I & & \\
k_3 \updownarrow k_{-3} & & k_4 \updownarrow k_{-4} & & \\
EI + A & \underset{k_{-5}}{\overset{k_5}{\rightleftharpoons}} & EAI & \xrightarrow{k_6} & EI + P
\end{array}
\tag{2.63}
$$

The reaction mechanism contains two rate constants, k_2 for the uninhibited reaction, and k_6 for the EAI complex, and two dissociation constants each for the substrate A and for the inhibitor I:

$$K_A = \frac{k_{-1}}{k_1} = \frac{[E][A]}{[EA]} \; ; \quad K_{Ai} = \frac{k_{-5}}{k_5} = \frac{[EI][A]}{[EAI]} \; ;$$

$$K_{ic} = \frac{k_{-3}}{k_3} = \frac{[E][I]}{[EI]} \; ; \quad K_{iu} = \frac{k_{-4}}{k_4} = \frac{[EA][I]}{[EAI]} \; .$$

They are connected by the relationship

$$\frac{K_A}{K_{Ai}} = \frac{K_{ic}}{K_{iu}} , \tag{2.64}$$

so that the fourth constant can be calculated if the other three are known.

The differential equations for the free enzyme and for the enzyme–substrate complex can be taken as zero, assuming steady-state conditions:

$$\frac{d[E]}{dt} = (k_{-1} + k_2)[EA] - (k_1[A] + k_3[I])[E] + k_{-3}[EI] = 0 , \tag{2.65}$$

$$\frac{d[EA]}{dt} = -(k_{-1} + k_2 + k_4)[EA] + k_1[A][E] + k_{-4}[EAI] = 0 . \tag{2.66}$$

The equation for the reaction rate v is:

$$\frac{d[P]}{dt} = k_2[EA] + k_6[EAI] = v . \tag{2.67}$$

2 Enzyme Kinetics

The total amount of the enzyme is the sum of all possible enzyme complexes, whereby the terms for the enzyme–inhibitor complexes can be replaced by the inhibition constants:

$$[E]_0 = [E] + [EA] + [EI] + [EAI] = [E] + [EA] + \frac{[E][I]}{K_{ic}} + \frac{[EA][I]}{K_{iu}}. \tag{2.68}$$

From this an expression for [E] is obtained:

$$[E] = \frac{[E]_0 - [EA]\left(1 + \frac{[I]}{K_{iu}}\right)}{1 + \frac{[I]}{K_{ic}}}. \tag{2.69}$$

In Eq. (2.65) [EI] is replaced by $K_{ic} = k_{-3}/k_3$:

$$-(k_1[A] + k_3[I])[E] + (k_{-1} + k_2)[EA] + k_3[E][I] = 0$$

$$-k_1[A][E] + (k_{-1} + k_2)[EA] = 0$$

and [E] is eliminated applying Eq. (2.69):

$$\frac{(k_{-1} + k_2)[EA]}{k_1[A]} = \frac{[E]_0 - [EA]\left(1 + \frac{[I]}{K_{iu}}\right)}{1 + \frac{[I]}{K_{ic}}}.$$

This gives a term for [EA], setting $K_m = (k_{-1} + k_2)/k_1$:

$$[EA] = \frac{[E]_0}{\frac{K_m}{[A]}\left(1 + \frac{[I]}{K_{ic}}\right) + \left(1 + \frac{[I]}{K_{iu}}\right)}.$$

Inserting this term in the rate equation (2.67), applying $V_1 = k_2[E]_0$, $V_2 = k_6[E]_0$ and replacing [EAI] by K_{iu}, it follows:

$$v = \left(k_2 + \frac{k_6[I]}{K_{iu}}\right)[EA]$$

$$v = \frac{\left(V_1 + \frac{V_2[I]}{K_{iu}}\right)[A]}{K_m\left(1 + \frac{[I]}{K_{ic}}\right) + \left(1 + \frac{[I]}{K_{iu}}\right)[A]}. \tag{2.70}$$

This is the rate equation for the general inhibition scheme (2.63), which also holds for a distinct inhibition mechanism, the *partially non-competitive inhibition*, which will be discussed separately in Section 2.5.3.5. Figure 2.19 gives a survey

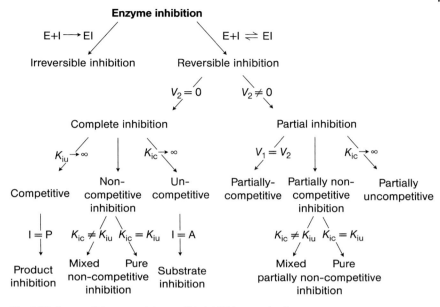

Fig. 2.19 Survey of the essential reversible inhibition mechanisms.

of all essential types of reversible inhibitions, which are included in this scheme. There is a branching point, where it is to be decided whether the enzyme is still active in the presence of inhibitor, such mechanisms are called *partial inhibition*. If binding of the inhibitor completely inactivates the enzyme, *complete* inhibitions result, EAI will be completely inactive ($k_6 \sim V_2 = 0$) and designated as a *dead-end complex*. Different inhibition mechanisms are obtained if the inhibitor is allowed to bind either to both, E and EA, or to only one enzyme form. For the non-binding forms infinite values for the respective binding (inhibition) constants are assumed. In the case of partial inhibitions also the ratios of k_2 to k_6, respectively V_1 to V_2, can vary.

2.5.3.2 Non-competitive Inhibition and Graphic Representation of Inhibition Data

If the EAI complex is assumed to be an inactive dead-end complex, k_6 resp. $V_2 = 0$, the reaction scheme Eq. (2.63) reduces to:

$$
\begin{array}{ccccc}
E + A & \underset{k_{-1}}{\overset{k_1}{\rightleftarrows}} & EA & \overset{k_2}{\rightarrow} & E + P \\
+ & & + & & \\
I & & I & & \\
k_3 \updownarrow k_{-3} & & k_4 \updownarrow k_{-4} & & \\
EI + A & \underset{k_{-5}}{\overset{k_5}{\rightleftarrows}} & EAI & &
\end{array}
\qquad (2.71)
$$

and Eq. (2.70) is simplified to ($V_1 = V$):

$$v = \frac{V[A]}{K_m\left(1 + \frac{[I]}{K_{ic}}\right) + \left(1 + \frac{[I]}{K_{iu}}\right)[A]} \quad . \tag{2.72}$$

The reciprocal form for the Lineweaver-Burk plot is:

$$\frac{1}{v} = \frac{1 + \frac{[I]}{K_{iu}}}{V} + \frac{K_m\left(1 + \frac{[I]}{K_{ic}}\right)}{V[A]} \quad . \tag{2.73}$$

Transformation according to the Hanes plot:

$$\frac{[A]}{v} = \frac{[A]\left(1 + \frac{[I]}{K_{iu}}\right)}{V} + \frac{K_m\left(1 + \frac{[I]}{K_{ic}}\right)}{V} \tag{2.74}$$

and the Eadie-Hofstee plot:

$$v = \frac{V}{1 + \frac{[I]}{K_{iu}}} - \frac{K_m\left(1 + \frac{[I]}{K_{ic}}\right)}{\left(1 + \frac{[I]}{K_{iu}}\right)} \cdot \frac{v}{[A]} \quad . \tag{2.75}$$

This general form of a complete inhibition is termed *non-competitive* (for the notations of inhibition types see Section 2.5.3.7). The apparent Michaelis constant and the maximum velocity are both affected by the inhibitor.

Non-competitive inhibition is the most important type of inhibition for the regulation of the cell metabolism. The enzyme activity can be affected by metabolites without any substrate analogy, as in the case of end product inhibition and with allosteric enzymes (Section 1.5.4). Due to cooperative effects these enzymes often show sigmoidal saturation curves, which cannot be linearized by the conventional plots. This complicates the analysis, the pattern of lines to identify the respective mechanism, i.e. joint intersects (see below) must be derived by extrapolation to infinite substrate concentration. Typically, non-competitive inhibition appears with multiple substrate reactions. An analog of one substrate will act and inhibit the reaction by displacing just this substrate from its binding site (competitive inhibition, Section 2.5.3.3). However, it cannot displace the cosubstrate, which can still bind to the enzyme but, with the inactive analog bound to the first substrate site, the reaction cannot proceed. So the analog will be a non-competitive inhibitor with respect to the cosubstrate. A similar situation arises with product inhibition, where competition is usually observed with the substrate from which the product is directly derived, while it is non-competitive with respect to the cosubstrate (competitive product inhibition, Section 2.4.3). This should only exemplify the principle of non-competitive inhibition. Actually, the situation is more complicated, dependent on the respective multisubstrate mechanisms (see Section 2.6.3).

The different inhibition types, such as non-competitive inhibition, can be identified by an experimental approach, where at first the turnover rates are determined at varying substrate concentrations. From this test series K_m and V will be obtained according to the Michaelis-Menten equation. In a second series the enzyme is tested with the same substrate concentrations, but now in the presence of a constant amount of the inhibitor. Two or three further series are performed in the same way, changing only the inhibitor concentration, which must always be kept constant within the same test series. Substrate and inhibitor concentrations are varied around their respective Michaelis and inhibition constant (a power of 10 below and above). For a normal inhibition mechanism each test series yields a hyperbolic curve by plotting the velocity against the substrate concentration (Fig. 2.20A). In the presence of the inhibitor the curves will become flatter, and even at high substrate concentrations the level of the maximum velocity of the uninhibited curve will not be reached. The relationship between the different curves becomes clearer in the linear diagrams. Even in the presence of the inhibitor linearity is conserved. A joint intercept left of the ordinate in the double-reciprocal plot is indicative of non-competitive inhibition. The lines differ in slope and ordinate intercept, because the inhibitor influences both the apparent Michaelis constant and the maximum velocity (Fig. 2.20B). The relative position of the common intercept of the lines gives additional information about the mechanism. If $K_{ic} = K_{iu}$ (in this case also $K_A = K_{Ai}$) the inhibitor binds with equal affinity to both the free enzyme and the enzyme–substrate complex, i.e. binding of substrate does not influence the inhibitor binding and, vice versa, inhibitor binding will not influence substrate binding. In this case the intercept of all lines is at the abscissa, K_m is not changed in this mechanism. For $K_{ic} < K_{iu}$, the common intercept is located above the abscissa. In this frequently occurring case the inhibitor binds with higher affinity to the free enzyme than to the enzyme–substrate complex, the already bound substrate impedes the binding of the inhibitor, e.g. by steric or electrostatic hindrance. Consequently, the binding of the substrate will also be affected similarly by the already bound inhibitor ($K_A < K_{Ai}$). In more rare cases already bound substrate or inhibitor facilitates the binding of the respective other compound, $K_{ic} > K_{iu}$ and $K_A > K_{Ai}$. The straight line intercept is now located below the abscissa.

Different terms are frequently used for these three cases of non-competitive inhibition. Completely independent binding of substrate and inhibitor with a common abscissa intercept is designated as *pure non-competitive inhibition*, while the other cases are called *mixed inhibition*, because that one with the intercept above the abscissa directs toward (or is "mixed with") the competitive inhibition with a joint ordinate intercept, that with the intercept below the abscissa approaches to (or is "mixed with") the uncompetitive inhibitions, where parallel lines are observed (see Sections 2.5.3.3 and 2.5.3.4). The term mixed inhibition is not quite adequate, because all three cases are really non-competitive inhibitions obeying the reaction mechanism according to Eq. (2.64), while there is a clear distinction between the other inhibition types.

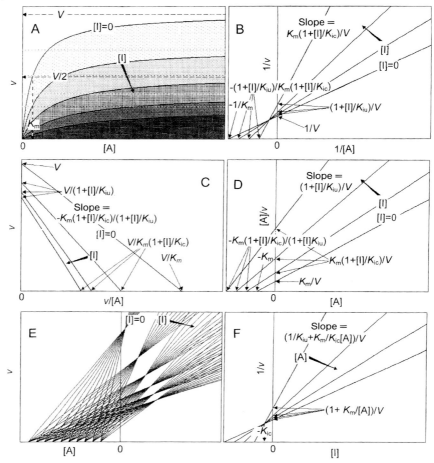

Fig. 2.20 Non-competitive inhibition in different representations. (A) Direct plot, (B) double-reciprocal plot, (C) Eadie-Hofstee plot, (D) Hanes plot, (E) direct linear plot, (F) Dixon inhibition plot (1953). The determinations of kinetic constants are indicated.

The Hanes plot (Fig. 2.20 D) shows the same straight line pattern as the Lineweaver-Burk plot, only the location of the intercepts above or below the abscissa is reversed. In the Eadie-Hofstee plot (Fig. 2.20 C) the straight line intercept for $K_i < K_{iu}$ is located in the upper left, that for $K_{ic} > K_{iu}$ in the lower right quadrant, $K_{ic} = K_{iu}$ yields parallels.

Besides identifying the respective inhibition mechanisms, the linear diagrams can be used to determine the kinetic constants. While K_m and V can be obtained from the uninhibited reactions, the inhibition constants are involved in any inhibitor-dependent change of the lines. From Eq. (2.73) it follows that K_{ic}

enters the slope, and K_{iu} the ordinate intercept of the double-reciprocal diagram. Similar relationships hold for the other diagrams (Eqs. (2.73)–(2.75)). *Secondary plots* (replots) are recommended as a convenient procedure not only for graphical determination of the constants, but also for a further confirmation of the inhibition mechanism. These diagrams can be derived from all three linearization methods, although it is demonstrated here only with the example of the double-reciprocal plot. The features changed by inhibition in the primary plot, e.g. slopes and ordinate intercepts, are plotted against the respective inhibitor concentrations. The inhibition constants can be taken from the resulting straight line. Straight lines, however, are only obtained with complete inhibitions with an inactive dead-end complex, partial inhibitions show non-linear dependences, so that both these reversible inhibition types can be distinguished.

The expression for the slope Sl in Eq. (2.73) for the Lineweaver-Burk diagram is:

$$Sl = \frac{K_m}{V} + \frac{[I]K_m}{K_{ic}V}, \tag{2.76}$$

and for the ordinate intercept Or:

$$Or = \frac{1}{V} + \frac{[I]}{K_{iu}V}. \tag{2.77}$$

Plotting the slopes or the ordinate intercepts of the respective Lineweaver-Burk diagram against the inhibitor concentrations used for each test series should yield straight lines with the abscissa intercepts $-K_{ic}$ and $-K_{iu}$, respectively (Fig. 2.21).

The direct linear plot is an alternative presentation of the enzyme kinetic data, where each pair of variables is drawn as a straight line instead of a point and the constants are obtained from the joint intercept of all lines. A similar

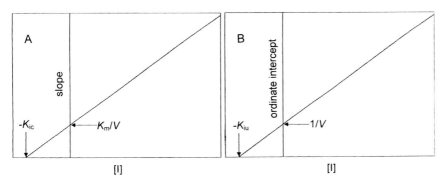

Fig. 2.21 Secondary plots of slopes (A) and ordinate intercepts (B) from a double-reciprocal plot for a non-competitive inhibition.

pattern is obtained in the presence of the inhibitor, only the intercept moves, in a direction characteristic for the respective inhibition mechanism. For the non-competitive inhibition it moves transversely downwards, as can be seen from Fig. 2.20 E.

Transformation of Eq. (2.73)

$$\frac{1}{v} = \frac{1}{V}\left(1 + \frac{K_m}{[A]}\right) + \frac{[I]}{V}\left(\frac{1}{K_{iu}} + \frac{K_m}{K_{ic}[A]}\right) \tag{2.78}$$

reveals a linear dependence of $1/v$ on [I]. For this plot, suggested by Dixon (1953), the inhibitor concentration is varied at several, constant substrate concentrations. This plot serves also to discriminate between complete and partial inhibitions, the former yielding straight lines, the latter non-linear dependences. For the non-competitive inhibition the lines intersect left of the ordinate, for $K_{ic} < K_{iu}$ above, $K_{ic} > K_{iu}$ below, and $K_{ic} = K_{iu}$ at the abscissa, similarly to the double-reciprocal plot. The X-coordinate of the intercept is K_{ic} (Fig. 2.20 F). A disadvantage of this plot is, that both non-competitive (for the case of $K_{ic} < K_{iu}$) and competitive inhibition show the same pattern of an intercept in the second quadrant and cannot be distinguished.

Another method of Dixon (1972), suitable also for K_m determination, only requires the dependence of the turnover rate on the inhibitor concentration at a single (saturating) substrate amount. For this procedure, the respective inhibition type must be known. V' is the uninhibited turnover rate at $[I]=0$ (not identical with the maximum velocity V at infinite substrate concentration). With increasing inhibitor the activity declines in a hyperbolic manner and runs towards the base line at complete inhibition (Fig. 2.22 A). Connecting lines are drawn between V' and the points on the curve $v' = V'(n-1)/n$ (for $n = 1, 2, 3$, etc., i.e. $V'/2, 2V'/3, 3V'/4, 4V'/5$, etc.). They cut the base line at equal distances, with the value of

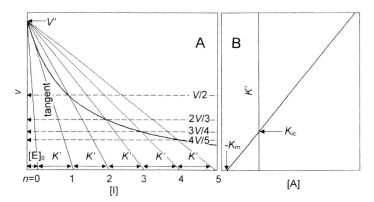

Fig. 2.22 Dixon method for the determination of inhibition and Michaelis constants (A). (B) Secondary plot of the apparent K' values against the substrate concentrations.

K_{ic} in a non-competitive inhibition (referring to the concentration scale of the abscissa). The line for $n=1$ is equal to the source tangent to the curve. If the distance for K_{ic} is once again entered to the left of its intercept, the line for $n=0$ is obtained. The distance from there to the ordinate ($[I]=0$) corresponds to the applied enzyme concentration. For competitive inhibition the intercept distances of the connecting lines with the base line depend on the substrate concentration. In this case the experiment must be performed at varying substrate concentrations. Plotting the distances against substrate amounts results in a straight line intercepting the ordinate ($[A]=0$) at K_{ic} and the abscissa at $-K_m$ (Fig. 2.22 B).

Alternatively inhibition mechanisms can be analyzed by applying the integrated Michaelis-Menten equation. For non-competitive inhibition integration of Eq. (2.72) gives:

$$\frac{K_m\left(1+\dfrac{[I]}{K_{ic}}\right)}{1+\dfrac{[I]}{K_{iu}}} \ln \frac{[A]_0}{[A]} + [A]_0 - [A] = \frac{Vt}{1+\dfrac{[I]}{K_{iu}}}. \tag{2.79}$$

As already described in Section 2.3.2.1 three linear forms can be derived from the integrated Michaelis-Menten equation, also in the extended form of Eq. (2.79). Here only one of them is presented:

$$\frac{[A]_0 - [A]}{t} = \frac{V}{1+\dfrac{[I]}{K_{iu}}} - \frac{K_m\left(1+\dfrac{[I]}{K_{ic}}\right)}{\left(1+\dfrac{[I]}{K_{iu}}\right)} \cdot \frac{\ln \dfrac{[A]_0}{[A]}}{t}. \tag{2.80}$$

Progress curves with a distinct initial substrate concentration and varying inhibitor concentrations yield groups of straight lines with a joint intercept. The ratio of inhibition constants can be read from its location: in the second quadrant $K_{ic} < K_{iu}$ (Fig. 2.23 A), in the fourth quadrant for $K_{ic} > K_{iu}$ (Fig. 2.23 C), and parallels for $K_{ic} = K_{iu}$ (Fig. 2.23 B). The inhibition constants can be obtained from secondary plots (Fig. 2.23 D and E).

2.5.3.3 Competitive Inhibition

This inhibition mechanism assumes that the inhibitor binds exclusively to the free enzyme. The reaction scheme (Eq. (2.71)) for the non-competitive inhibition is reduced correspondingly:

$$\begin{array}{c} E + A \underset{k_{-1}}{\overset{k_1}{\rightleftarrows}} EA \xrightarrow{k_2} E + P \\ + \\ I \\ k_3 \updownarrow k_{-3} \\ EI \end{array} \tag{2.81}$$

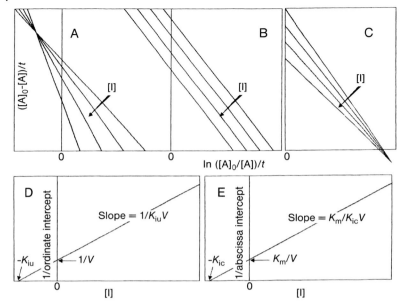

Fig. 2.23 Non-competitive inhibition in linearized plots of the integrated Michaelis-Menten equation. (A) $K_{ic} < K_{iu}$, (B) $K_{ic} = K_{iu}$, (C) $K_{ic} > K_{iu}$, (D) secondary plot of reciprocal ordinate intercepts against inhibitor concentration, (E) secondary plot of reciprocal abscissa intercepts against inhibitor concentration.

The rate equation for non-competitive inhibition (Eq. (2.72)) is simplified by considering $K_{iu} \to \infty$, $[I]/K_{iu}=0$, so that only one inhibition constant, K_{ic}, remains:

$$v = \frac{V[A]}{K_m\left(1 + \dfrac{[I]}{K_{ic}}\right) + [A]} \quad . \tag{2.82}$$

Linearization of Eq. (2.82) according to Lineweaver-Burk gives:

$$\frac{1}{v} = \frac{1}{V} + \frac{K_m\left(1 + \dfrac{[I]}{K_{ic}}\right)}{V[A]} \quad , \tag{2.83}$$

according to Hanes:

$$\frac{[A]}{v} = \frac{[A]}{V} + \frac{K_m\left(1 + \dfrac{[I]}{K_{ic}}\right)}{V} \quad , \tag{2.84}$$

according to Eadie-Hofstee:

$$v = V - K_m \left(1 + \frac{[I]}{K_{ic}}\right) \cdot \frac{v}{[A]} \tag{2.85}$$

and according to Dixon:

$$\frac{1}{v} = \frac{1}{V}\left(1 + \frac{K_m}{[A]}\right) + \frac{[I]K_m}{VK_{ic}[A]} . \tag{2.86}$$

As a consequence of the competition mechanism a great surplus of the substrate will displace the inhibitor from its binding site and, therefore, the maximum velocity V, defined for $[A] \to \infty$, will not be changed in this inhibition. Conversely, large quantities of inhibitor displace the substrate. This feature, noticeable in the double-reciprocal plot by a joint ordinate intercept of all straight lines at $1/V$ (Fig. 2.24 B), is indicative of this type of inhibition. Correspondingly, the hyperbolae in the direct diagram tend towards the same saturation level (Fig. 2.24 A). Only the apparent Michaelis constant is affected by the inhibition and, consequently, those parameters containing K_m are altered in the linearized plots (Fig. 2.24 B–D), e.g. slopes and abscissa intercepts in the Lineweaver-Burk plot. The binding constant for the competitive inhibitor, the *competitive inhibition constant* K_{ic} can be obtained from the abscissa intercept of a secondary plot of the slopes against the inhibitor concentration. In the Dixon plot the straight lines meet in a joint intercept in the second quadrant with $-K_{ic}$ as the X-coordinate (Fig. 2.24 F). The joint intercept of the straight lines in the direct linear plot moves straight towards the right (Fig. 2.24 E).

The integrated Michaelis-Menten equation for a competitive inhibition can be deduced from Eq. (2.72) for the non-competitive inhibition disregarding the terms containing K_{iu}:

$$K_m \left(1 + \frac{[I]}{K_{ic}}\right) \ln \frac{[A]_0}{[A]} + [A]_0 - [A] = Vt . \tag{2.87}$$

In the linear plot obeying equation:

$$\frac{[A]_0 - [A]}{t} = V - K_m\left(1 + \frac{[I]}{K_{ic}}\right) \cdot \frac{\ln \frac{[A]_0}{[A]}}{t} \tag{2.88}$$

straight lines with a common intercept at V are obtained. Plotting the slopes or the reciprocal abscissa intercepts against the inhibitor concentrations in a secondary plot yields $-K_{ic}$ from the abscissa intercept (see Fig. 2.23 E).

The competitive inhibition mechanism presumes alternative binding of substrate and inhibitor, both cannot bind simultaneously. The common interpretation of this mechanism is a competition of both compounds for the same site, e.g. the catalytic center, one displacing the other. The study of competitive inhi-

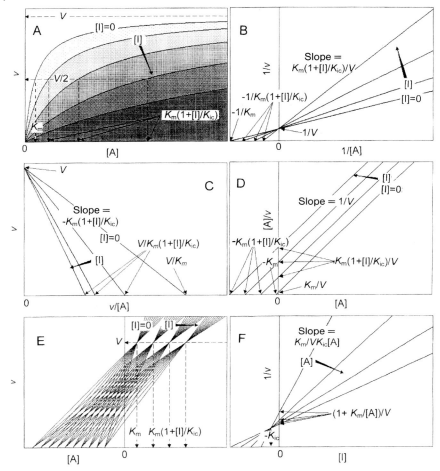

Fig. 2.24 Competitive inhibition in various representations. (A) Direct diagram, (B) Lineweaver-Burk plot, (C) Eadie-Hofstee plot, (D) Hanes plot, (E) direct linear plot, (F) Dixon inhibition plot. Modes of determination of kinetic constants are indicated.

bition with different substrate analogs gives valuable information about enzyme specificity. Competitive inhibitors (antagonists) also serve as targeted blockers of enzyme reactions in therapy. In the regulation of the metabolism competitive inhibition occurs mostly as product inhibition (Section 2.4.3). This prevents accumulation of higher amounts of product, even in the presence of a surplus of substrate.

Although competitive inhibition is usually assumed to require structural analogy between substrate and inhibitor, this inhibition type is sometimes found in apparently unrelated compounds. Cibracon dyes, although structurally completely different, compete for NAD in dehydrogenases. These dyes serve as ligands for the purification of the respective enzymes by affinity chromatography. Conversely, a competitive inhibition pattern is not in any case conclusive for competition. O-phenanthroline shows the pattern of competitive inhibition with respect to NAD in alcohol dehydrogenase, but the inhibition is actually based on complexing essential zinc ions (Bowie and Branden 1977). Interaction of the inhibitor with a regulatory site and strict prevention of substrate binding, caused by a conformational change, can also be described by this mechanism, but this is rather rare. Finally, *partially competitive inhibition* shows the same inhibition pattern in linearized diagrams as true competitive inhibition, although it is based on a completely different mechanism (Section 2.5.3.7).

2.5.3.4 Uncompetitive Inhibition

In this rare inhibition type the inhibitor binds exclusively to the enzyme–substrate complex. Such a mechanism will be realized when the binding site for the inhibitor is only formed in interaction with substrate:

$$E + A \underset{k_{-1}}{\overset{k_1}{\rightleftarrows}} EA \xrightarrow{k_2} E + P$$
$$+$$
$$I$$
$$k_4 \updownarrow k_{-4}$$
$$EAI \tag{2.89}$$

In Eq. (2.72) for the non-competitive inhibition $K_{ic} \to \infty$, $[I]/K_{ic}=0$, so that K_{iu} remains as a single inhibition constant:

$$v = \frac{V[A]}{K_m + \left(1 + \frac{[I]}{K_{iu}}\right)[A]} \, . \tag{2.90}$$

The equation according to Lineweaver-Burk is:

$$\frac{1}{v} = \frac{1 + \frac{[I]}{K_{iu}}}{V} + \frac{K_m}{V[A]} \, , \tag{2.91}$$

according to Hanes:

$$\frac{[A]}{v} = \frac{[A]\left(1 + \frac{[I]}{K_{iu}}\right)}{V} + \frac{K_m}{V} \, , \tag{2.92}$$

according to Eadie-Hofstee:

$$v = \frac{V}{1+\frac{[I]}{K_{iu}}} - \frac{K_m}{\left(1+\frac{[I]}{K_{iu}}\right)} \cdot \frac{v}{[A]} ,\qquad(2.93)$$

and according to Dixon:

$$\frac{1}{v} = \frac{1}{V}\left(1+\frac{K_m}{[A]}\right) + \frac{[I]}{VK_{iu}} .\qquad(2.94)$$

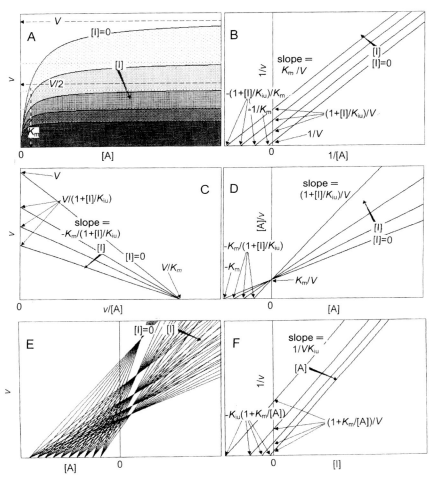

Fig. 2.25 Uncompetitive inhibition in different representations. (A) Direct plot, (B) double-reciprocal plot, (C) Eadie-Hofstee plot, (D) Hanes plot, (E) direct linear plot, (F) Dixon inhibition plot. Modes of determination of the kinetic constants are indicated.

The rise of hyperbolae at low substrate concentrations is similar both in the absence and presence of the inhibitor, at higher concentrations the curves move towards differing saturation values (Fig. 2.25 A). The double-reciprocal and the Dixon plots both yield parallels (Fig. 2.25 B and F), the Eadie-Hofstee plot has a common abscissa intercept (Fig. 2.25 C) and a common ordinate intercept appears in the Hanes plot (Fig. 2.25 D). The *uncompetitive inhibition constant* K_{iu} is obtained from a secondary plot of the ordinate sections (Fig. 2.21 B). In the direct linear plot the intercepts tend straight downwards (Fig. 2.25 E). The integrated Michaelis-Menten equation for this inhibition type is:

$$\frac{K_m}{\left(1+\frac{[I]}{K_{iu}}\right)} \ln \frac{[A]_0}{[A]} + [A]_0 - [A] = \frac{Vt}{\left(1+\frac{[I]}{K_{iu}}\right)} \quad (2.95)$$

and linearized:

$$\frac{[A]_0 - [A]}{t} = \frac{V}{1+\frac{[I]}{K_{iu}}} - \frac{K_m \ln \frac{[A]_0}{[A]}}{\left(1+\frac{[I]}{K_{iu}}\right)t} . \quad (2.96)$$

A joint abscissa intercept is obtained. A secondary plot of the reciprocal ordinate sections against the inhibitor concentrations yields $-K_{iu}$ as an abscissa section.

2.5.3.5 Partially Non-competitive Inhibition

The mechanisms of partial inhibition are characterized by a still active EAI complex. Its activity can either be unchanged with regard to the EA complex, or it can be changed in a negative or a positive manner. Thus, in contrast to complete inhibition, activation can also occur.

The reaction scheme (Eq. (2.63)) and the rate equation (Eq. (2.70)) for this type of inhibition have already been presented (Section 2.5.3.1). The double-reciprocal form reads:

$$\frac{1}{v} = \frac{K_m\left(1+\frac{[I]}{K_{ic}}\right)}{\left(V_1+\frac{V_2[I]}{K_{iu}}\right)[A]} + \frac{1+\frac{[I]}{K_{iu}}}{V_1+\frac{V_2[I]}{K_{iu}}} , \quad (2.97)$$

the Hanes equation:

$$\frac{[A]}{v} = \frac{K_m\left(1+\frac{[I]}{K_{ic}}\right)}{V_1+\frac{V_2[I]}{K_{iu}}} + \frac{\left(1+\frac{[I]}{K_{iu}}\right)[A]}{V_1+\frac{V_2[I]}{K_{iu}}} , \quad (2.98)$$

and the Eadie-Hofstee equation:

$$v = \frac{V_1 + \dfrac{V_2[I]}{K_{iu}}}{1 + \dfrac{[I]}{K_{iu}}} - \frac{K_m \left(1 + \dfrac{[I]}{K_{ic}}\right)}{\left(1 + \dfrac{[I]}{K_{iu}}\right)} \cdot \frac{v}{[A]} . \tag{2.99}$$

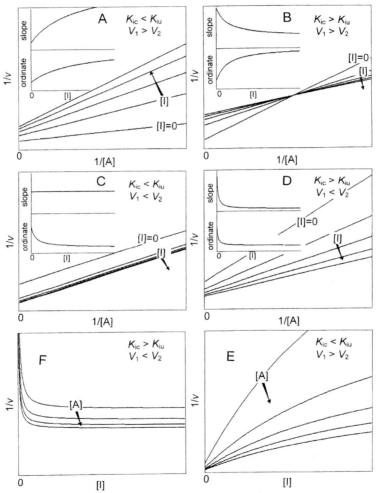

Fig. 2.26 Partially non-competitive inhibition in the double-reciprocal plot (A–D) and in the Dixon plot (E, F) in various combinations of inhibition constants and maximum velocities V_1 and V_2. The inserts show secondary plots of slopes and ordinate intercepts set against inhibition concentrations.

Linear dependences are also observed in all three diagrams for partial inhibitions and the relative positions of the straight lines are often similar to the respective complete inhibition types. However, a larger variety of curve patterns can be derived, since besides the inhibition constants the turnover rate constant k_6 for the EAI complex can change. This complicates the identification and analysis of this inhibition type. Figure 2.26 shows some of such combinations in the Lineweaver-Burk plot (A–D) and the Dixon inhibition diagram (E, F). The possible location of a common intercept in the first quadrant of the double reciprocal-plot is remarkable (Fig. 2.26 B). Characteristic for partial inhibitions and a criterion for discrimination from complete inhibitions is the fact that all plots with the inhibitor as variable, i.e. secondary and Dixon plots, do not yield straight lines (Fig. 2.26 E, F and inserts in A–D). A further peculiarity of partial inhibition mechanisms is that they can describe activations in two different ways. If $k_6 > k_2$ the EAI complex becomes more active than the EA complex. Activation will also be observed if $K_{ic} > K_{iu} \approx K_A > K_{Ai}$ and then the equilibrium shifts from the inactive E and EI forms to the active EA and EAI complexes.

2.5.3.6 Partially Uncompetitive Inhibition

This inhibition type is a special form of uncompetitive inhibition when the ternary EAI complex is still active:

$$
\begin{array}{c}
E + A \underset{k_{-1}}{\overset{k_1}{\rightleftharpoons}} EA \xrightarrow{k_2} E + P \\
+ \\
I \\
k_4 \updownarrow k_{-4} \\
EAI \xrightarrow{k_6} EI + P
\end{array}
\qquad (2.100)
$$

The rate equation for this inhibition type is:

$$
v = \frac{\left(V_1 + \dfrac{V_2[I]}{K_{iu}}\right)[A]}{K_m + \left(\dfrac{1 + [I]}{K_{iu}}\right)[A]}, \qquad (2.101)
$$

in the double-reciprocal form:

$$
\frac{1}{v} = \frac{K_m}{\left(V_1 + \dfrac{V_2[I]}{K_{iu}}\right)[A]} + \frac{1 + \dfrac{[I]}{K_{iu}}}{V_1 + \dfrac{V_2[I]}{K_{iu}}}, \qquad (2.102)
$$

after Hanes:

$$\frac{[A]}{v} = \frac{K_m}{V_1 + \dfrac{V_2[I]}{K_{iu}}} + \frac{\left(1 + \dfrac{[I]}{K_{iu}}\right)[A]}{V_1 + \dfrac{V_2[I]}{K_{iu}}} \ , \qquad (2.103)$$

after Eadie-Hofstee:

$$v = \frac{V_1 + \dfrac{V_2[I]}{K_{iu}}}{1 + \dfrac{[I]}{K_{iu}}} - \frac{K_m}{\left(1 + \dfrac{[I]}{K_{iu}}\right)} \cdot \frac{v}{[A]} \ . \qquad (2.104)$$

The graphics can be similar to those of the completely uncompetitive inhibition, but, dependent on the respective constants, deviations may occur and the lines in the double-reciprocal plot need no longer be parallel (Fig. 2.27 A, B). Dixon and secondary plots are non-linear (Fig. 2.27 A–D). Activation is also possible with this inhibition mechanism (Fig. 2.27 B, D).

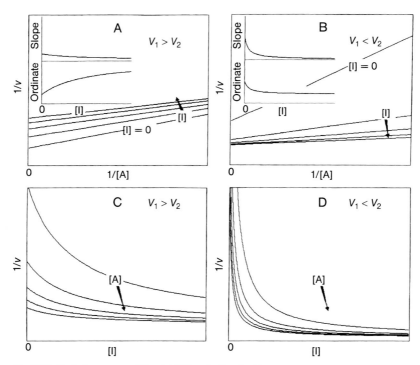

Fig. 2.27 Partially uncompetitive inhibition in the double-reciprocal plot (A, B) and in the Dixon plot (C, D) for various combinations of maximum velocities V_1 and V_2. The inserts show secondary plots of slopes and ordinate intercepts set against inhibition concentrations.

2.5.3.7 Partially Competitive Inhibition

Partial inhibition, where the EAI complex is still active, is possible with both non-competitive and uncompetitive inhibition, but a partially competitive inhibition should not be possible because an EAI complex cannot exist in a competitive mechanism. Actually, this inhibition type is not defined by its reaction mechanism, but because its pattern of curves in the linearized diagrams is quite similar to, and thus confusable with, those of complete competitive inhibition (Fig. 2.28A). Because of this formal coincidence this mechanism is called *partially competitive inhibition*, although its mechanism is completely different. It is principally a partially non-competitive inhibition obeying the same reaction scheme (Eq. (2.63)) with the only exception being that $k_2 = k_6$, the inhibitor does not affect the turnover rate:

$$
\begin{array}{ccccc}
E + A & \underset{k_{-1}}{\overset{k_1}{\rightleftarrows}} & EA & \xrightarrow{k_2} & E + P \\
+ & & + & & \\
I & & I & & \\
k_3 \updownarrow k_{-3} & & k_4 \updownarrow k_{-4} & & \\
EI + A & \underset{k_{-5}}{\overset{k_5}{\rightleftarrows}} & EAI & \xrightarrow{k_2} & EI + P
\end{array}
\tag{2.105}
$$

The rate equation for this inhibition type is:

$$
v = \frac{V[A]}{K_m \cdot \dfrac{\left(1 + \dfrac{[I]}{K_{ic}}\right)}{\left(1 + \dfrac{[I]}{K_{iu}}\right)} + [A]}, \tag{2.106}
$$

after Lineweaver-Burk:

$$
\frac{1}{v} = \frac{1}{V} + \frac{K_m \left(1 + \dfrac{[I]}{K_{ic}}\right)}{V \left(1 + \dfrac{[I]}{K_{iu}}\right)} \cdot \frac{1}{[A]}, \tag{2.107}
$$

after Hanes:

$$
\frac{[A]}{v} = \frac{[A]}{V} + \frac{K_m \left(1 + \dfrac{[I]}{K_{ic}}\right)}{V \left(1 + \dfrac{[I]}{K_{iu}}\right)}, \tag{2.108}
$$

and after Eadie-Hofstee:

$$
v = V - \frac{K_m \left(1 + \dfrac{[I]}{K_{ic}}\right)}{\left(1 + \dfrac{[I]}{K_{iu}}\right)} \cdot \frac{v}{[A]}. \tag{2.109}
$$

Although the same pattern of curves as for the true competitive mechanism is observed in the primary plots (Fig. 2.28 A), the two mechanisms can be simply differentiated by the non-linear secondary plots (Fig. 2.28 A, B) and the Dixon plots (Fig. 2.28 C, D). Also with this mechanism activation can be observed, if $K_{ic} > K_{iu}$ (Fig. 2.28 B, D).

It is remarkable that all inhibition types discussed so far contain the term *competitive*, although it is valid only for one of these inhibitions. Originally, inhibitions were classified according to the straight line patterns in the linearized diagrams, e.g. the double-reciprocal plot. According to this, competitive inhibition is characterized by a straight line intercept on the ordinate, while an intercept on the abscissa is characteristic of *non*-competitive inhibition. Parallel lines are neither of the above, thus they are *un*competitive. If the intercept is located between the axes the inhibition is *mixed*.

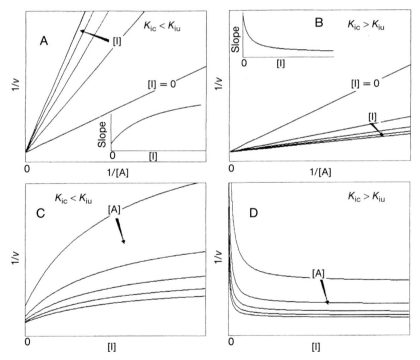

Fig. 2.28 Partially competitive inhibition in the double-reciprocal plots (A, B) and in the Dixon inhibition plot (C, D), in various combinations of inhibition constants, causing inhibition in (A) and (C), and activation in (B) and (D). The inserts show secondary plots of slopes set against inhibition concentrations.

2.5.3.8 Noncompetitive and Uncompetitive Product Inhibition

Inhibition of enzymes by the product is competitive for enzyme reactions with only one substrate (see Section 2.4.3) and mostly also for multi-substrate reactions with respect to the substrate from which the product is directly derived. However, other inhibition types can also occur. Especially with respect to co-substrates non-competitive or uncompetitive product inhibitions are observed. The rate equations for the respective inhibition mechanism can easily be converted to the equations for product inhibition by replacing the term [I] by [P] (not shown here). The reverse reaction, which may occur in the presence of product, will be neglected for simplicity. The respective straight line patterns follow the corresponding inhibition type. The inhibition constants have the meaning of product binding constants.

Integration of Eq. (2.72) for the case of a non-competitive product inhibition results in the following term, setting $[P] = [A]_0 - [A]$:

$$K_m \left(1 + \frac{[A]_0}{K_{ic}}\right) \cdot \ln \frac{[A]_0}{[A]} + \left(1 - \frac{K_m}{K_{ic}}\right)[P] + \frac{[P]^2}{2K_{iu}} = Vt. \qquad (2.110)$$

Because of the quadratic term this relationship cannot be converted into a linear equation according to the linearization methods of the integrated Michaelis-Menten equation, as demonstrated by plotting $([A]_0 - [A])/t$ against $\ln([A]_0/[A])/t$:

$$\frac{[A]_0 - [A]}{t} = \frac{V}{1 + \frac{[A]_0 - [A]}{2K_{iu}} - \frac{K_m}{K_{ic}}} - \frac{K_m\left(1 + \frac{[A]_0}{K_{ic}}\right)}{1 + \frac{[A]_0 - [A]}{2K_{iu}} - \frac{K_m}{K_{ic}}} \cdot \frac{\ln \frac{[A]_0}{[A]}}{t}. \qquad (2.111)$$

By adding a constant amount of product [P] from the start of the reaction, Eq. (2.111) becomes:

$$\frac{[A]_0 - [A]}{t} = \frac{V}{1 + \frac{[P]_0}{K_{iu}} + \frac{[A]_0 - [A]}{2K_{iu}} - \frac{K_m}{K_{ic}}} - \frac{K_m\left(1 + \frac{[A]_0}{K_{ic}}\right)}{1 + \frac{[P]_0}{K_{iu}} + \frac{[A]_0 - [A]}{2K_{iu}} - \frac{K_m}{K_{ic}}} \cdot \frac{\ln \frac{[A]_0}{[A]}}{t}. \qquad (2.112)$$

As non-competitive inhibition also includes competitive and uncompetitive inhibition, their equations can be obtained by reduction of Eqs. (2.110)–(2.112), setting $K_{iu} = \infty$ for competitive, $K_{ic} = \infty$ for uncompetitive inhibition. An evaluation of progress curves for the identification of the inhibition type can be performed with the Foster and Niemann method described in Section 2.4.3.

2.5.3.9 Substrate Inhibition

This frequently observed inhibition type, also called *substrate surplus inhibition*, is identified when, at high substrate concentrations, the reaction rate decreases instead of tending towards the maximum velocity. This inhibition is caused by a second substrate molecule, which binds to the active EA complex and inhibits the reaction. This second binding site can be the product binding site of the enzyme. If an additional substrate molecule binds to this site, product can no longer be formed from the first substrate molecule. It must be assumed that the first binding substrate molecule will occupy the genuine substrate binding site (otherwise the enzyme will not be functional). The inhibitory substrate molecule can only bind to the EA complex and acts as an uncompetitive inhibitor:

$$E + A \underset{k_{-1}}{\overset{k_1}{\rightleftarrows}} EA \overset{k_2}{\rightarrow} E + P$$
$$+$$
$$A$$
$$k_4 \updownarrow k_{-4}$$
$$EAA \tag{2.113}$$

Eqs. (2.90)–(2.94) for uncompetitive inhibition apply by substituting [I] by [A]:

$$v = \frac{V[A]}{K_m + \left(1 + \frac{[A]}{K_{iu}}\right)[A]}, \tag{2.114}$$

Lineweaver-Burk equation:

$$\frac{1}{v} = \frac{1 + \frac{[A]}{K_{iu}}}{V} + \frac{K_m}{V[A]}, \tag{2.115}$$

Hanes equation:

$$\frac{[A]}{v} = \frac{[A]\left(1 + \frac{[A]}{K_{iu}}\right)}{V} + \frac{K_m}{V}, \tag{2.116}$$

Eadie-Hofstee equation:

$$v = \frac{V}{1 + \frac{[A]}{K_{iu}}} - \frac{K_m}{\left(1 + \frac{[A]}{K_{iu}}\right)} \cdot \frac{v}{[A]}, \tag{2.117}$$

Dixon equation:

$$\frac{1}{v} = \frac{1}{V}\left(1 + \frac{K_m}{[A]}\right) + \frac{[A]}{VK_{iu}}. \tag{2.118}$$

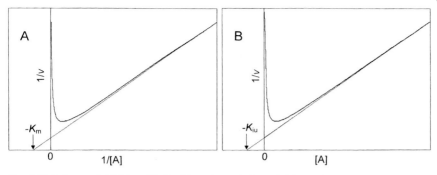

Fig. 2.29 Substrate inhibition. (A) Double-reciprocal plot, (B) Dixon plot.

As the same compound acts at the same time as both substrate and inhibitor, the two opposing effects cannot be separated, the linearized plots do not yield straight lines and their shape identifies the substrate inhibition. In the double-reciprocal plot the curve bends upwards towards the ordinate (Fig. 2.29 A). This makes the determination of the constants more difficult, because the Michaelis constant cannot be determined in the absence of inhibitor. Approximately, it may be assumed that the inhibitory effect is negligible at low substrate concentrations and from an asymptote to the curve in this region K_m and V can be estimated (Fig. 2.29 A). The Dixon plot, where the substrate (now regarded as inhibitor) is directly plotted against the reciprocal velocity, shows a similar deviation to that of the double-reciprocal plot and, here, the inhibition constant for the substrate can be extrapolated in the same manner from the abscissa intercept with an asymptote to the high substrate region, where the inhibitory effect predominates.

Progress curves can also not be linearized by integration of Eq. (2.114):

$$\frac{[A]_0 - [A]}{t} = V - \frac{[A]_0^2 - [A]^2}{2K_{iu}t} - K_m \frac{\ln \frac{[A]_0}{[A]}}{t} \, . \tag{2.119}$$

2.5.4
Enzyme Reactions with Two Competing Substrates

Substrate analogs are mostly inactive and inhibit the enzyme competitively, as discussed in Section 2.5.3.3. Sometimes, however, the analog can still be active, or the enzyme can accept a homologous substrate, as for example, the liver alcohol dehydrogenase which also oxidizes higher alcohols. If both, A_1 and A_2 are simultaneously present, only one at a time can be accepted and converted to product:

2 Enzyme Kinetics

$$\begin{array}{c} A_1 \xrightleftharpoons[k_{-1}]{k_1} EA_1 \xrightarrow{k_2} E + P_1 \\ + \\ E \\ + \\ A_2 \xrightleftharpoons[k_{-3}]{k_3} EA_2 \xrightarrow{k_4} E + P_2 \, . \end{array} \qquad (2.120)$$

One of the substrates, usually the physiological one, will be more efficient, while the poorer substrate has a retarding influence, slowing down the overall turnover rate. The situation is complicated by the fact that, for estimation of the substrate efficiency, distinction must be made between binding and catalysis. One substrate can be superior to the other in both respects, but they can also differ and both effects may counteract each other, one substrate may bind with higher affinity but evolve a slower turnover than the second one, and vice versa.

The derivation of a rate equation for the simultaneous presence of two active substrates follows the steady-state rules:

$$\frac{d[EA_1]}{dt} = k_1[E][A_1] - (k_{-1} + k_2)[EA_1] = 0 \, , \qquad (2.121)$$

$$\frac{d[EA_2]}{dt} = k_3[E][A_2] - (k_{-3} + k_4)[EA_2] = 0 \, , \qquad (2.122)$$

$$v = k_2[EA_1] + k_4[EA_2] \, , \qquad (2.123)$$

$$[E]_0 = [E] + [EA_1] + [EA_2] \, . \qquad (2.124)$$

Replacement of [E] from Eq. (2.124) in Eq. (2.121) and Eq. (2.122) yields:

$$[EA_2] = [E]_0 - [EA_1]\left(1 + \frac{k_{-1} + k_2}{k_1[A_1]}\right) \qquad (2.125)$$

$$[EA_1] = [E]_0 - [EA_2]\left(1 + \frac{k_{-3} + k_4}{k_3[A_2]}\right) \, . \qquad (2.126)$$

Equation (2.125) is inserted into Eq. (2.126), considering $K_{m1} = (k_{-1} + k_2)/k_1$ and $K_{m2} = (k_{-3} + k_4)/k_3$:

$$[EA_1] = \frac{K_{m2}[E]_0[A_1]}{K_{m2}[A_1] + K_{m1}[A_2] + K_{m1}K_{m2}} \, , \qquad (2.127\,a)$$

$$[EA_2] = \frac{K_{m1}[E]_0[A_2]}{K_{m2}[A_1] + K_{m1}[A_2] + K_{m1}K_{m2}} \, . \qquad (2.127\,b)$$

Inserting Eq. (2.127a) and Eq. (2.127b) into Eq. (2.123) results in the rate equation for a reaction with two alternate substrates, setting $V_1 = k_2[E]_0$ and $V_2 = k_4[E]_0$:

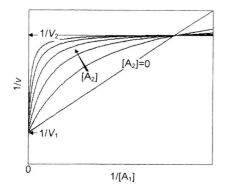

Fig. 2.30 Reaction with two different substrates in the double-reciprocal diagram. The maximum velocity V_1 of the varying substrate A_1 is higher than V_2 of the constant substrate A_2. Equal Michaelis constants are taken for both substrates.

$$v = \frac{V_1 K_{m2}[A_1] + V_2 K_{m1}[A_2]}{K_{m2}[A_1] + K_{m1}[A_2] + K_{m1} K_{m2}} . \tag{2.128}$$

Figure 2.30 shows an example of a reaction with two competing substrates. The linear dependence in the presence of only one substrate A_1 changes with both substrates into a biphasic characteristic, one phase approaching the maximum velocity V_1 of the first substrate, and the other the maximum velocity V_2 of the second substrate A_2. For $V_1 > V_2$, as depicted in Fig. 2.30, the curves bend upwards from the straight line for $[A_2] = 0$, in the reverse case, for $V_1 < V_2$, they deviate downwards and tend towards the, then lower, value $1/V_2$.

2.5.5
Different Enzymes Catalyzing the Same Reaction

There exist a large number of examples of enzymes carrying out the same reaction. One classical example is hexokinase, an enzyme occurring in all organs and catalyzing the first step in glycolysis. In the liver there is another enzyme with similar specificity, but a higher K_m value, glucokinase. At low concentrations glucose will preferentially be fed into the glycolytic pathway for energy generation in all organs, while at higher concentrations the liver will skim the glucose surplus for storage as glycogen. Lactate dehydrogenase, a tetrameric enzyme occurs in two forms (isoenzymes, isozymes), the M-type preferentially in skeletal muscle, the H-type in the heart. Compared with the M-type H has a higher affinity for its substrate and becomes inhibited by high amounts of pyruvate. Besides the pure H_4 and M_4 forms there exist also hybrids containing subunits from both types.

If enzymes (1, 2 … n), differing in their efficiencies, catalyze the same reaction, their velocities will behave additively:

$$r = \frac{V_1[A]}{K_{m1} + [A]} + \frac{V_2[A]}{K_{m2} + [A]} + \cdots \frac{V_n[A]}{K_{mn} + [A]} . \tag{2.129}$$

A conformable behavior has already been discussed for the case of binding of the same ligand to non-identical binding sites, differing in their affinities (Section 1.4) and similar saturation curves will be obtained (see Fig. 1.6 A). In the direct diagram the separate hyperbolic saturation curves of each enzyme type will produce a curve starting in the low substrate range according to the enzyme species with the smaller Michaelis constant and passing continuously into the curve belonging to the enzyme species with the higher K_m value. Finally, the curve will strive for a saturation value according to the maximum velocities, which depend both on the respective catalytic constants k_{cat1}, k_{cat2} ... and on the actual amount of the enzyme species. It must be considered that, for binding of ligands to non-identical binding sites, as discussed in Section 1.4, the binding sites are located on the same enzyme or macromolecule in a fixed ratio, while in the case of isoenzymes their respective ratio depends on their actual amounts in the reaction mixture, e.g. on the particular isolation procedure. Even when the two different catalytic sites are located on the same enzyme molecule, as in the case of hybrid lactate dehydrogenases, the ratios of the respective catalytic constants do not give an integer as in the case of a defined number of binding sites. Therefore, the characteristic of this effect, i.e. the extent of deviation from a simple hyperbolic or linear shape, can vary and may be insignificant if one species is clearly dominant.

In the linearized plots (Fig. 1.6 B–D) biphasic (or multiple-phasic, if $n > 2$) curves are obtained, from which the individual phases may be derived by computer analysis, as described in Section 1.4. The pattern of deviation from linearity in all these plots resembles the case of negative cooperativity (see Section 1.5.6), which must, therefore, be excluded.

2.6
Multi-substrate Reactions

2.6.1
Nomenclature

The Michaelis-Menten equation was originally derived for an irreversible enzyme reaction with only one substrate, but it has been demonstrated that it remains valid also for reversible reactions (Section 2.4.1) and in the presence of inhibitors (Section 2.5.3). However, to be regarded as the fundamental relationship of enzyme kinetics the Michaelis-Menten equation must prove to be valid also for reactions with two or more substrates, which, indeed, represent the majority of enzyme reactions. In a simple approach such reactions can be treated under conditions where all components, i.e. cosubstrates, cofactors, or coenzymes, are kept at constant, saturating surplus and only one substrate will be varied as in the case of a one-substrate reaction and, formally, the constants (K_m and V) for just this substrate may be obtained. However, such a procedure will reveal nothing about the special mechanism and the interactions between the multiple substrates. For this a more detailed analysis must be undertaken, considering all components involved in the reaction, and this must be based on adequate equations. Treatment of multi-substrate reactions according to the steady-state rules as discussed so far, however, will be a difficult and laborious undertaking and, therefore, a particular approach is required. Theoretical treatments of multi-substrate reactions have been reported by R. A. Alberty (1959), K. Dalziel (1957), and especially by W. Wallace Cleland (1963).

The following discussion is based on the concise rules and nomenclature introduced by Cleland. Substrates are termed in the sequence of their binding to the enzymes as A, B, C, products in the sequence of their release P, Q, R. Different enzyme states are named E, F, G. With substrates and products the enzyme forms *transitory complexes*, EA, EP, etc., that decay in a unimolecular step releasing substrate or product. Catalytic reactions occur from *central complexes* which are set in parentheses, e.g. (EAB), to distinguish them from transitory complexes. Central complexes cannot bind further substrate or product as all sites are already occupied, rather, they release substrate or product in unimolecular steps. As steady-state kinetics provides no information on the catalytic conversion of substrate into product (and vice versa) on the enzyme molecule, only one central complex for both states before and after catalysis (EAB–EPQ) is defined. The number of substrates participating in the forward and of products participating in the reverse reaction will be expressed by the terms *uni, bi, ter, quad* for 1, 2, 3, and 4, respectively, and these terms are added to the name of the reaction mechanisms. Two substrates being condensed into one product is a *bi uni* mechanism. Mechanisms where all substrates must bind before product can be released are called *sequential* mechanisms. The binding of substrates may be either *random* or *ordered*. In a *ping-pong* mechanism, product is already released before all substrates are bound. The enzyme here exists in two or more forms, modified by groups of the substrate. In *iso mechanisms* the enzyme isomerizes into two or more stable conformations.

The reaction equations are presented in a schematic form. The progressing reaction is symbolised by a horizontal line, the reaction coordinate. The various enzyme states appearing during the reaction course are indicated below this line. Substrate binding is indicated by vertical arrows on the reaction coordinate, dissociation of product by vertical arrows from it. Rate constants are ascribed on the sites of the arrows, the forward rate constant on the left and the reverse on the right. Alternatively, the binding constant is indicated. The reaction schema for an *ordered bi bi mechanism* in the conventional style:

$$E + A \underset{k_{-1}}{\overset{k_1}{\rightleftarrows}} EA$$

$$EA + B \underset{k_{-2}}{\overset{k_2}{\rightleftarrows}} EAB\text{-}EPQ \underset{k_{-3}}{\overset{k_3}{\rightleftarrows}} EQ + P$$

$$EQ \underset{k_{-4}}{\overset{k_4}{\rightleftarrows}} E + Q$$

will be transformed accordingly:

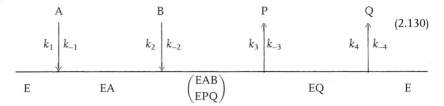

(2.130)

In the following the most frequent multi-substrate mechanisms are presented and methods to analyse and distinguish between them will be discussed. Finally methods for deriving complex rate equations are introduced.

2.6.2
Random Mechanism

Typical representatives of this multi-substrate mechanism are kinases or phosphorylase B. It is assumed that binding of substrates and products occurs in a random sequence. This is the case when the components bind completely independently from one another without any mutual interaction. The reaction schema for a random bi bi mechanism with two substrates and two products is:

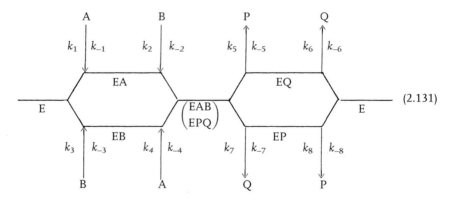

(2.131)

Compared to the one-substrate mechanism, for which the Michaelis-Menten equation has been derived, a set of new constants must be considered. Besides two maximum velocities, V_1 for the forward reaction and V_2 for the reverse reaction, each substrate and product has its own Michaelis constant, K_{mA}, K_{mB}, K_{mC} ... and K_{mP}, K_{mQ}, K_{mR} ..., respectively. As with the simple Michaelis-Menten equation the Michaelis constants represent the interaction of the substrate (or product in the reverse reaction) with the enzyme to form an active (central) enzyme substrate complex. In the case of multi-substrate reactions, however, the first binding step of the substrate to the free enzyme will not create an active enzyme complex, rather, a transitory complex, as long as the cosubstrate is not bound. Therefore the first binding step is characterized by a pure dissociation constant, which can differ from the Michaelis constant for the same substrate. To discriminate from the Michaelis constants the binding constants of substrate or products to the free enzyme are designated as K_{iA}, K_{iB}, K_{iC} ..., K_{iP}, K_{iQ}, K_{iR}, where 'i' means 'inhibition'. As will be discussed later (Section 2.6.5) this first binding step can be used to study multi-substrate mechanisms by analysing the inhibitory effect of the respective compound in the reverse reaction (product inhibition). Therefore, these constants are called *inhibition constants*. Accordingly, for a random bi bi mechanism there exist ten kinetic constants: two maximum velocities, four inhibition constants and four Michaelis constants for substrates and products. These constants cannot be obtained by simple kinetic treatment as described before and need a more detailed analysis. The procedure is similar to that already described for the analysis of enzyme inhibition (indeed, the random mechanism has far-reaching similarities with the non-competitive inhibition, see Section 2.5.3.2). In one test series one substrate (e.g. A) will be varied in the presence of a constant amount of the second substrate (B), and several test series are performed changing the second substrate.

Because of its alternate reaction pathways the random mechanism in its general form is described by a complex rate equation, which yields no hyperbolic substrate saturation curves (or linear curves in the respective diagrams) as long as the substrates are not present in saturating amounts. A significant simplification is achieved by the assumption of a rapid equilibrium compared with a relatively slow conversion of the central ternary complexes (EAB) and (EPQ) (*rapid equilibrium-random mechanism*). The rate equation for the random bi bi mechanism then is:

$$v = \frac{V_1 V_2 \left([A][B] - \frac{[P][Q]}{K_e}\right)}{K_{iA} K_{mB} V_2 + K_{mB} V_2 [A] + K_{mA} V_2 [B] + \frac{K_{mQ} V_1 [P]}{K_e} + \frac{K_{mP} V_1 [Q]}{K_e} + V_2 [A][B] + \frac{V_1 [P][Q]}{K_e}},$$
(2.132)

where K_e is the equilibrium constant of the overall reaction. In a closed reaction cycle, as for the random bi bi mechanism (in similarity to the non-competitive inhibition) the constants are related: $K_{iA}/K_{iB} = K_{mA}/K_{mB}$ and $K_{iP}/K_{iQ} = K_{mP}/K_{mQ}$, so that only three of the four constants must be determined. Eq. (2.132) can be further simplified by regarding only initial velocities in one reaction, e.g. the forward reaction for $[P]=[Q]=0$:

$$v = \frac{V_1[A][B]}{K_{iA}K_{mB} + K_{mB}[A] + K_{mA}[B] + [A][B]}.$$
(2.133)

If one of the two substrates is kept constant, Eq. (2.133) reduces to the ordinary Michaelis-Menten equation and by variation of the other substrate hyperbolic dependences result, becoming linear on applying the double-reciprocal equation:

$$\frac{1}{v} = \frac{K_{iA} K_{mB}}{V_1 [A][B]} + \frac{K_{mA}}{V_1 [A]} + \frac{K_{mB}}{V_1 [B]} + \frac{1}{V_1} . \qquad (2.134)$$

If one substrate, e.g. [A], is varied while [B] is constant then a straight line is obtained and if, in a second series, [A] is varied with a different constant amount of [B] a second line results, and so on, then all the lines meet at a common intercept left of the ordinate in the double-reciprocal plot (Fig. 2.31 B). The relative position of this intercept gives further information on the mechanism in a similar manner as for non-competitive inhibition. For a strict random mechanism, binding of the same substrate (i.e. A) to the free enzyme E, as well as to the enzyme–cosubstrate complex EB, should be completely equal, i.e. $K_{iA} = K_{mA}$ and $K_{iB} = K_{mB}$. In this case an intercept on the abscissa with the value $-1/K_{mA}$ is obtained. Accordingly, upon variation of B the abscissa intercept has the meaning $1/K_{mB}$. If the reaction is reversible, the Michaelis constants for the products can be obtained correspondingly, thus for this case all kinetic constants are easily accessible.

If already bound cosubstrate impedes the binding of the other substrate, binding to the free enzyme will occur with higher affinity, thus $K_{iA} < K_{mA}$, and $K_{iB} < K_{mB}$. A common intercept above the abscissa identifies this mechanism. Conversely, an intercept below the abscissa is indicative of an improvement of binding of one substrate by the already bound other substrate, and $K_{iA} > K_{mA}$, $K_{iB} > K_{mB}$. Similar patterns of lines are obtained regardless of whether [A] or [B] is varied in the presence of constant amounts of the respective other substrate. The Michaelis and inhibition constants can be determined from the slopes and coordinate intercepts of the lines, as shown in Fig. 2.31. However, the secondary plot method already discussed in inhibition kinetics can be recommended. Linearity in these plots is an additional test for the assumed mechanism (Table 2.2). The slope Sl_A of the straight line in the primary plot, according to Eq. (2.134), is:

$$Sl_A = \frac{K_{iA} K_{mB}}{V_1 [B]} + \frac{K_{mA}}{V_1} . \qquad (2.135)$$

A straight line with an abscissa intercept $-K_{mA}/K_{iA}K_{mB}$ (or $-1/K_{mB}$ for $K_{mA} = K_{iA}$) results by plotting the slope against $1/[B]$.

The ordinate intercepts Or_A of the primary plot

$$Or_A = \frac{K_{mB}}{V_1 [B]} + \frac{1}{V_1} \qquad (2.136)$$

plotted against $1/[B]$ also result in a straight line, cutting the abscissa at $-1/K_{mB}$.

Variation of [B] in the primary plot results in a straight line in a secondary plot from slopes Sl_B plotted against $1/[A]$ with an abscissa intercept at $-1/K_{iA}$:

$$Sl_B = \frac{K_{iA} K_{mB}}{V_1 [A]} + \frac{K_{mB}}{V_1} . \qquad (2.137)$$

2.6 Multi-substrate Reactions | 129

An abscissa intercept of $-1/K_{mA}$ is yielded from plotting the ordinate intercepts Or_B against $1/[A]$:

$$Or_B = \frac{K_{mA}}{V_1[A]} + \frac{1}{V_1} . \tag{2.138}$$

Similar analysis can be performed applying the other linearization methods. According to Hanes $[A]/v$ is plotted against $[A]$, or $[B]/v$ against $[B]$, respectively:

$$\frac{[A]}{v} = \frac{1}{V_1}\left(K_{mA} + \frac{K_{iA}K_{mB}}{[B]}\right) + \frac{[A]}{V_1}\left(1 + \frac{K_{mB}}{[B]}\right) \tag{2.139 a}$$

$$\frac{[B]}{v} = \frac{K_{mB}}{V_1}\left(1 + \frac{K_{iA}}{[A]}\right) + \frac{[B]}{V_1}\left(1 + \frac{K_{mA}}{[A]}\right) . \tag{2.139 b}$$

A common intercept left of the ordinate is revealed (Fig. 2.31 C). Kinetic constants can be obtained either from the slopes and coordinate intersections of the primary plot or from secondary plots of the slopes or ordinate intercepts plotted against the reciprocal cosubstrate concentrations (Table 2.2).

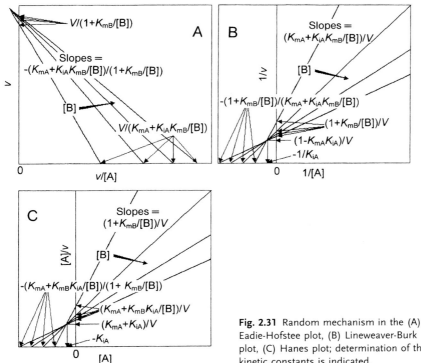

Fig. 2.31 Random mechanism in the (A) Eadie-Hofstee plot, (B) Lineweaver-Burk plot, (C) Hanes plot; determination of the kinetic constants is indicated.

2 Enzyme Kinetics

Table 2.2 Abscissa and ordinate intercepts of secondary plots for bisubstrate reactions, Sl_A, Sl_B are the slopes, Or_A, Or_B the ordinate

Primary plot		Secondary plot					
Designation of axis		Designation of axis		Intercepts			
Y	X	Y	X	Random/ordered mechanism		Ping-pong mechanism	
				Ordinate	Abscissa	Ordinate	Abscissa
$1/v$	$1/[A]$	Sl_A	$1/[B]$	K_{mA}/V	$-K_{mA}/K_{iA}K_{mB}$		
		Ab_A	$1/[B]$			$1/K_{mA}$	$-1/K_{mB}$
		Or_A	$1/[B]$	$1/V$	$-1/K_{mB}$	$1/V$	$-1/K_{mB}$
$1/v$	$1/[B]$	Sl_B	$1/[A]$	K_{mB}/V	$-1/K_{iA}$		
		Ab_B	$1/[A]$			$1/K_{mB}$	$-1/K_{mA}$
		Or_B	$1/[A]$	$1/V$	$-1/K_{mA}$	$1/V$	$-1/K_{mA}$
$[A]/v$	$[A]$	Sl_A	$1/[B]$	$1/V$	$-1/K_{mB}$	$1/V$	$-1/K_{mB}$
		$1/Ab_A$	$1/[B]$			K_{mA}	$-1/K_{mB}$
		Or_A	$1/[B]$	K_{mA}/V	$-K_{mA}/K_{iA}K_{mB}$		
$[B]/v$	$[B]$	Sl_B	$1/[A]$	$1/V$	$-1/K_{mA}$	$1/V$	$-1/K_{mA}$
		$1/Ab_B$	$1/[A]$			K_{mB}	$-1/K_{mA}$
		Or_B	$1/[A]$	K_{mB}/V	$-1/K_{iA}$		
$v/[A]$	v	Sl_A	$1/[B]$			$-1/K_{mA}$	$-1/K_{mB}$
		$1/Ab_A$	$1/[B]$	$1/V$	$-1/K_{mB}$	$1/V$	$-1/K_{mB}$
		Or_A	$1/[B]$	K_{mA}/V	$-K_{mA}/K_{iA}K_{mB}$		
$v/[B]$	v	Sl_B	$1/[A]$			$-1/K_{mB}$	$-1/K_{mA}$
		$1/Ab_B$	$1/[A]$	$1/V$	$-1/K_{mA}$	$1/V$	$-1/K_{mA}$
		Or_B	$1/[A]$	K_{mB}/V	$1/K_{iA}$		

Applying the Eadie-Hofstee diagram $v/[A]$, respectively $v/[B]$, is plotted against v:

$$v = \frac{V_1}{1+\frac{K_{mB}}{[B]}} - \frac{v}{[A]} \cdot \frac{K_{mA} + \frac{K_{iA}K_{mB}}{[B]}}{1+\frac{K_{mB}}{[B]}} \tag{2.140 a}$$

$$v = \frac{V_1}{1+\frac{K_{mA}}{[A]}} - \frac{v}{[B]} \cdot \frac{K_{mB}\left(1+\frac{K_{iA}}{[A]}\right)}{1+\frac{K_{mA}}{[A]}} \tag{2.140 b}$$

A joint intercept is found left of the ordinate for $K_{iA} > K_{mA}$ (Fig. 2.31 A), and to the right of the ordinate, below the abscissa, for $K_{iA} < K_{mA}$. If $K_{iA} = K_{mA}$ parallel lines are obtained. In the secondary plots the reciprocal abscissa or ordinate intercepts are plotted against the corresponding cosubstrate concentrations (Table 2.2).

2.6.3
Ordered Mechanism

According to its designation, this mechanism, which was already presented in the reaction schema (2.130), is characterized by a strict sequence of substrate binding. The detailed rate equation of an ordered bi bi mechanism is derived in Section 2.7.1, Eq. (2.170). Principally, the ordered mechanism is related to the random mechanism and the same rate equations can be used. In the case $K_{iA} < K_{mA}$ the substrate A will preferentially bind to the free enzyme, for $K_{iA} > K_{mA}$ it will preferentially bind to the enzyme cosubstrate complex. In both cases an order of binding is determined, while for $K_{iA} = K_{mA}$ no preference is given. Thus, the latter case represents a pure random mechanism, while the first two cases tend to an ordered mechanism. The more the constants differ, the clearer the order of binding. Therefore, both can be described as borderline cases of the same general mechanism, the pure random mechanism for no interaction and the ordered mechanism for strong substrate interaction. Equation (2.133), already introduced for the rapid equilibrium *random bi bi* mechanism, can also be taken for the ordered bi bi mechanism and both mechanisms may only be distinguished by the differences between their inhibition and Michaelis constants or, in the linearized diagrams, by the relative position of the common straight line intercepts.

Ordered mechanisms are frequently observed in dehydrogenases. In alcohol dehydrogenase there is the exceptional case of the *Theorell-Chance mechanism*, in which the central complex decomposes so rapidly that its stationary concentration is negligible and can be disregarded in the rate equation:

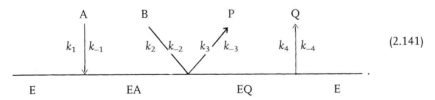

(2.141)

The rate equation (2.133) for the forward reaction of the normal ordered mechanisms also holds for this mechanism, so that no distinction can be made by graphic analysis. However, in the absence of the central complex the product inhibition pattern of both mechanisms differs (Table 2.3).

Generally, mechanisms showing modifications of the normal mechanisms are called *iso-mechanisms*. An *iso-ordered mechanism* exists when the enzyme isomerizes in the central complex from EAB to FPQ:

Table 2.3 Product inhibition pattern in bisubstrate mechanisms (after Cleland 1963). C, competitive; NC, non-competitive; UC, uncompetitive; nI, no inhibition

Mechanism	Inhibiting product	Variable substrate			
		A		B	
		Not saturated	Saturated with B	Not saturated	Saturated with A
Ordered *bi-bi*	P	NC	UC	NC	NC
	Q	C	C	NC	nI
Theorell Chance	P	NC	nI	C	C
	Q	C	C	NC	nI
Iso ordered *bi-bi*	P	NC	UC	NC	NC
	Q	NC	NC	NC	UC
Random *bi-bi* Rapid equilibrium	P or Q	C	nI	C	nI
Ping-pong *bi-bi*	P	NC	nI	C	C
	Q	C	C	NC	nI
Iso ping-pong *bi-bi* (isomerisation of the enzyme)	P	NC	nI	C	C
	Q	NC	NC	NC	NC

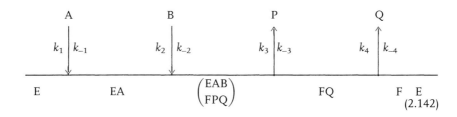

$$\text{(2.142)}$$

The complete rate equation for this mechanism contains additional links, which, however, are irrelevant for the forward reaction, so that Eq. (2.133) is still valid, but product inhibition is influenced (Table 2.3).

2.6.4
Ping-pong Mechanism

The name illustrates the alternate binding of substrates and release of products characteristic of this mechanism:

2.6 Multi-substrate Reactions

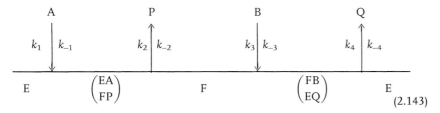

(2.143)

After binding of the first substrate, the first product is released, followed by binding of the second substrate and release of the second product. A stringent feature of this mechanism is the formation of an intermediary enzyme form in the reaction with the first substrate, usually by transferring a reactive group. The second substrate will remove this group to form the second product. Aminotransferase (transaminase) reactions are typical examples of this mechanism. An amino acid, e.g. aspartate, transfers its amino group to the pyridoxal phosphate cofactor of the enzyme and is released as the α-oxo acid (oxalacetate). Pyridoxamine phosphate as a covalent enzyme intermediate is formed. A second α-oxo acid (α-oxoglutarate) accepts the amino group from the cofactor and is converted into an amino acid (glutamate). A *multi-site ping-pong mechanism* is found in fatty acid synthase. The substrate is passed on over seven catalytic centers by the growing fatty acid chain bound to the central pantetheine residue.

The general rate equation for the *ping-pong bi bi mechanism* is derived in Section 2.7.3. The reduced equation for the forward reaction is:

$$v = \frac{V_1[A][B]}{K_{mB}[A] + K_{mA}[B] + [A][B]} .$$ (2.144)

The ping-pong mechanism can be differentiated from other multi-substrate mechanisms by the parallel lines resulting in the Lineweaver-Burk plot (Fig. 2.32 B):

$$\frac{1}{v} = \frac{K_{mA}}{V_1[A]} + \frac{K_{mB}}{V_1[B]} + \frac{1}{V_1} .$$ (2.145)

Secondary plots of the ordinate and abscissa intercepts, plotted against the reciprocal cosubstrate concentrations, can be drawn (Table 2.2).

In the Hanes plot

$$\frac{[A]}{v} = \frac{K_{mA}}{V_1} + \frac{[A]}{V_1}\left(1 + \frac{K_{mB}}{[B]}\right)$$ (2.146 a)

$$\frac{[B]}{v} = \frac{K_{mB}}{V_1} + \frac{[B]}{V_1}\left(1 + \frac{K_{mA}}{[A]}\right) .$$ (2.146 b)

The straight lines intersect at K_{mA}/V_1 and K_{mB}/V_1 (Fig. 2.32 C). Secondary plots are obtained by plotting the slopes or the abscissa intercepts, respectively, against the reciprocal cosubstrate concentrations (Table 2.2).

The Eadie-Hofstee plot also yields a joint straight line intercept on the abscissa (Fig. 2.32 A):

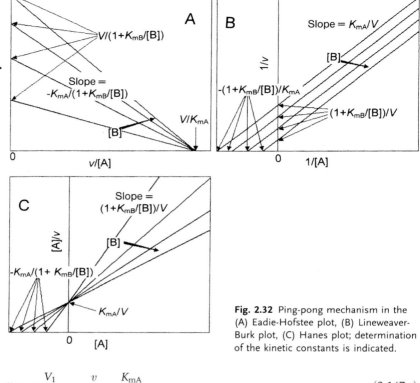

Fig. 2.32 Ping-pong mechanism in the (A) Eadie-Hofstee plot, (B) Lineweaver-Burk plot, (C) Hanes plot; determination of the kinetic constants is indicated.

$$v = \frac{V_1}{1+\frac{K_{mB}}{[B]}} - \frac{v}{[A]} \cdot \frac{K_{mA}}{1+\frac{K_{mB}}{[B]}} \qquad (2.147\,\text{a})$$

$$v = \frac{V_1}{1+\frac{K_{mA}}{[A]}} - \frac{v}{[B]} \cdot \frac{K_{mB}}{1+\frac{K_{mA}}{[A]}} \qquad (2.147\,\text{b})$$

Secondary plots are obtained from the slopes and the reciprocal abscissa intercepts (Table 2.2).

In an *iso ping-pong mechanism* the enzyme isomerizes into form F and further into form G:

$$
\begin{array}{ccccccc}
& A & & P & & B & & Q \\
& \downarrow k_1 \; k_{-1} & & \uparrow k_2 \; k_{-2} & & \downarrow k_3 \; k_{-3} & & \uparrow k_4 \; k_{-4} \\
E & & \binom{EA}{FP} & & F & & \binom{FB}{GQ} & & G \quad E
\end{array} \qquad (2.148)
$$

This mechanism can be identified in the product inhibition pattern (Table 2.3).

2.6.5
Product Inhibition in Multi-substrate Reactions

The various multi-substrate mechanisms including their iso-mechanisms and especially those involving three substrates cannot be identified only by graphic analysis, the unavoidable scattering of data makes it more difficult to decide about the relative position of joint intercepts or the prevalence of parallel lines. An additional method for analysis of multi-substrate reactions is the study of product inhibition patterns. As described in Section 2.4.3, product is assumed to inhibit competitively with respect to its substrate. In multi-substrate mechanisms two or more substrates are involved and competition may be expected for the substrate from which the respective product is directly derived, e.g. for lactate dehydrogenase, pyruvate will compete with lactate but not with NAD as cosubstrate. If lactate is displaced by pyruvate, NAD, added even in high amounts, cannot reactivate the enzyme and promote the reaction. Thus, with respect to NAD non-competitive inhibition patterns should be observed. The actual conditions for multi-substrate reactions are even more complex and depend on the respective mechanism and the degree of saturation with the cosubstrate. Each substrate can be tested with respect to its inhibition mechanism with its own product, as well as with the coproducts. For two substrates there result 4 inhibition types, for 3 substrates (and 3 products) even 9 types (Table 2.3). A further extension of possibilities is the test in the presence of limiting or saturating amounts, respectively, of the cosubstrate. For a distinct multi-substrate mechanism the pattern of inhibition types is always the same and, therefore, provides a further criterion for the identification of multi-substrate mechanisms besides the graphic methods (Table 2.3).

2.6.6
Haldane Relationships in Multi-substrate Reactions

Haldane relationships describe the correlations between kinetic constants in enzyme reactions after attaining the equilibrium state. Correspondingly, as discussed already for one-substrate reactions in Section 2.4.2, Haldane relationships can also be derived for other kinetic mechanisms. In the general form they read:

$$K_e = \left(\frac{V_1}{V_2}\right)^n \frac{K_P K_Q K_R}{K_A K_B K_C} \tag{2.149}$$

where K_e is the equilibrium constant of the overall reaction. The numerator contains the maximum velocity for the forward reaction and the constants for all products, the denominator is composed of the maximum velocity for the reverse reaction and the constants for all substrates. The constants are, depending on the respective mechanism, either Michaelis or inhibition constants. The exponent n assumes mostly the values 0, 1, or 2, but there is always at least one

Haldane relationship for $n=1$ which serves to eliminate K_e from the rate equation. This can be obtained from the constant term in the denominator of the respective rate reaction (e.g. $K_{iA}K_{mB}V_2$ in Eq. (2.132)). If the numerator and denominator of Eq. (2.132) are extended by K_e/V_1, the constant term becomes $K_{iA}K_{mB}V_2 \cdot K_e/V_1$. Expressed in the coefficient form with combination of the rate constants (see Eq. (2.165); for the definition of coefficients see Section 2.7.1) yields the following equation:

$$K_{iA} K_{mB} V_2 K_e \left(\frac{1}{V_1}\right) = \frac{Co}{CoA} \cdot \frac{CoA}{CoAB} \cdot \frac{N_2}{CoPQ} \cdot \frac{N_1}{N_2} \cdot \frac{CoAB}{N_1}. \qquad (2.150)$$

After reduction and extension with CoQ, the remaining coefficients can be converted into kinetic constants:

$$\frac{K_{iA} K_{mB} V_2 K_e}{V_1} = \frac{Co}{CoQ} \cdot \frac{CoQ}{CoPQ} = K_{iQ} K_{mP}. \qquad (2.151)$$

From this the following Haldane relationship is obtained:

$$K_e = \frac{V_1 K_{iQ} K_{mP}}{V_2 K_{iA} K_{mB}}. \qquad (2.152)$$

Other Haldane relationships can be derived with the coefficient form of the rate equation and with K_e from the denominator terms in a similar way.

2.6.7
Mechanisms with more than Two Substrates

Principally all multi-substrate reactions, including those with three or four substrates, can be reduced to the three main mechanisms: random, ordered, and ping-pong. An *ordered ter ter* mechanism

$$
\begin{array}{cccccc}
\quad A & B & C & P & Q & R \\
k_1 \downarrow k_{-1} & k_2 \downarrow k_{-2} & k_3 \downarrow k_{-3} & k_4 \uparrow k_{-4} & k_5 \uparrow k_{-5} & k_6 \uparrow k_{-6} \\
E & EA & EAB & \begin{pmatrix} EABC \\ EPQR \end{pmatrix} & EQR & ER \qquad E
\end{array} \qquad (2.153)
$$

obeys the following equation for the forward reaction:

$$v = \frac{V_1[A][B][C]}{K_{iA}K_{iB}K_{mC} + K_{iB}K_{mC}[A] + K_{iA}K_{mB}[C] + K_{mC}[A][B] + K_{mB}[A][C]} \\ \frac{V_1[A][B][C]}{+K_{mA}[B][C] + [A][B][C]}. \qquad (2.154)$$

As described already for bisubstrate reactions one distinct substrate is varied at fixed concentrations of a cosubstrate, which will be changed in different test series. The third (and fourth) substrate remains constant during the whole process. A similar procedure can be carried out by variation of the second and, finally, the third (and fourth) substrate. The double-reciprocal plot yields intercepts left of the ordinate and corresponding secondary plots can be derived from Eq. (2.154).

In the *hexa uni ping-pong* mechanism all substrates and products bind alternately and three central complexes are formed:

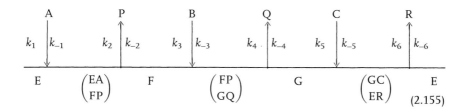
(2.155)

The rate equation for the forward reaction is:

$$v = \frac{V_1[A][B][C]}{K_{mC}[A][B] + K_{mB}[A][C] + K_{mA}[B][C] + [A][B][C]} \quad . \tag{2.156}$$

Parallel lines are obtained in the double-reciprocal plot upon variation of each of the three substrates against each cosubstrate, respectively.

With three and four substrates involved, combinations of the three main mechanisms are observed, like hybrid *ping-pong ordered* or *ping-pong random* mechanisms. Four *ping-pong* mechanisms with two ligands binding or detaching in ordered sequence, and the third in a ping-pong sequence exist, i.e. the *bi uni uni bi ping-pong* mechanism:

```
       A          B         P          C         Q          R
       ↓          ↓         ↑          ↓         ↑          ↑
   k₁ |k₋₁    k₂ |k₋₂   k₃ |k₋₃    k₄ |k₋₄   k₅ |k₋₅    k₆ |k₋₆
       ↓          ↓         ↓          ↓         ↓          ↓
   E        EA       (EAB)        F        (FC)       ER         E
                     (FP )                 (EQR)
```
(2.157)

the *uni bi bi uni*, the *bi bi uni uni*, and the *uni uni bi bi* mechanism. The latter two have the same reverse reaction. All four mechanisms have two central complexes and they share the same rate equation for the forward reaction:

$$v = \frac{V_1[A][B][C]}{K_{iA}K_{mB}[C] + K_{mC}[A][B] + K_{mB}[A][C] + K_{mA}[B][C] + [A][B][C]} \tag{2.158}$$

and thus cannot be distinguished by graphic representations alone, only by including the analysis of product inhibition. The double-reciprocal plot yields parallel lines. However, according to the ordered part of the mechanism, with variation of [A] against different amounts of [B], or of [B] against several concentrations of [A], the straight lines will intersect in a point left of the ordinate.

Quad mechanisms with four participating substrates are seldom found, one example is the carbamoylphosphate synthetase.

2.6.8
Other Nomenclatures for Multi-substrate Reactions

Besides the widely used nomenclature for multi-substrate reactions introduced by Cleland there exist also notations suggested by other authors. The one by Dalziel (1957) deviates farthest, as V is inserted into other kinetic constants. The rate reactions can be converted into the Dalziel notation by multiplying numerator and denominator by the reciprocal catalytic constants $\Phi_0 = 1/k_{cat}$. Eq. (2.133) for the forward reaction of a random mechanism is converted into the following form:

$$v = \frac{[E]_0 [S_1][S_2]}{\Phi_{12} + \Phi_2[S_1] + \Phi_1[S_2] + \Phi_0[S_1][S_2]} \cdot \qquad (2.133\,a)$$

The Dalziel coefficients are defined as: $\Phi_1 = K_{mA}/k_{cat}$ and $\Phi_2 = K_{mB}/k_{cat}$; $\Phi_{12}/\Phi_2 = K_{iA}$. The notation of Alberty (1953) corresponds essentially to the Cleland nomenclature ($K_A = K_{mA}$, etc.), but for product $K_{iA}K_{mB}$ in Eq. (2.133) a common constant $K_{AB} = K_{iA}K_{mB}$ is introduced.

2.7
Derivation of Rate Equations of Complex Enzyme Mechanisms

2.7.1
King-Altmann Method

It is evident that the derivation of rate reactions for complicated enzyme mechanisms from differential equations, following the rules of steady-state kinetics, frequently leads to complex relationships, which are difficult to resolve. Various approaches have been proposed in order to simplify the derivation of rate equations for complicated enzyme mechanisms without excessive mathematical effort. The method of E. L. King and C. Altmann (1956) has widely been accepted. It is demonstrated here by the example of an *ordered bi bi* mechanism.

Step 1: Construction of a polygon. The enzyme reaction is written in the shape of a polygon, the corners formed by the enzyme forms occurring in the mechanism. Double arrows connect the enzyme forms and above and below the arrows the respective rate constants are indicated, multiplied (if occurring) by entering ligands (substrates or products). The release of ligands is not indicated.

2.7 Derivation of Rate Equations of Complex Enzyme Mechanisms

The reaction scheme must always yield a completely closed figure, including all possible pathways and side reactions. Central complexes of substrates and products are combined. The polygon for the ordered bi bi mechanism reads:

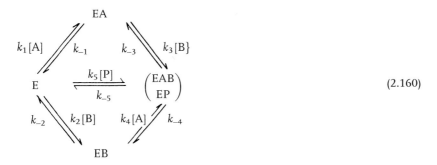 (2.159)

For comparison the polygon for a *random bi uni* mechanism (two substrates forming one product):

 (2.160)

Step 2: Derivation of patterns: From the polygon all possible patterns are drawn which connect all enzyme forms by single lines. These patterns must be open, closed figures must be avoided. Eight such patterns can be drawn for the *random bi uni* mechanism:

and four for the ordered bi bi mechanism:

Generally, the number of possible patterns is $m!/(n-1)!(m-n+1)!$ for mechanisms with only one reaction cycle; n is the number of different enzyme forms and m the number of connections between the enzyme forms in the original reaction schema. In the case of multiple cycles the combinations forming closed patterns have to be deducted for each cycle. They have to be calculated for each cycle according to $(m-r)!(n-r-1)!(m-n+1)!$, r being the number of connections in a closed cycle. For the *random bi uni* mechanism there would be $5!/(4-1)!(5-4+1)! = 10$ patterns, for the two cycles each with three connections $(5-3)!/(4-3-1)!(5-4+1) = 1$ must be deducted twice, resulting in eight open patterns.

Step 3: Deriving the expressions for the enzyme forms: One after the other enzyme forms are assigned in their respective positions into each pattern, e.g. E is marked into the four patterns of the ordered bi bi mechanism:

$k_{-1}k_{-2}k_{-3}[P]$ $k_2 k_3 k_4 [B]$ $k_{-1}k_3 k_4$ $k_{-1}k_{-2}k_4$.

All arrows pointing to this enzyme form are drawn in each pattern together with the respective rate constants and concentration terms, according to the original polygon (Eq. (2.159)). The rate constants together with the concentration terms for each pattern are written as a product. Since, for the ordered bi bi mechanisms, four patterns are present, correspondingly, four product terms are obtained for one enzyme form. These are summarized to give the expression for this enzyme form:

$$N_E = \frac{k_{-1}k_{-2}k_{-3}[P] + k_2 k_3 k_4 [B] + k_{-1}k_3 k_4 + k_{-1}k_{-2}k_4}{D} \qquad (2.161\,a)$$

The same procedure is accomplished for the other enzyme forms and the respective expressions are:

$$N_{EA} = \frac{k_1 k_{-2} k_{-3}[A][P] + k_{-2}k_{-3}k_{-4}[P][Q] + k_1 k_2 k_3 [A] + k_1 k_{-2} k_4 [A]}{D} \qquad (2.161\,b)$$

$$N_{EAB} = \frac{k_1 k_2 k_{-3}[A][B][P] + k_2 k_{-3}k_{-4}[B][P][Q] + k_{-1}k_{-3}k_{-4}[P][Q] + k_1 k_2 k_4 [A][B]}{D}$$

$$(2.161\,c)$$

$$N_{EQ} = \frac{k_1 k_2 k_3 [A][B] + k_2 k_3 k_{-4}[B][Q] + k_{-1}k_3 k_{-4}[Q] + k_{-1}k_{-2}k_{-4}[Q]}{D} \qquad (2.161\,d)$$

In the final rate equation, the sum of all *N*-terms forms the denominator *D*:

2.7 Derivation of Rate Equations of Complex Enzyme Mechanisms | 141

$$D = N_E + N_{EA} + N_{EAB} + N_{EQ}$$
$$= k_{-1}k_{-2}k_{-3}[P] + k_2k_3k_4[B] + k_{-1}k_3k_4 + k_{-1}k_{-2}k_4 + k_1k_{-2}k_{-3}[A][P] + k_{-2}k_{-3}k_{-4}[P][Q]$$
$$+ k_1k_3k_4[A] + k_1k_{-2}k_4[A] + k_1k_2k_{-3}[A][B][P] + k_2k_{-3}k_{-4}[B][P][Q] + k_{-1}k_{-3}k_{-4}[P][Q]$$
$$+ k_1k_2k_4[A][B] + k_1k_2k_3[A][B] + k_2k_3k_{-4}[B][Q] + k_{-1}k_{-3}k_{-4}[Q] + k_{-1}k_{-2}k_{-4}[Q]$$
(2.162)

while the numerator is the difference between a product formed by a numerator coefficient N_1 for the forward reaction and all substrate concentrations and a numerator coefficient N_2 for the backward reaction and all product concentrations. The numerator coefficients include the total enzyme concentration and all rate constants of the forward reaction, $N_1 = [E]_0 k_1 k_2 k_3 \ldots$, of the reverse reaction $N_2 = [E]_0 k_{-1} k_{-2} k_{-3} \ldots$, respectively.

$$v = \frac{(N_1[A][B][C]\ldots - N_2[P][Q][R]\ldots)}{D} \quad . \tag{2.163}$$

The rate equation for the ordered mechanism is now obtained in the form of rate constants:

$$v = \frac{k_1k_2k_3k_4[A][B] - k_{-1}k_{-2}k_{-3}k_{-4}[P][Q]}{k_{-1}k_4(k_{-2}+k_3) + k_1k_4(k_{-2}+k_3)[A] + k_2k_3k_4[B] + k_{-1}k_{-2}k_{-3}[P] + k_{-1}k_{-4}(k_{-2}+k_3)[Q]}$$
$$\overline{+k_1k_2(k_3+k_4)[A][B] + k_1k_{-2}k_{-3}[A][P] + k_2k_3k_{-4}[B][Q] + k_{-3}k_{-4}(k_{-1}+k_{-2})[P][Q]}$$
$$\overline{+k_1k_2k_{-3}[A][B][P] + k_2k_{-3}k_{-4}[B][P][Q]}$$
(2.164)

For this the denominator is rearranged according to the concentration terms.

Step 4: Transformation of the rate equation into the coefficient form.

All rate constants with the same concentration term are expressed as coefficient Co of the respective concentration term, e.g. $k_1k_4(k_{-2}+k_3)[A] = CoA$. The complete denominator term in the coefficient form reads:

$$D = Co + CoA[A] + CoB[B] + CoP[P] + CoQ[Q] + CoAB[A][B] + CoAP[A][P] +$$
$$CoBQ[B][Q] + CoPQ[P][Q] + CoABP[A][B][P] + CoBPQ[B][P][Q] \quad . \tag{2.165}$$

The kinetic constants are now defined by coefficients. The maximum velocity of the forward reaction V_1 is the quotient of the numerator coefficient N_1 and the coefficient of all substrates. The maximum velocity of the reverse reaction V_2 is the quotient of N_2 and the coefficient of all products:

$$V_1 = \frac{N_1}{CoABC\ldots} \; ; \; V_2 = \frac{N_2}{CoPQR\ldots} \quad . \tag{2.166}$$

For the Michaelis constants the ratio is formed between the respective coefficient of all substrates (all products) except the variable one and the coefficient of all substrates (or products):

$$K_{mA} = \frac{C_oCB\ldots}{C_oABC\ldots}\ ;\ K_{mB} = \frac{C_oAC\ldots}{C_oABC\ldots}$$
$$K_{mP} = \frac{C_oQR\ldots}{C_oPQR\ldots}\ ;\ K_{mQ} = \frac{C_oPR\ldots}{C_oPQR\ldots}\ . \tag{2.167}$$

K_e, the equilibrium constant of the overall reaction is the ratio of the numerator coefficients, the ratio of the rate constants of the forward reaction to those of the backward reaction:

$$K_e = \frac{N_1}{N_2} = \frac{k_1 k_2 k_3 \ldots}{k_{-1} k_{-2} k_{-3} \ldots}\ . \tag{2.168}$$

The respective constants for the *ordered bi bi* mechanism are given by:

$$V_1 = \frac{k_1 k_2 k_3 k_4 [E]_0}{k_1 k_2 (k_3 + k_4)} = \frac{k_3 k_4 [E]_0}{k_3 + k_4}$$

$$V_2 = \frac{k_{-1} k_{-2} k_{-3} k_{-4} [E]_0}{k_{-3} k_{-4} (k_{-1} + k_{-2})} = \frac{k_{-1} k_{-2} [E]_0}{k_{-1} + k_{-2}}$$

$$K_{mA} = \frac{k_2 k_3 k_4}{k_1 k_2 (k_3 + k_4)} = \frac{k_3 k_4}{k_1 (k_3 + k_4)}$$

$$K_{mB} = \frac{k_1 k_4 (k_{-2} + k_3)}{k_1 k_2 (k_3 + k_4)} = \frac{k_4 (k_{-2} + k_3)}{k_2 (k_3 + k_4)}$$

$$K_{mP} = \frac{k_{-1} k_{-4} (k_{-2} + k_3)}{k_{-3} k_{-4} (k_{-1} + k_{-2})} = \frac{k_{-1} (k_{-2} + k_3)}{k_{-3} (k_{-1} + k_{-2})}$$

$$K_{mQ} = \frac{k_{-1} k_{-2} k_{-3}}{k_{-3} k_{-4} (k_{-1} + k_{-2})} = \frac{k_{-1} k_{-2}}{k_{-4} (k_{-1} + k_{-2})}$$

$$K_e = \frac{k_1 k_2 k_3 k_4}{k_{-1} k_{-2} k_{-3} k_{-4}}\ .$$

Step 5: Transformation of the rate equation into the form of kinetic constants: The numerator and denominator of the rate equation in its coefficient form are multiplied by the constant factor $N_2/(C_oABC\ldots C_oPQR\ldots)$ and, term by term, the coefficients should be replaced by kinetic constants according to the above definitions. Correspondingly, the numerator changes into the general form:

$$V_1 V_2 \left([A][B][C]\ldots - \frac{[P][Q][R]\ldots}{K_e} \right)\ . \tag{2.169}$$

The conversion of the denominator is, at first sight, not so obvious. Table 2.4 shows the transformation of all denominator terms of Eq. (2.165) for the ordered bi bi mechanism. In the first column the single terms, extended by the constant factor (in bold letters) are indicated. For the second term, taken as an example, the conversion is quite obvious, the coefficient *CoA* (lacking B) divided by *CoAB* is defined as K_{mB} and N_2 divided by *CoPQ* becomes the maximum velocity in the reverse reaction. Accordingly, the coefficients in the third term can be replaced by

K_{mA} and V_2. In the fourth or fifth term the coefficients cannot be directly converted into kinetic constants, this can, however, easily be achieved by multiplying numerator and denominator by a suitable factor, i.e. N_1 in both cases. Other terms, like the first one, however, cannot be resolved by appropriate extensions. In fact, the definitions for the constants discussed above is incomplete, the constants for the binding of ligands to the free enzyme, the inhibition constants, are not yet considered. The reason for this omission is the fact that they do not fit easily into a unified scheme, but can be derived from the actual case. Principally, the definition for the Michaelis constant also holds for the inhibition constants, i.e. that numerator and denominator coefficients include the same ligands but the variable ligand may not appear in the numerator coefficient. In contrast to the Michaelis constants, the coefficients can comprise both substrates and products. Accordingly, in the first term Co/CoA can be taken for K_{iA}. If the coefficients are replaced by the respective rate constants the expression k_{-1}/k_1 results. This is just the definition of the dissociation constant for A, confirming the correctness of the definition of the inhibition constant. In term seven $CoP/CoAP$ appears, which can equally be defined as the inhibition constant K_{iA}. This seems inconsistent, but when the coefficients are expressed by the respective rate constants then again k_{-1}/k_1 is obtained, proving the assumption to be correct. Obviously different definitions for the same constant are possible.

Finally the rate equation in the form of the kinetic constants for the *ordered bi bi* mechanism is:

Table 2.4 Transformation of the denominator from Eq. (2.165) from the coefficient form into the one of kinetic constants. The constant factor by which all denominator terms are multiplied, is marked in the first column in bold letters

Denominator terms in coefficients form	Extension	Denominator terms in form of constants	Definition of inhibition constants
$\frac{N_2 Co}{CoAB \cdot CoPQ}$	CoA	$K_{iA} K_{mB} V_2$	$K_{iA} = \frac{Co}{CoA} = \frac{k_{-1}}{k_1}$
$\frac{N_2 CoA[A]}{CoAB \cdot CoPQ}$		$K_{mB} V_2 [A]$	
$\frac{N_2 CoB[B]}{CoAB \cdot CoPQ}$		$K_{mA} V_2 [B]$	
$\frac{N_2 CoP[P]}{CoAB \cdot CoPQ}$	N_1	$\frac{K_{mQ} V_1 [P]}{K_e}$	
$\frac{N_2 CoQ[Q]}{CoAB \cdot CoPQ}$	N_1	$\frac{K_{mP} V_1 [Q]}{K_e}$	
$\frac{N_2 CoAB[A][B]}{CoAB \cdot CoPQ}$		$V_2 [A][B]$	
$\frac{N_2 CoAP[A][P]}{CoAB \cdot CoPQ}$	$CoP\ N_1$	$\frac{K_{mQ} V_1 [A][P]}{K_{iA} K_e}$	$K_{iA} = \frac{CoP}{CoAP} = \frac{k_{-1}}{k_1}$
$\frac{N_2 CoBQ[B][Q]}{CoAB \cdot CoPQ}$	CoB	$\frac{K_{mA} V_1 [B][Q]}{K_{iQ}}$	$K_{iP} = \frac{CoB}{CoBQ} = \frac{k_4}{k_{-4}}$
$\frac{N_2 CoPQ[P][Q]}{CoAB \cdot CoPQ}$	N_1	$\frac{V_1 [P][Q]}{K_e}$	
$\frac{N_2 CoABP[A][B][P]}{CoAB \cdot CoPQ}$		$\frac{V_2 [A][B][P]}{K_{iP}}$	$K_{iP} = \frac{CoAB}{CoABP} = \frac{k_3 + k_4}{k_{-3}}$
$\frac{N_2 CoBPQ[B][P][Q]}{CoAB \cdot CoPQ}$	N_1	$\frac{V_1 [B][P][Q]}{K_{iB} K_e}$	$K_{iB} = \frac{CoPQ}{CoBPQ} = \frac{k_{-1} + k_{-2}}{k_2}$

$$v = \frac{V_1 V_2 \left([A][B] - \dfrac{[P][Q]}{K_e}\right)}{K_{iA} K_{mB} V_2 + K_{mB} V_2 [A] + K_{mA} V_2 [B] + \dfrac{K_{mQ} V_1 [P]}{K_e} + \dfrac{K_{mP} V_1 [Q]}{K_e} + V_2 [A][B] + \dfrac{K_{mQ} V_1 [A][P]}{K_{iA} K_e}}$$

$$+ \frac{K_{mA} V_2 [B][Q]}{K_{iQ}} + \frac{V_1 [P][Q]}{K_e} + \frac{V_2 [A][B][P]}{K_{iP}} + \frac{V_1 [B][P][Q]}{K_{iB} K_e} \quad . \tag{2.170}$$

This still very complicated equation is reduced to the already mentioned Eq. (2.133) if initial rates for the forward reaction are applied, assuming $[P]=[Q]=0$:

$$v = \frac{V_1 [A][B]}{K_{iA} K_{mB} + K_{mB} [A] + K_{mA} [B] + [A][B]} \tag{2.133 a}$$

and for the reverse reaction if $[A]=[B]=0$:

$$v = \frac{V_2 [P][Q]}{K_{iA} K_{mB} + K_{mQ} [P] + K_{mP} [Q] + [P][Q]} \quad . \tag{2.133 b}$$

2.7.2
Simplified Derivations Applying Graph Theory

The King-Altman method represents a considerable simplification compared with the derivation of rate equations applying steady-state rules, but for more complex mechanisms it is still very complicated. Applying the graph theory developed originally for electronic signals Volkenstein and Goldstein (1966) reduced the King-Altman method. A further reduction, requiring "only a rudimentary understanding of algebra", was suggested by Fromm (1970).

Closed reaction schemes, as in the King-Altman method, have to be composed for the respective mechanism. Individual enzyme forms are taken as junction points and are numbered consecutively:

$$\begin{array}{c}
\overset{①}{E} \underset{k_{-1}}{\overset{k_1 [A]}{\rightleftharpoons}} \overset{②}{EA} \\
k_4 \updownarrow k_{-4}[Q] \qquad k_{-2} \updownarrow k_2 [B] \\
\underset{④}{EQ} \underset{k_3}{\overset{k_{-3}[P]}{\rightleftharpoons}} \begin{pmatrix} EAB \\ EPQ \end{pmatrix} ③
\end{array} \tag{2.171}$$

The determinant of one junction, e.g. 1 ($\approx$[E]) is composed of two parts. First, the constants or terms of the shortest one-step pathways leading to this junction are taken, hence $2 \to 1 \approx k_{-1}$ and $4 \to 1 \approx k_4$ for this example. To this the numbers of the junction points not touched by the respective one-step pathways are written:

$$[E] = (2 \rightarrow 1)(3)(4) + (4 \rightarrow 1)(2)(3) \ .$$

Each of these numbers is substituted by expressions of those arrows which direct away from the respective junction:

$$[E] = k_{-1}(k_{-2} + k_3)(k_{-3}[P] + k_4) + k_4(k_{-1} + k_2[B])(k_3 + k_{-2}) \ .$$

Multiplying of the terms yields:

$$[E] = k_{-1}k_{-2}k_{-3}[P] + k_{-1}k_3k_{-3}[P] + k_{-1}k_{-2}k_4 + k_{-1}k_3k_4 + k_{-1}k_3k_4 + k_2k_3k_4[B]$$
$$+ k_{-1}k_{-2}k_4 + k_2k_{-2}k_4[B] \ .$$

'Forbidden' terms containing rate constants for both forward and reverse reactions of the same step, i.e. $k_{-1}k_3k_{-3}[P]$ and $k_2k_{-2}k_4[B]$, are omitted. In redundant terms ($k_{-1}k_{-2}k_4$, $k_{-1}k_3k_4$) only one expression is left so that the final relationship for [E] becomes identical to that of the King-Altman method:

$$[E] = k_{-1}k_{-2}k_{-3}[P] + k_{-1}k_{-2}k_4 + k_{-1}k_3k_4 + k_2k_3k_4[B] \ .$$

This procedure performed for all enzyme forms results in a rate equation according to the King-Altman method.

2.7.3
Combination of Equilibrium and Steady State Approach

A further simplification is based on the original assumption of Michaelis and Menten that equilibria are attained much faster than the catalytic turnover (Cha 1968). More complex mechanisms are divided into several segments in which the individual reaction steps are in rapid equilibrium with each other. These segments are separated by slow reaction steps. A fractional concentration factor f_i gives the ratio of the concentration of a certain enzyme form (E_i) to the sum of the concentrations of all enzyme forms of the respective rapid equilibrium segment:

$$f_i = \frac{[E_i]}{\sum_{i=1}^{n}[E_i]} \tag{2.172}$$

This procedure is demonstrated by the example of a *ping-pong bi bi* mechanism, assuming slow catalytic equilibria compared to rapid binding steps:

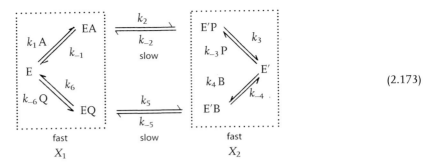
(2.173)

The enzyme forms of both segments $X_1=[E]+[EA]+[EQ]$ and $X_2=[E']+[E'B]+[E'P]$ are in rapid equilibrium with each other. The rate equation according to the general King-Altman form is:

$$v = \frac{(k_2 k_5 f_2 f_5 - k_{-2} k_{-5} f_{-2} f_{-5})[E]_0}{k_2 f_2 + k_{-2} f_{-2} + k_5 f_5 + k_{-5} f_{-5}} \quad (2.174)$$

The fractional concentration factors for the respective enzyme forms are defined by taking one enzyme form, e.g. E or E′, as reference and setting it as 1. The other enzyme forms are replaced by the concentration variables of the corresponding ligands and the rate and dissociation constants. If the ligand binds to the respective enzyme form, the concentration variable stands on the fraction bar and the dissociation constant below. If it dissociates, the fraction is reversed (hereafter the rate constants are given in place of the dissociation constants):

$$f_{EA} = \frac{[EA]}{X_1} = \frac{\frac{k_1[A]}{k_{-1}}}{1 + \frac{k_1[A]}{k_{-1}} + \frac{k_{-6}[Q]}{k_6}} = \frac{k_1 k_6 [A]}{k_{-1} k_6 + k_1 k_6 [A] + k_{-1} k_{-6} [Q]},$$

$$f_{E'P} = \frac{[E'P]}{X_2} = \frac{\frac{k_{-3}[P]}{k_3}}{1 + \frac{k_4[B]}{k_{-4}} + \frac{k_{-3}[P]}{k_3}} = \frac{k_{-3} k_{-4} [P]}{k_3 k_{-4} + k_3 k_4 [A] + k_{-3} k_{-4} [P]},$$

$$f_{E'B} = \frac{[E'B]}{X_2} = \frac{\frac{k_4[B]}{k_{-4}}}{1 + \frac{k_4[B]}{k_{-4}} + \frac{k_{-3}[P]}{k_3}} = \frac{k_3 k_4 [B]}{k_3 k_{-4} + k_3 k_4 [A] + k_{-3} k_{-4} [P]},$$

$$f_{EQ} = \frac{[EQ]}{X_1} = \frac{\frac{k_{-6}[Q]}{k_6}}{1 + \frac{k_1[A]}{k_{-1}} + \frac{k_{-6}[Q]}{k_6}} = \frac{k_{-1} k_{-6} [Q]}{k_{-1} k_6 + k_1 k_6 [A] + k_{-1} k_{-6} [Q]}.$$

If EQ is taken instead of E as reference and set as 1, the fractional concentration factor f_{EA} is:

$$f_{EA} = \frac{[EA]}{X_1} = \frac{\dfrac{k_1[A]k_6}{k_{-1}k_{-6}[Q]}}{\dfrac{k_6}{k_{-6}[Q]} + \dfrac{k_1[A]k_6}{k_{-1}k_{-6}[Q]}} = \frac{k_1 k_6 [A]}{k_{-1}k_6 + k_1 k_6 [A] + k_{-1}k_{-6}[Q]},$$

yielding the same expression. After replacing the fractional concentration factors in Eq. (2.174) the rate equation for the ping-pong mechanism in the form of the rate constants is obtained. Compared with an equation derived with the King-Altman scheme, the constants k_2, k_{-2}, k_5 and k_{-5} for the slow step are partially disregarded:

$$v = \frac{(k_1 k_2 k_3 k_4 k_5 k_6 [A][B] + k_{-1} k_{-2} k_{-3} k_{-4} k_{-5} k_{-6} [P][Q])[E]_0}{\begin{array}{l} k_1 k_2 k_3 k_{-4} k_{-6}[A] + k_{-1} k_3 k_4 k_5 k_6 [B] + k_1 k_3 k_4 k_6 (k_2 + k_5)[A][B] + k_{-1} k_{-2} k_{-3} k_{-4} k_6 [P] \\ + k_{-1} k_3 k_{-4} k_5 k_{-5} k_{-6}[Q] + k_{-1} k_{-3} k_{-4} k_{-6}(k_{-2} + k_{-5})[P][Q] \\ + k_1 k_{-3} k_{-4} k_6 (k_2 + k_{-2})[A][P] + k_{-1} k_3 k_4 k_{-6}(k_5 + k_{-5})[B][Q] \end{array}}. \quad (2.175)$$

2.8
Kinetic Treatment of Allosteric Enzymes

Allosteric enzymes have been discussed extensively in Section 1.5.4 where the fundamental features of these enzymes, cooperativity and allostery are described on the basis of equilibrium processes. Two essential models, the symmetry and the sequence models, are likewise suitable to explain allosteric behavior. Thus to investigate allosteric enzymes direct determination of ligand binding will be the most appropriate procedure. However, due to the difficulties of binding measurements, especially the requirement of high amounts of the enzyme, kinetic methods, which need only catalytic enzyme amounts, are often preferred. In most cases this is justified and the results obtained are principally comparable to those received from binding measurements, because the reaction rate v is proportional to the amount of active enzyme–substrate complex and can be a direct indication for the prevalence of the active enzyme form, i.e. the relaxed R-form.

There are, however, some restrictions to be considered. The ligand responsible for the cooperative effect is, in many cases, but not inevitably, the substrate. There are cases where an effector molecule shows a cooperative effect, while substrate saturation obeys normal hyperbolic kinetics. This can also be detected by kinetic studies, when changing the effector concentration at constant amounts of the substrate. Another aspect is the fact that cooperativity may either be caused by a modification of the affinity for the ligand (K-systems), or of the catalytic efficiency (V-systems) between both (T and R) states of the enzyme. A combination of binding and kinetic methods can resolve this question.

Analysis of allosteric enzymes by kinetic methods can be performed in a similar way as with binding methods, using the reaction velocity v instead of the saturation function [A]$_{bound}$, r=[A]$_{bound}$/[E]$_0$. The same diagrams can be used, with the difference that n, the number of binding sites on the enzyme, cannot be obtained directly, as is the case with binding measurements. The Hill plot as a special diagram for the study of cooperative phenomena (see Fig. 1.9) can also be used, the x-coordinate remains log [A] but for the ordinate log $v/(V-v)$ must be used instead of log $\overline{Y}/(1-\overline{Y})$.

Therefore, a detailed treatment of allosteric enzymes with regard to kinetic aspects is not necessary and the following discussion will be limited to the special phenomenon of *kinetic cooperativity*.

2.8.1
Hysteretic Enzymes

Several enzymes, like phosphofructokinase and the thiamine diphosphate dependent pyruvate dehydrogenase complex, show a behavior which cannot simply be interpreted by the steady-state theory. According to this theory an enzyme reaction should proceed, at least in its initial phase, in a linear, zero order manner. With this enzyme class, however, when substrate is added and all conditions for the enzyme test are fulfilled at time $t=0$, the reaction starts with a very slow, often zero, turnover, but without any further influence the rate increases time dependently until a constant turnover is reached and the steady state phase is attained. Finally, when the substrate becomes depleted, the reaction rate declines, similarly to normal enzyme reactions (Fig. 2.33 A). This initial lag phase may last a few seconds up to several minutes, depending on the special enzyme and the test conditions, such as concentrations of components, pH or temperature. For the different enzymes showing this phenomenon, different mechanisms have been discussed. Obviously the enzyme exists in an inactive state of rest when there is no substrate present. In the presence of substrate it will turn into an active state; this transformation, however, appears to be very slow. There has been speculation about the possible reasons for such slow activation processes. To elucidate this problem it may be considered that a bacteria, like *Escherichia coli* with an average lifetime of 20 min will need about 10% of its total lifetime for full activation of one of its enzymes. In analogy to magnetism, Carl Frieden created the term *hysteretic enzymes* for this kind of enzyme. As these enzymes do not react immediately, but remember already passed processes, they show a kind of enzymatic memory (*mnemonic enzymes*). They have a dampening effect on metabolic changes, short-time impulses are ignored, fluctuations are equalized.

The duration of the *lag phase* τ is estimated by extrapolation of the linear steady-state region to the abscissa or the ordinate (Fig. 2.33 A). In a semi-logarithmic diagram of the turnover rate ln v_i against time t the course of the lag phase can be linearized and the value τ can be obtained from the slope (Fig. 2.33 B). For this v_i must be reduced by the steady-state rate v_{ss} (determined from tangents on distinct points of the progress curves). Linearity in this plot serves as control of a pseudo-first order process of the lag phase. It must, how-

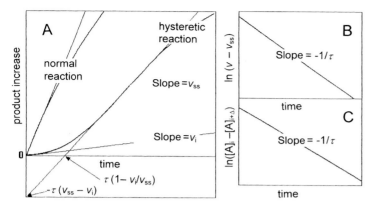

Fig. 2.33 Progression of a hysteretic reaction compared to a normal enzyme reaction of zero order. (A) Progress curve, determination of the rate in the initial (v_i) and the steady-state phase (v_{ss}) and length of the lag phase τ are indicated; (B) semi-logarithmic plot, (C) Guggenheim plot.

ever, be taken into consideration that the deviation from the linear steady-state range due to substrate depletion, as shown in Fig. 2.33A, may occur relatively early. With normal progress curves which show the steady-state range, and thus also the fastest turnover already at the start of the reaction, this is no problem. In the presence of a lag phase, however, substrate depletion may occur already before the steady-state range is actually attained, the increasing velocity during the lag phase will convert directly into the depletion range and thus the real steady-state rate will never be reached. Moreover, the direct transition from the lag to the depletion phase may give the impression at the turning point of a linear steady-state range. In this case τ can be obtained from the slope of a Guggenheim plot (Fig. 2.33C). The difference between two substrate concentrations $[A]_i$ and $[A]_{i+\Delta}$ separated by a constant time interval Δt is plotted logarithmically against time. The slope of the straight line has the value $1/\tau$. The relationship for this plot is derived from Eq. (2.3) for a first order reaction, where $e^{\Delta t/\tau}$ remains constant:

$$[A]_i - [A]_{i+\Delta} = [A]_0 \, e^{-t/\tau}(1 - e^{\Delta t/\tau}) \,. \tag{2.176}$$

2.8.2
Kinetic Cooperativity, the Slow Transition Model

Slow activation processes in enzymatic reactions observed with hysteretic enzymes can be responsible for deviations from the Michaelis-Menten kinetics in the sense of positive or negative cooperativity, without requiring interactions between subunits (*kinetic cooperativity*). Indeed this mechanism was originally discovered when

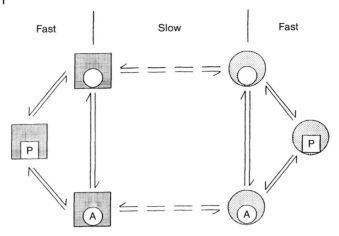

Fig. 2.34 Scheme of the slow transition model for a monomeric enzyme.

sigmoidal saturation behavior was observed with RNAase, a monomeric enzyme. According to the dogma of allosteric enzymes, sigmoidal kinetics as an expression of cooperativity should only be explainable by the interaction of (identical) subunits. The lack of any subunit interactions challenged the development of a new theory. A further contribution to this theory was revealed from the observation of slow activation processes. It is assumed that the enzyme persists in a low active form E in the absence of substrate. Substrate binding converts the enzyme into the highly active state E'. As a prerequisite of this theory the conversion from E to E' must be slow compared with the binding and the catalytic process. Therefore the theory, which is presented schematically in Fig. 2.34, is called the *slow transition model*. Sigmoidal saturation behavior of the substrate is observed if the turnover rate is measured in dependence on the substrate concentration. It can be imagined that at low amounts of substrate the enzyme will be converted to the more active E' form upon substrate binding and will return to the less active E state before a new substrate molecule binds. At high substrate concentrations, however, the next substrate molecule will bind immediately after release of the product and there will not be enough time to return to the E state. Thus, the enzyme becomes activated by increasing substrate concentrations. With this model the Hill coefficient can reach a maximum value of $n_h = 2$ as the reaction of substrate is of second order and the substrate reacts in two different places (EA and E'A). Sigmoidal saturation behavior will only be observed when the reaction rate is measured, while binding of the substrate obeys normal hyperbolic saturation behavior. Therefore the slow transition model can easily be differentiated from the symmetry and sequential model for allosteric enzymes by comparing kinetic and binding methods.

As the kinetic cooperativity was first discovered with monomeric enzymes, like RNAase and wheat germ hexokinase, the presence of only one subunit for an enzyme showing sigmoidal saturation behavior can be taken as a strong indication for the validity of this model. On the other hand, the presence of more identical subunits is not an unequivocal prove for subunit–subunit interactions. The bacterial pyruvate dehydrogenase complex as the largest soluble enzyme aggregate of the cell with an assembly of 24 protomers is also a hysteretic enzyme obeying the slow transition model (Bisswanger 1984).

2.9
pH and Temperature Dependence of Enzymes

2.9.1
pH Optimum and Determination of pK Values

Enzymes react very sensitively to pH changes. If the activity of an enzyme is tested in dependence on the pH mostly a bell-shaped curve is obtained (Fig. 2.35 A). Its maximum (*pH optimum*) coincides frequently with the physiological pH value (about pH 7.4) and the activity decreases significantly at both the acid and the alkaline site and drops finally to zero towards the more extreme pH ranges at both sites. Actually the shape of the curve is determined by at least two effects, the participation of ionic groups in the catalytic mechanism and the involvement of charged groups for stabilization of the protein structure.

For determination of the pH dependence of an enzyme reaction some experimental aspects must be considered. Saturating conditions with respect to all components, like substrates or cofactors are required. Since saturation can only be attained at infinite concentrations, measurements are performed at a high (but not completely saturating) substrate concentration and, therefore, only an apparent maximum velocity V_{app} instead of the true V (which can only be obtained by extrapolation) will be obtained. To vary the pH in an enzyme assay, the pH of the buffer will be changed. It must, however, be considered, that the components in the assay mixture can influence the pH. With the reactions of several enzymes, such as lipase and cholinesterase, the pH will change during the reaction due to the release of acid components. Therefore the pH should be measured directly in the assay mixture before and at the end of the reaction. It must further be borne in mind that the capacity of buffer systems is limited to one pH unit below and above of its pK_a value, at most. Therefore a pH dependence of an enzyme cannot be determined with only one buffer system. However, care must be taken when combining different buffer systems, because the enzyme activity depends also on the type of the buffer, the nature of the ions and the ion strength. Different buffers must at least be compared in overlapping regions. The use of universal buffers, covering a broad pH range, like the Teorel-Stenhagen buffer, is strongly recommended. It must further be considered that essential components in the test assay, substrates, coenzymes (NAD) or helper enzymes in coupled tests also possess their own pH dependences, which may deviate severely from that of the enzyme and may cause misinterpretations.

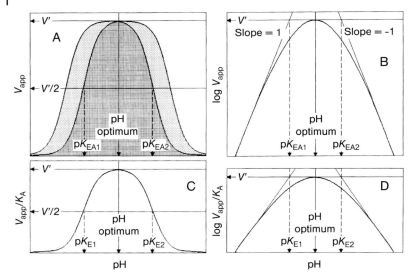

Fig. 2.35 pH behavior of enzymes. (A) Direct plot of the apparent maximum velocity V_{app} against pH. The inner curve shows an ideal pH optimum curve with two titratable essential groups in the catalytic center, while the outer curve shows a pH stability curve. Modes of determination of the pK values of the EA complex (B) and of the free enzyme E according to Dixon and Webb (1973) are shown in (C) and (D).

Ionic groups are frequently involved in enzyme catalysis, e.g. as acid–base catalysts, and the state of protonation is essential for the reaction. A prominent example is the catalytic triade of serine proteases, like chymotrypsin, where the hydroxy group of serine, the imidazole group of histidine, and the carboxy group of aspartate act together and the transfer of a proton from the serine hydroxy group to an imidazol nitrogen is an essential step for the cleavage of the peptide bond. A pH changes influence the protonated state and impairs the catalytic mechanism. In a simple diprotic system the pH optimum curve is composed of two titration curves, an increasing and a decreasing one, and the pK_{EA} values for both groups in the enzyme substrate complex corresponds to the pH values of the respective inflexion points (Fig. 2.35 A).

According to Dixon and Webb (1974) a logarithmic scale is applied for V_{app} (Fig. 2.35 B). Normal titration curves will rise, with slopes of 1 and −1, respectively, at both sites of the optimum curve. The respective tangents meet a horizontal line through the maximum of the curve at the positions of the pK_{EA} values. If protonation at the active site of the enzyme is altered by substrate binding, the pK_{EA} values change to pK_E values and the apparent substrate binding constant K_A becomes dependent on the protonation of the respective groups and, thus, on the pH value (while intrinsic K_A-values are pH independent). The pK_E values of the free enzyme are derived from plots of V_{app}/K_A (or V_{app}/K_m) or log V_{app}/K_A against the pH values in the same manner as the pK_{EA} values (Fig. 2.35 C and D).

Ionization constants can also be derived from secondary plots of the Lineweaver-Burk diagram. The dependence of the velocity v on the substrate concentration is measured at different pH values. Plotting $1/v$ against $1/[A]$ results in a series of straight lines with a common intercept left of the ordinate. A secondary plot of the slopes of these lines against $1/[H^+]$ yields a straight line intersecting the abscissa at $-1/K_{EA}$. However, the values of pK_1 and pK_2 must differ by more than 3.5 pH units to be resolved by this plotting method. If there are more than two ionic groups involved deviating curves will be obtained.

If only one single group is to be protonated (or the process of protonation is only relevant for enzyme activities within a certain pH range, either acid or basic), a pH optimum curve with only one flank will arise. With more than two groups the titration curves superimpose and may become flattened or steps will appear in the flanks.

The pK_a values derived from the pH optimum curve can give an indication of the type of the groups involved in the catalytic process, e.g. for the special amino acid residues like aspartate ($pK_a = 3.86$), histidine (6.09), cysteine (8.3), serine (9.15), tyrosine (10.11), and arginine (12.28). However, the integration of these groups into the protein structure can modify these values even by some pH units.

2.9.2
pH Stability

Besides the specific effects of groups directly involved in the catalytic mechanism, it must be considered that an enzyme carries a lot of charged groups inside and on the surface. These are important for the enzyme function, for the balance between flexibility significant for the catalytic mechanism, like formation of the transition state, and rigidity to maintain the three-dimensional structure or for executing controlled conformational changes. While protonation of essential residues in the active center is predominantly a reversible process, changes in the charge of structurally important groups often cause irreversible damage to the native structure. Reversible and irreversible pH effects can be differentiated by comparing pH stability curves with pH optimum curves. For optimum curves the respective pH is directly adjusted in the test assay and the activity is measured immediately after addition of the enzyme. For pH stability curves the enzyme is pre-incubated alone (or in combination with distinct components) at the respective pH for a defined time interval (e.g. 1 h) and thereafter tested at the optimum pH in the normal enzyme assay. Reversible processes should not influence the enzyme activity. Therefore, in the optimum range, a broad plateau will frequently be observed, and also the reduction in activity occurs more at extreme pH values (Fig. 2.35 A). The stability test should also be performed with pH-sensitive components in the test assay.

Enzymes tend to destabilization especially at their isoelectric point pI, where positive and negative surplus charges are just balanced and the solubility of the enzyme becomes strongly reduced.

2.9.3
Temperature Dependence

Enzymes are very sensitive to temperature changes. Enzyme catalyzed reactions obey the same rules as normal chemical reactions, which increase their velocity according to the van 't Hoff rule by a factor of 2–3 per each 10 °C. Theoretically there is no upper limit for this increase, however, above a distinct temperature the rate of enzyme catalysis reactions decreases to zero (Fig. 2.36 A). This decrease is due to the limited stability, a general feature of any protein. Therefore the temperature behavior of an enzyme is composed of two different and independent processes. Although the curves obtained when the enzyme activity is tested in dependence on temperature resemble the bell-shaped pH optimum curves with an increasing and decreasing branch, they cannot be interpreted in a similar manner. Due to the decline in the reaction rate a maximum with highest activity is passed but, unlike the pH dependence, the enzyme is not at its optimum, as the maximum indicates the beginning of an irreversible denaturation process. Therefore, the term "temperature optimum" for this maximum should be avoided. Optimal temperatures are not those with the highest activity, rather they must clearly be below the range of inactivation. Moreover, protein denaturation is not only a temperature-dependent, but also a time-dependent process. Incubation of the enzyme at a high temperature for a longer time will cause larger inactivation and the temperature maximum will shift towards lower temperatures. Because of this the temperature maximum of a distinct enzyme cannot be regarded as a constant value, in contrast to the pH optimum.

Most enzymes remain active up to the physiological temperature (37 °C), but become unstable at only a few degrees beyond it, between 40 °C and 50 °C. Very temperature sensitive enzymes suffer inactivation at even lower temperatures, like the alcohol dehydrogenase which becomes instable shortly above 30 °C, while the closely related lactate dehydrogenase denatures beyond 50 °C. Thermo-

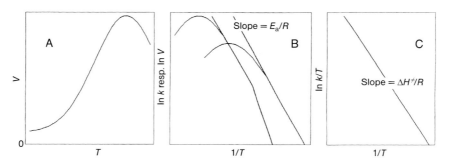

Fig. 2.36 Temperature behaviour of enzymes. (A) Direct diagram of the apparent maximum velocity V_{app} against the absolute temperature T (K); (B) Arrhenius plot; (C) diagram for the determination of the reaction enthalpy $\Delta H^{\neq}$ of the transition state.

philic microorganisms living in hot springs survive even in boiling water and it may be assumed that all of their enzymes can resist such high temperatures. This is indeed the case for many, but not for all of these enzymes, some become inactivated significantly below the growing temperature of the organism, so that there must exist stabilizing factors in the cell. High concentrations of proteins and other cell components are known to convey a stabilizing effect on enzymes. It is, anyway, an interesting fact that enzymes from thermophilic organisms are structurally closely related to those found in mesophilic organisms and even in the amino acid composition no essential differences can be detected, beside some increase in hydrophobic and charged residues, causing a higher rigidity of the proteins. So proteins need not necessarily be temperature sensitive and considering that the primeval atmosphere in which the first organism lived was considerably hot and became colder during evolution, it may be assumed that enzymes from mesophilic organisms and those with a constant body temperature lost their stability because they no longer need it.

The temperature behavior of enzymes can be studied by two different procedures. The enzyme may either be tested directly in an assay adjusted to the respective temperature, or it may be pre-incubated at a distinct temperature and tested after different time intervals at the normal assay temperature. While the first procedure considers directly the temperature dependence of the activity, the second procedure observes the time-dependence and thus the long-term stability of the enzyme. In the first case the reaction is tested immediately after addition of the enzyme. Longer pre-incubations should be avoided to exclude the time-dependent aspect, which should be studied separately. In the lower temperature range the enzyme should remain stable, while at higher temperatures inactivation starts. Therefore, to test the enzyme stability as a control a lower pre-incubation temperature should be chosen to be compared with the inactivation process observed at higher temperatures (Fig. 2.37 A). It must be remembered that enzymes can lose their activity due to other processes, like proteolytic attack, or oxidative processes. The thermal instability of other compounds, substrates, coenzymes or helper enzymes in coupled assays must also be considered. The time course for thermal inactivation of an enzyme follows frequently a first order reaction. Actually, denaturation of an enzyme is a complicated process, passing through several steps, but it can be assumed that one single step will be responsible for the loss of the catalytic activity. If the denaturation curves (Fig. 2.37 A) are plotted in a semi-logarithmic manner, straight lines may be expected and rate constants for the denaturation at the respective temperature can be calculated from the slopes (Fig. 2.37 B).

For plotting the temperature curves the catalytic constant, k_{cat} should be taken, but mostly the reaction velocity v is used, which is directly related to k_{cat} as long as the enzyme concentration can be taken as constant. Saturating concentrations should be applied for all components, especially substrates and cofactors, so that v (respectively V_{app}) approaches V (see Section 2.9.1). The curvature obtained by direct plotting (Fig. 2.36 A) can be analyzed by a diagram based on the empirical equation of Svante Arrhenius (1889):

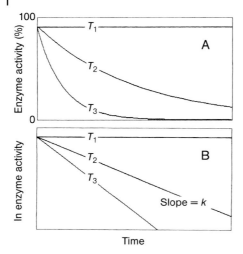

Fig. 2.37 Thermostability of enzymes in the direct (A) and semi-logarithmic plot (B). The enzyme is pre-incubated before determination of its activity at distinct temperatures, T_1, below, T_2, within, T_3 above the temperature maximum.

$$k = A \cdot e^{-E_a/RT}, \qquad (2.177)$$

which by logarithmic conversion yields the linear relationship:

$$\ln k = \ln A - \frac{E_a}{RT} \quad \text{or} \quad \log k = \log A - \frac{E_a}{2.3RT} \qquad (2.178)$$

where E_a is the activation energy of the transition state within the catalytic mechanism. It can be derived from the slope of the Arrhenius diagram. The constant A represents the probability of the reaction and contains components for the collision frequency and the orientation of the colliding particles. R is the gas constant, T is the absolute temperature (Kelvin, $0\,°C = 273.15$ K). The Arrhenius equation describes the increasing part of the temperature function without considering the thermal inactivation of the protein and this area should become linear (Fig. 2.36 B, right curve). Thus, deviation in the upper temperature range due to commencing inactivation can easily be detected.

Deviations are sometimes also observed in the lower temperature range and can be due to conformational changes of the respective enzyme. Especially, thermophilic enzymes, which cover a broad temperature area, often show inhomogeneities in the upper linear range, which can be interpreted as an adaptation to the higher temperature range (Fig. 2.36 B, left curve). Further reasons can be a mixture of isoenzymes or coupled enzyme assays containing two or more enzyme species differing in their temperature behavior. In membrane-bound enzymes transitions of the membrane can be transmitted to the enzyme. The activation energy is a complex

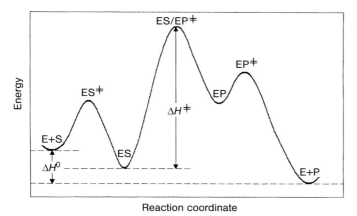
Fig. 2.38 Energy profile of an enzyme-catalyzed reaction.

constant which includes several processes occurring during the catalytic mechanism, each possessing its own temperature correlation, so that the rate determining step can change during the temperature increase. The values for E_a range mostly between 40 and 50 kJ mol^{-1} and can be estimated from the slope of the Arrhenius plot or by measuring v at two different temperatures T_1 and T_2 according to Eq. (2.179), which is obtained from Eq. (2.177) by definite integration:

$$\log \frac{k_2}{k_1} = \frac{E_a}{2.3R} \left(\frac{T_2 - T_1}{T_1 T_2} \right). \tag{2.179}$$

The theory of Henry Eyring (1953) gives a description of the principle of enzyme catalysis. Upon binding to the enzyme the substrate forms a transition state which reduces the activation energy necessary to surmount the energy barrier for conversion to the product. Only substrates disposing of sufficient energy are able to overcome the energy barrier of the transition state. The lower the barrier the higher the share of substrate molecules to become converted to product. Fig. 2.38 shows the energy profile of an enzyme-catalyzed reaction, assuming transition states for binding of the reactants from both directions, as well as for the effective conversion process. For several enzyme reactions compounds (*transition state analogs*) have been created which mimic the transition state. They bind by more strongly, by several orders of magnitude, to the enzyme than the intrinsic substrate and, although binding non-covalently, can hardly be removed from the enzyme (see Section 2.5.2.3).

Arrhenius developed his equation in analogy to the van't Hoff reaction isobar, describing the dependence of the dissociation constant K_d on the temperature at the reaction equilibrium:

$$\left(\frac{d \ln K_d}{dT} \right)_P = \frac{\Delta H^0}{RT^2}. \tag{2.180}$$

The standard reaction enthalpy ΔH^0 can be taken as temperature independent within narrow ranges, so that integration

$$\ln K_a = -\frac{\Delta H^0}{RT} + C \qquad (2.181)$$

yields a linear relationship between the logarithm of the dissociation constant and $1/T$. ΔH^0 can be derived from the slope. The integration constant includes the reaction entropy ΔS^0:

$$\ln K_a = -\frac{\Delta H^0}{RT} + \frac{\Delta S^0}{R}, \qquad (2.182)$$

which can be obtained from the ordinate intercept. Eq. (2.180) is derived from the relationship for the Gibbs free standard energy $\Delta G^0 = \Delta H^0 + T\Delta S^0$. The Gibbs free energy is related to the dissociation constant: $\Delta G^0 = RT \ln K_d$. For the free energy of the transition state $\Delta G^{\ddagger}$ the expression

$$\Delta G^{\ddagger} = -RT \ln K^{\ddagger} = \Delta H^{\ddagger} - T\Delta S^{\ddagger} \qquad (2.183)$$

is obtained. According to quantum mechanics the rate constant $k^{\ddagger}$ for the formation of the transition state is related to the equilibrium constant of the transition state $K^{\ddagger}$: $k^{\ddagger} = K^{\ddagger}(RT/N_A h)$. N_A is the Avogadro constant, h is Planck's constant. From Eq. (2.183) the relationship for the transition state is obtained:

$$\log \frac{k}{T} = -\frac{\Delta H^{\ddagger}}{2.3RT} + \frac{\Delta S^{\ddagger}}{2.3R} + \log \frac{R}{N_A h}. \qquad (2.184)$$

A linear function should result by plotting $\log k/T$ against $1/T$ and $\Delta H^{\ddagger}$ can be derived from the slope (Fig. 2.36C). In contrast to the Arrhenius plot absolute values must be taken for $k^{\ddagger}$. The Arrhenius activation energy is related to the enthalpy of the transition state by the relation: $E_a = \Delta H^{\ddagger} + RT$.

2.10
Isotope Exchange

Isotopes are valuable tools for the study of enzyme reactions. Two frequent applications are described here: isotope exchange kinetics, providing valuable information, especially for multi-substrate reactions, and the kinetic isotope effect, providing information about certain mechanisms of enzyme catalysis.

2.10.1
Isotope Exchange Kinetics

Isotope exchange kinetics can be described under two aspects: the system, consisting of substrate, product and enzyme, may be either in or out of equilibrium. Systems in equilibrium allow simple and unique interpretations, and only these will be discussed here. More extensive descriptions are to be found, e.g., in Fromm (1973), Huang (1979), and Purich and Allison (1980).

In the presence of enzyme the system is allowed to attain equilibrium. Substrates and products are already added in their equilibrium concentrations according to the equilibrium constant $K_e = [P][Q]\ldots/[A][B]\ldots$ to speed up the process. The equilibrium constant can be obtained by setting a fixed P/A ratio as shown in Fig. 2.39. After adding the enzyme, the size and direction of the shifts $\Delta[A]$ or $\Delta[P]$ are determined. The intercept of the resulting curve at $\Delta[A]$ or $\Delta[P] = 0$ equals the position of equilibrium. A particular substrate/product pair, e.g., B/P in a bisubstrate reaction, is varied at several concentration levels, the change lying in the ratio of the equilibrium concentration. The complementary pair A/Q remains constant. A small quantity of a component (e.g., A* as a radioactive isotope), not affecting the equilibrium, is added. A↔Q exchange as the result of changes in the B/P pair is recorded by time-dependent measurement.

Derivation of the rate equations of the exchange reaction R for different multi-substrate reactions is complex (see Huang, 1979; Purich and Allison, 1980) and not very informative, as kinetic constants can hardly be obtained from the isotope exchange itself. Rather, the profiles of the exchange rate are indicative of particular reaction types, and one can distinguish between hyperbolic (H), hyperbolic with complete depression (HCD) and partial depression (HPD),

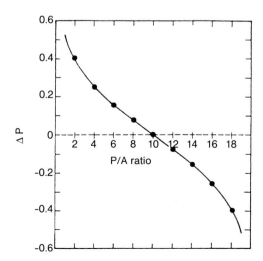

Fig. 2.39 Determination of the equilibrium constant of a reaction by approximation of the substrate/product ratio P/A. Enzyme is added to a fixed mixture of A and P and the direction of the change is monitored. Equilibrium is reached at ΔA or $\Delta P = 0$ (Purich and Allison, 1980).

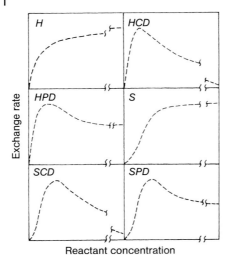

Fig. 2.40 Profiles of isotope exchanges for various enzyme reaction mechanisms. H, hyperbolic; HCD, hyperbolic with complete depression; HPD, hyperbolic with partial depression; S, sigmoidal; SCD, sigmoidal with complete depression; SPD, sigmoidal with partial depression (Purich and Allison, 1980).

Table 2.5 Profiles of isotope exchanges in single- and multi-substrate reaction H, hyperbolic; HCD, hyperbolic with complete depression; L, linear (according to Purich and Allison, 1980)

Mechanism	Exchange	Variable substrate-product pair			
		A-P	B-P	A-Q	B-Q
Uni-uni	A ↔ P	H			
Ordered bi-uni	A ↔ B	H	HCD		
	B ↔ P	H	H		
Random uni-bi	all	H	H		
Rapid equilibrium					
Ordered bi-bi	A ↔ P	H	HCD	H	HCD
	B ↔ P	H	H	H	H
	A ↔ Q	HCD	HCD	H	HCD
	B ↔ Q	HCD	HCD	H	H
Theorell-Chance bi-bi	A ↔ P	H	H	H	HCD
	B ↔ P	H	L	H	H
	A ↔ Q	HCD	H	H	HCD
	B ↔ Q	HCD	H	H	H
Random bi-bi	all	H	H	H	H

sigmoidal (S), and sigmoidal with complete depression (SCD) and partial depression (SPD). Depression is defined as a decreasing curve at higher concentrations of the varied substrate/product pair (Fig. 2.40, Table 2.5). Sigmoidal curves can also be indicative of cooperative effects.

2.10 Isotope Exchange

Especially, the A↔Q exchange rate is diagnostic for an obligatory *ordered bi-bi mechanism*, where A precedes binding of B, and P is released before Q (Scheme 2.130). A↔Q exchange requires an enzyme species that can bind to the labeled compound A*. This is the case at medium concentrations of the B/P pair. At first the A↔Q exchange rate increases but, at high B/P amounts, the enzyme species accessible for A* becomes depleted and the A↔Q exchange rate decreases (Fig. 2.40 C, D). A normal hyperbolic exchange profile is achieved for the BP exchange by increase in the A/Q pair, as this will create predominance of the enzyme species EA and EQ, required for the combination with B* and P* (Fig. 2.40 H).

Because of alternative pathways, where A* can bind either to E or to EB, no decrease in the A↔Q exchange rate is observed on increasing the B/P pair in the *random bi-bi mechanism*. For symmetry reasons, this statement is valid for all kinds of exchange in this mechanism. If in a random mechanism the conversion of the ternary complexes is rate limiting, the velocity for each exchange will be identical. Otherwise exchange rates will differ.

A speciality of the *ping-pong mechanism* is that the exchange is already possible with half of the reaction, e.g., an AP exchange can be performed in the absence of the B/Q pair (and *vice versa*):

$$E + A \underset{k_{-1}}{\overset{k_1}{\rightleftharpoons}} EA \underset{k_{-2}}{\overset{k_2}{\rightleftharpoons}} E' + P, \tag{2.185 a}$$

$$E' + B \underset{k_{-3}}{\overset{k_3}{\rightleftharpoons}} E'B \underset{k_{-4}}{\overset{k_4}{\rightleftharpoons}} E + Q. \tag{2.185 b}$$

According to steady-state rules the following equation is valid for the exchange rate of the partial reaction (Eq. (2.185 a)) (A* is the marked compound, only the initial reaction is recorded, the reverse reaction of P* to EA* is disregarded):

$$\frac{d[EA^*]}{dt} = k_1[E][A^*] - (k_{-1} + k_2)[EA^*] = 0, \tag{2.186}$$

$$[EA^*] = \frac{k_1[E][A^*]}{k_{-1} + k_2}. \tag{2.187}$$

The velocity of the exchange reaction is:

$$v^* = k_2[EA^*]. \tag{2.188}$$

Inserting for [EA*] gives:

$$v^* = \frac{k_1 k_2[E][A^*]}{k_{-1} + k_2} \tag{2.189}$$

If the exchange takes place in the absence of B and Q, $[E]_0 = [E] + [E'] + [EA]$:

$$[EA] = \frac{k_1[A][E]}{k_{-1}}, \quad [E'] = \frac{k_1 k_2 [A][E]}{k_{-1} k_{-2}},$$

$$[E]_0 = [E]\left(1 + \frac{k_1[A]}{k_{-1}} + \frac{k_1 k_2 [A]}{k_{-1} k_{-2}}\right) \tag{2.190}$$

$$v^* = \frac{k_2 [E]_0 [A^*]}{\dfrac{(k_{-1} + k_2)}{k_1}\left(1 + \dfrac{k_1[A]}{k_{-1}} + \dfrac{k_1 k_2 [A]}{k_{-1} k_{-2} [P]}\right)}. \tag{2.191}$$

At similar specific radioactivity [A*] and [A] may be set equal. The reciprocal formula is:

$$\frac{1}{v^*} = \frac{k_{-1} + k_2}{k_1 k_2 [E]_0}\left(\frac{1}{[A]} + \frac{k_1}{k_{-1}} + \frac{k_1 k_2}{k_{-1} k_{-2} [P]}\right), \tag{2.192}$$

and in the form of the kinetic constant:

$$\frac{1}{v^*} = \frac{K_{mA}}{V_1 [A]} + \frac{K_{mA}}{V_1 K_{iA}}\left(1 + \frac{K_{iP}}{[P]}\right). \tag{2.193}$$

From a plot of $1/v^*$ against $1/[A]$ with different amounts of [P], parallels are obtained, whose ordinate intercepts, plotted against $1/[P]$ in a secondary diagram, yield a straight line with the ordinate intercept of a reciprocal maximum exchange velocity: $V^* = k_{-1} k_2 [E]_0 / (k_{-1} + k_2)$. This differs from the maximum velocity of the forward reaction of the ping-pong mechanism ($V_1 = k_2 k_4 [E]_0 / (k_2 + k_4)$), the ratio between k_{-1} and k_4 being the determining factor. If both rate constants are equal then both maximum velocities are also identical. If k_{-1} is larger than k_4, the exchange rate $V^*_{A \leftrightarrow P} > V_1$, and vice versa. Beyond this there exist no further relationships between these two terms.

Between the maximum initial and exchange velocities the following relationship holds:

$$\frac{1}{V^*_{A \leftrightarrow P}} + \frac{1}{V^*_{B \leftrightarrow Q}} = \frac{1}{V_1} + \frac{1}{V_2}, \tag{2.194}$$

in the form of the rate constants,

$$\frac{k_{-1} + k_2}{k_{-1} k_2} + \frac{k_{-3} + k_4}{k_{-3} k_4} = \frac{k_2 + k_4}{k_{-2} k_4} + \frac{k_{-1} + k_{-3}}{k_{-1} k_{-3}} \tag{2.195}$$

i.e., all four parameters must be known in order to evaluate the relationship between exchange rate and initial velocity. This also demonstrates that it is practically impossible to determine kinetic parameters from exchange experiments. The ping-pong mechanism is one of the few exceptions. For the partial reaction of the A↔P exchange in the absence of B and Q the slope of the straight line in the double-reciprocal plot has the value K_{mA}/V_1, and the ordinate intercept $K_{mA}/V_1 K_{iA}(1+K_{iP}/[P])$. The ordinate intercepts, plotted against $1/[P]$, yield $K_{mA}/V_1 K_{iA}$ as ordinate, and $-1/K_{iP}$ as abscissa intercepts. Conversely, the constants K_{mB}, K_{iB}, and K_{iQ} are obtained from the B↔Q exchange in the absence of A and P.

Abortive complexes, i.e., non-productive enzyme species seriously influence the analysis of exchange data. Such abortive complexes can evolve when ligands bind to the enzyme under conditions that prevent the enzyme from performing the catalysis. Binding of pyruvate and NAD^+ by lactate dehydrogenase is an example. Both ligands are already oxidized and hydrogen transfer is not possible.

2.10.2
Isotope Effects

2.10.2.1 Primary Kinetic Isotope Effect

Isotope effects rest on the fact that the altered mass of the isotope influences the turnover rate of the reaction. In most cases the mass differences are minimal, e.g., only 8% between ^{13}C and ^{12}C. The maximum isotope effects observed are $^{13}C/^{12}C = 1.07$, $^{15}N/^{14}N = 1.04$, and $^{18}O/^{16}O = 1.06$. This requires high precision of the respective analytic method, e.g., mass spectroscopy. Mass differences of 100 or 200%, however, exist between deuterium (D) or tritium (T) and hydrogen (H). In reactions where proton transfer is rate-determining, a significant reduction in the reaction rate is observed in D_2O. The ratio of the rate constants k_H/k_D ranges between 2 and 15. The effect is even more pronounced in tritium with a relationship of $\log k_H/k_T = 1.44 \log k_H/k_D$. However, only D_2O is available in 100% concentration, so that each molecule reacts in a homogeneous population, while in reactions with T_2O only 1 in 10^{10} molecules exists as ^3H. Macroscopically, a reduction in the turnover rate cannot be realized under this condition and a different method must be applied to detect the effect. If, e.g., ^3H is released into water in the LDH reaction, ^3H_2O is formed in a non-rate-determining proton transfer ($k_H/k_T = 1$) with the same specific radioactivity as the substrate, at $k_H/k_T = 10$, however, only with 1/10. If 1 mmol lactate is converted into pyruvate, only 100 µmol ^3H_2O is formed. Water can be distinguished from substrate by its volatility. Thus, the reaction discriminates ^3H-marked molecules so that the specific radioactivity of the remaining substrate molecules increases.

The isotope effect is based on the difference in energy of zero point oscillations. While heavy C atoms are fixed, the frequency of the extended oscillation depends on the mass difference between D and H. The energy of the basic state for C–D binding is lower than that of C–H binding, while both possess the same energy in the transition state. Thus the energy difference between basic and transition states for C–D binding is higher than for C–H binding. The activation energy for the cleavage of a C–D bond is 4.8 kJ mol^{-1} higher than for C–H binding. This corresponds to an almost 7-fold difference in velocity.

It is generally assumed that with a reduction of the velocity by a factor of 2–15 a primary kinetic isotope effect is exhibited, and the cleavage of a C–H bond is rate determining. Conversely, the absence of an isotope effect indicates that the cleavage of a C–H bond is not rate limiting, even if it occurs during the total reaction process. For a reduction of the velocity by less than two, the proton transfer is only partially rate determining. There either exist two or more comparatively slow steps, or a secondary isotope effect may be effective (see Section 2.10.2.3).

2.10.2.2 Influence of the Kinetic Isotope Effect on V and K_m

The isotope effect is manifested in the reduction of proton transfer velocity. Thus, in an enzymatic reaction the maximum velocity is affected, and not so much the Michaelis constant. This can be tested by a double-reciprocal plot of $1/v$ against $1/[A]$, the turnover rate having been measured in dependence on the substrate concentration with hydrogen, as well as in the presence of the isotope. With normal behavior, straight lines are obtained that display different ordinate intercepts because of different maximum velocities, but show a common abscissa intercept due to the same Michaelis constant. For $V_H/V_D > 1$, V is partially or completely controlled by a single step including a C–H bond cleavage. High V_H/V_D ratios of up to 8 are rarely observed; the V_H/V_D ratio ranges mostly between 1.5 and 2.0. These relatively small isotope effects are scarcely determined by a single step. A high energy barrier may be substituted by several small barriers, each step being partially rate determining.

If $K_m = K_d$ no isotope effect is to be expected for the Michaelis constant, as the physical binding step is not sensitive to isotope exchanges. Because of the different size of D and H there might be a steric isotope effect when the active center is of limited size. Also, bindings with deuterium are more difficult to polarize than those with hydrogen. If the formation of a transition complex includes removal of a proton, K_m may also be affected.

For the description of the isotope effect the conversion from A to P and the release of product, which are both combined in the simple Michaelis-Menten equation, have to be differentiated:

$$E + A \underset{k_{-1}}{\overset{k_1}{\rightleftarrows}} EA \underset{k_{-2}}{\overset{k_2}{\rightleftarrows}} EP \underset{k_{-3}}{\overset{k_3}{\rightleftarrows}} E + P. \quad (2.196)$$

According to the steady-state rules (under conditions of the initial rate k_{-2} and k_{-3} are not considered) the following equation may be derived:

$$v = \frac{\frac{k_2 k_3 [E]_0 [A]}{k_2 + k_3}}{\frac{(k_{-1} + k_2) k_3}{k_1 (k_2 + k_3)} + [A]} \quad (2.197)$$

where $V = k_2 k_3 [E]_0/(k_2 + k_3)$ and $K_m = (k_{-1} + k_2) k_3 / k_1(k_2 + k_3)$. The influence of the isotope effect depends on the ratio between k_2 and k_3. For $k_2/k_3 < 1$ or $k_2 \ll k_3$ V becomes $V = k_2[E]_0$, or $V_H/V_D = k_{2(H)}/k_{2(D)}$, respectively. The observed isotope effect approximates the true isotope effect. For $k_2/k_3 > 1$ or $k_2 k_3$, respectively, the release of product becomes rate determining, and because of

$$\frac{V_H}{V_D} = \frac{k_{2(H)}(k_{2(D)} + k_3)}{k_{2(D)}(k_{2(H)} + k_3)} \to 1$$

an existing isotope effect is suppressed.

The relationship of the kinetic constants according to Eq. (2.197) is:

$$\frac{V}{K_m} = \frac{k_1 k_2 [E]_0}{k_{-1} + k_2} . \qquad (2.198)$$

Inserting this into the relation of the isotopes gives:

$$\frac{\left(\dfrac{V}{K_m}\right)_H}{\left(\dfrac{v}{K_m}\right)_D} = \frac{k_{2(H)}(k_{-1} + k_{2(D)})}{k_{2(D)}(k_{-1} + k_{2(H)})} = \frac{k_{2(H)}}{k_{2(D)}} \left(\frac{1 + \dfrac{k_{2(D)}}{k_{-1}}}{1 + \dfrac{k_{2(H)}}{k_{-1}}} \right). \qquad (2.199)$$

The apparent isotope effect is reciprocally dependent on the ratio of k_2/k_{-1}. For $k_2 \ll k_{-1}$, Eq. (2.199) approximates $k_{2(H)}/k_{2(D)}$, the isotope effect is fully expressed. For $k_2 \gg k_{-1}$ the isotope effect is suppressed, as $(k_{2(H)}/k_{2(D)})/(k_{2(D)}/k_{2(H)}) = 1$. Under this condition catalysis is significantly faster than decomposition of the ES complex into E + S so the isotope effect cannot manifest itself.

In many enzymatic reactions the effects on V and V/K_m are identical. In these cases K_m is identical for unlabeled and for deuterized substrates. Only if K_m has been altered for the deuterized substrate are the isotope effects on V and V/K_m different.

2.10.2.3 Other Isotope Effects

A *secondary kinetic isotope effect* occurs if a reaction is affected by an isotope-substituted C–H bond in the a-position that is not itself cleaved during the reaction process. The reason for this effect is a change in hybridization. A basically tetrahedral, sp^3-hybridized carbon atom transforms into a transition state conforming to the carbonium ion, with a planar sp^2 arrangement. The substitution of a C–H bond by a C–D bond reduces the frequency of deformation vibrations. It is easier for the substrate to form the sp^2 intermediate with a C–H bond than with a C–D bond. A ratio of $k_H/k_D = 1.38$ is to be expected; ratios from 1.02 to 1.40 are observed. The secondary isotope effect is much less pronounced than the primary and is easy to identify. A secondary isotope effect was observed, e.g., for the dehydration of malate in the fumarase reaction.

D_2O can also affect the enzyme reaction with its properties as a solvent. The change of proton concentration in D_2O compared to H_2O and thus the altered ionization in substrates and enzymes may affect functionality. In D_2O the pH value adjusted by standard buffer is changed: pD=pH+0.4. Most acids are 3–5 times weaker in D_2O than in water, corresponding to a pK difference of 0.5–0.7. The number and strength of hydrogen bonds and hydrophobic interactions are also altered. D_2O is 23% more viscous than H_2O, the O–D bond is 0.004 nm shorter than the O–H bond. Thus there are changes in the degree of polarization and in the solvent structure, which may also affect the enzyme structure.

2.11
Special Enzyme Mechanisms

2.11.1
Ribozymes

For a period of about 100 years it was a strict dogma that enzymes must in any case be proteins, at the most possessing a non-proteinogenic component, a prosthetic group or a cofactor, supporting the catalytic mechanism, when in 1982 RNA molecules with catalytic activity were detected. This appeared to be an exciting discovery which supported a previous hypothesis, that RNA may be predecessors of proteins in the development of organisms. On the other hand, although an increasing number of examples of ribozymes with new functions have been described, their role within the cell metabolism remains limited and so the general importance of protein enzymes as main catalysts and regulators of the cell metabolism need not be revised by this discovery.

Principally, two types of RNA enzyme must be differentiated. One type, including hairpin and hammerhead ribozymes and hepatitis delta virus, group I and group II introns, perform only single turnover reactions during which they become altered, e.g. by intramolecular self-cleavage reactions, resulting in a modified RNA product. This is, however, not in a strict sense a catalytic reaction, where the catalyst should not become altered during the reaction. In some cases the ribozyme can be manipulated to provide true catalytic cleavage with an external RNA substrate, establishing the potential catalytic character of these ribozymes. Ribozymes of the second type perform multiple turnovers and thus are true catalysts, like RNase P and 23S rRNA. All ribozymes depend strictly on divalent metal ions, usually Mg^{2+}. The activity of ribozymes increases over a broad range from pH 3 to 7, linearly with the pH; it is assumed that protonation of some residues will inhibit the active ribozyme form. For true catalytic units principally the rules of enzyme kinetics apply. Ribozymes follow ordinary Michaelis-Menten kinetics and a Michaelis constant and maximum velocity can be obtained (Symons 1992).

Single turnover mechanisms like the self-cleavage of native RNA into P follow a first order process

$$v = \frac{d[P]}{dt} = -\frac{d[RNA]}{dt} = k_1[RNA] \qquad (2.1\,a)$$

and can be treated as already described in Section 2.1.1. Complication can arise from an equilibrium between an inactive protonated and an active deprotonated ribozyme form preceding the catalytic reaction, as discussed by Bevilacqua et al. (2003).

2.11.2
Polymer Substrates

Enzymes cleaving polymer substrates with numerous identical bonds, such as starch, cellulose, or chitin, do not obey the normal Michaelis-Menten relationship. K_m increases with decreasing degree of polymerization or molecular mass, while V decreases, i.e., the kinetic constants change during the progress of the enzymatic cleavage reaction. Assuming all reactive bonds to be equal, the following relationship according to Chetkarov and Kolev (1984) can be applied:

$$v = \frac{k_2' \dfrac{M_A - M_{A\infty}}{N_A m_1}\left(1 + \dfrac{M_A}{M_E}\right)[E][A]}{\dfrac{k_{-1}' - k_2'}{k_1'} \cdot \dfrac{a}{(C_b^a)^\alpha} \cdot \dfrac{M_A}{N_A} + [A]} . \qquad (2.200)$$

K_m and k_2 are the kinetic constants for the reactive bonds, M_A and $M_{A\infty}$ the substrate molecular mass before and after infinite enzyme reaction. M_E is the molecular mass of the enzyme, m_1 that of the monomer unit of the polymer (e.g., glucose). N_A is the Avogadro constant, a the number of active centers per enzyme molecule, b the number of reactive bonds per substrate molecule, and b_∞ the number of uncleaved bonds after the enzyme reaction. $(C_b^a)^\alpha$ stands for the effective number of possible combinations between the active centers of the enzyme and the reactive bonds of the substrate. For most enzymes $a=1$ and $C_b^1 = b$; $\alpha = \sigma_A/\sigma_E$ is the ratio of the effective cross-section of the reactive substrate bond σ_A to the cross-section of the active site σ_E. Eq. (2.200) follows the Michaelis-Menten relationship, assuming $M_A \ll M_E$:

$$V = k_2[E] = k_2' \frac{M_A - M_{A\infty}}{N_A m_1}[E][A] = k_2'(b - b_\infty)[E][A] \qquad (2.201)$$

and

$$K_m(M) = \frac{K_m}{M_A} = \frac{k_{-1}' + k_2'}{k_1' N_A b^\alpha} = \frac{K_m'(M)}{b^\alpha} \qquad (2.202)$$

The corresponding constants can be obtained from plots after Eqs. (2.201) and (2.202), plotting the actually measured Michaelis constant and the maximum velocity, respectively, against the number of reactive centers of the substrate molecule (Fig. 2.41).

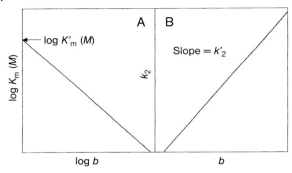

Fig. 2.41 Determination of the constants K'_m and K'_2 for the reactive bonds of a polymer substrate: (A) plotting the actual measured Michaelis constant K_m and (B) the maximum velocity k_2 against the number b of the reactive bonds of the substrate molecule.

2.11.3
Kinetics of Immobilized Enzymes

Immobilized enzymes are of increasing significance in biotechnological processes in enzyme reactors, in biosensors, and in medicine. In the cell, membrane-bound enzymes may be regarded as immobilized systems (see Section 2.11.5).

Kinetic treatment of an immobilized enzyme system depends on its specific structure, general rules can hardly be defined. Immobilization of enzymes is frequently achieved by covalent binding to solid surfaces, to a *matrix*, like dextran, agarose, synthetic polymers, glass or ceramics. As long as any effect by the matrix on the reactants, especially on the free diffusion of substrate or product, can be excluded, these systems may be treated kinetically like soluble enzymes. The matrix may, however, reject or attract substrates or products and thus affect their concentration in the vicinity of the immobilized enzyme, either negatively or positively. The (often hydrophobic) surface of the matrix may act as a barrier or boundary layer, impeding the passage of substrate to the catalytic center. This also applies to enzymes embedded in organelle membranes. Another principle of fixation is embedding the enzymes in a matrix permeable to substrate, e.g. agarose, polyacrylamide, or nylon beads. The enzyme is not modified by covalent fixation and retains its native structure to a large degree. The rules for enzymes in solution apply as long as the concentrations within the particle are equivalent to those of the surrounding solution. This requires free diffusion of all components, like substrates, products or ions (pH changes!). Deviations in the behavior of the immobilized enzyme will occur if diffusion is affected. If enzymatic transformation of the substrate is faster than its diffusion through the matrix, a depletion of substrate occurs in the area around the enzyme, the extent depending on substrate concentration in the circumfluent medium. In low substrate concentrations, with the enzyme reacting at maximum efficiency, substrate depletion is more pronounced, while being comparatively low near substrate saturation. Conversely, product accumulates with limited diffusion and is prevented from dissociating from the enzyme.

The kinetic models developed from such considerations only regard the interactions between the enzyme and its immediate surroundings on the matrix, especially the substrate. Special effects on individual enzymes cannot be considered. Immobilization of an enzyme by covalent modification of functional groups participating directly or indirectly in the catalytic mechanism can affect the reaction mechanism. Transitions of conformation into the active state or regulatory influences may be impeded or completely repressed by fixation to the matrix. The narrow attachment to the matrix may cause shielding of the active center.

Due to diffusion hindrance of the substrate, experimentally determined turnover rates v' of immobilised enzymes can differ considerably from that of the native enzyme in solution v_{kin}:

$$v' = \eta_e v_{kin} = \eta_e \frac{V[A]}{K_m + [A]} . \qquad (2.203)$$

The efficiency factor η_e is a function of the substrate concentration, for $\eta_e = 1$ the reaction is kinetically controlled. The reaction of the immobilized enzyme is equal to that of the free enzyme and behaves according to the Michaelis-Menten equation. The reaction becomes increasingly diffusion-controlled the lower the values of this factor ($\eta_e < 1$). The Michaelis-Menten equation is no longer valid, so that the usual linearization methods will yield no straight lines. Eq. (2.203) and the following considerations are based on a one-substrate reaction. As long as all other substrates and cofactors are present in saturating amounts, this treatment can also be applied for multi-substrate reactions.

Generally there exist two kinds of limited diffusion (Fig. 2.42). The external diffusion limitation is caused by a boundary layer between the matrix in which the enzyme is embedded and the circumfluent solution. The substrate has to overcome this barrier. In an internal diffusion limitation the matrix affects substrate diffusion.

2.11.3.1 External Diffusion Limitation

Substrate has to pass through a boundary layer to reach the catalytic center of an enzyme fixed to a solid liquid-impermeable surface. The processes of transport and catalysis take place successively. The flow of substrate J_A from the circumfluent solution with the substrate concentration $[A]_0$ to the active center on the surface with the substrate concentration $[A]$ is:

$$J_A = h_A([A]_0 - [A]) = \frac{D_A([A]_0 - [A])}{\delta} . \qquad (2.204)$$

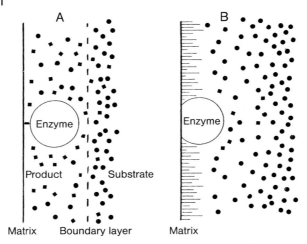

Fig. 2.42 Schematic representation of external (A) and internal (B) diffusion limitation; ●, substrate; ■, product.

$h_A = D_A/\delta$ is the transport coefficient and D_A the diffusion coefficient of the substrate, δ is the effective thickness of the boundary layer. D_A and h_A can be determined by methods like the radioactive tracer technique or diffusion cells (Rovito et al. 1973) or may be taken from the literature (Bird et al. 1960). The flow of substrate towards the active center and the enzymatic substrate turnover, usually following the Michaelis-Menten relationship, occur successively. Under steady-state conditions both processes take place with equal velocity:

$$h_A([A]_0 - [A]) = \frac{V[A]}{K_m + [A]} \ . \tag{2.205}$$

For substrate concentration, the non-dimensional term $a = [A]/K_m$ is set:

$$a_0 - a = \frac{Va}{h_A K_m (1+a)} = \mu \cdot \frac{a}{1+a} \ . \tag{2.206}$$

$\mu = V/h_A K_m$ is a non-dimensional substrate module, indicating the ratio between the reaction velocities and system transport. In the limiting case $K_m \gg [A]$, i.e., at very low substrate concentrations, Eq. (2.206) becomes:

$$h_A([A]_0 - [A]) = \frac{V[A]}{K_m} \ . \tag{2.207}$$

The overall reaction follows first order kinetics. The effective substrate concentration in the area around the active center is:

$$[A] = \frac{h_A [A]_0}{h_A + \dfrac{V}{K_m}} \,. \tag{2.208}$$

The experimentally determined turnover rate v' is:

$$v' = \frac{V[A]}{K_m} = \frac{\dfrac{V}{K_m} h_A [A]_0}{h_A + \dfrac{V}{K_m}} = \frac{[A]_0}{\dfrac{1}{h_A} + \dfrac{K_m}{V}} \,. \tag{2.209}$$

For $h_A \gg V/K_m$ the transport is faster than the kinetically controlled enzymatic reaction:

$$v' = v_{kin} = \frac{V[A]_0}{K_m} \,. \tag{2.210}$$

Vice versa, the reaction becomes diffusion-controlled at very slow transportation across the matrix, $h_A \ll V/K_m$:

$$v' = v_{diff} = h_A [A]_0 \,. \tag{2.211}$$

In the limiting case $K_m \ll [A]_0$, i.e., at saturating substrate concentration in the zero order region of the reaction, v' tends to the saturation value V (Eq. (2.205), Fig. 2.43 A). In the medium substrate range ($[A] \sim K_m$), depending on their respective size, either v_{kin} or v_{diff} contribute to the larger part of the reaction. A saturation function results, composed of parts of the transport process and the kinetic reaction. It displays, in accordance with the degree of diffusion limita-

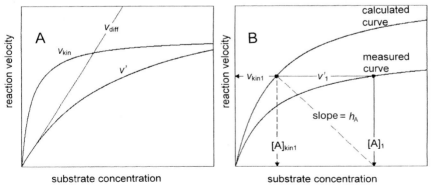

Fig. 2.43 Reaction of immobilized enzymes. (A) Comparison of measured (v') and diffusion controlled (v_{diff}) and kinetically controlled (v_{kin}) turnover rate. (B) Determination of v_{kin} and substrate concentration $[A]_{kin}$ at the active site of an immobilized enzyme.

tion, an increased apparent K_m value (obtained at half saturation) against the true K_m value of the enzyme reaction. The curve deviates with increasing diffusion limitation in the graphic linearization methods (e.g., in the double-reciprocal plot) from the linear behavior of the kinetically controlled reaction. A feature of the external diffusion limitation is the fact, that it can be influenced by stirring. Equilibration by diffusion between solution and immobilized enzyme is accelerated.

With a known transport coefficient h_A, the velocity of the kinetically controlled enzyme reaction v_{kin} and the corresponding substrate concentration $[A]_{kin}$ at the active center on the membrane surface can be determined graphically from the dependence of the measured velocity on the substrate concentration of the circumfluent solution (Fig. 2.43 B). A straight line with a slope h_A is drawn through a random point on the abscissa, corresponding to a fixed external substrate concentration $[A]_1$. The intercept of this straight line with the parallel to the abscissa at the point of the corresponding measured velocity v'_1 has the coordinates $[A]_{kin}$ and v_{kin1}. With this method characteristics of the kinetic reaction are obtained point by point, and the constants by the usual graphical methods.

2.11.3.2 Internal Diffusion Limitation

In contrast to external diffusion, internal diffusion runs parallel to the enzyme-catalyzed reaction. Due to substrate depletion, the velocity of the reaction declines with decreasing distance of the immobilized enzyme from the membrane surface, while product formation causes a local accumulation and the formation of a product gradient. The simultaneous processes of diffusion across the membrane and the kinetic reaction behave additively:

$$\frac{\partial [A]}{\partial t} = \left(\frac{\partial [A]}{\partial t}\right)_{diff} + \left(\frac{\partial [A]}{\partial t}\right)_{kin} . \tag{2.212}$$

For the diffusion, Fick's Second Law of Diffusion is applied, and the Michaelis-Menten relationship for the kinetic reaction. V''' is the intrinsic maximum velocity per volume unit of the porous medium or the membrane:

$$\frac{\delta [P]}{\delta t} = D_A \left(\frac{\delta^2 [A]}{\delta x^2}\right) - \frac{V'''[A]}{K_m + [A]} . \tag{2.213}$$

In a stationary state is $\delta[A]/\delta t = 0$:

$$D_A \left(\frac{\delta^2 [A]}{\delta x^2}\right) = \frac{V'''[A]}{K_m + [A]} . \tag{2.214}$$

The differential equation (2.214) can be resolved by numerical calculation. Non-dimensional terms are introduced for the substrate concentration $a = [A]/K_m$ and for the distance x from the surface, $l = x/L$, L being the thickness of the membrane (for a sphere with the particle radius r, L is substituted by $r/3$) and l the position within the membrane:

$$\frac{d^2 a}{dl^2} = \frac{L^2 V'''}{K_m D_A}\left(\frac{a}{1+a}\right) = \Phi_A^2 \left(\frac{a}{1+a}\right). \tag{2.215}$$

Φ_A is the substrate or Thiele module:

$$\Phi_A = L\sqrt{\frac{V'''}{K_m D_A}}. \tag{2.216}$$

The Thiele module contains three factors determining the substrate profile in the membrane: membrane thickness, diffusivity of substrate, and enzyme activity. With increasing Φ_A, the actual substrate concentration in the membrane decreases, the steepness of the substrate gradient in the membrane increases. The membrane is depleted of substrate and the enzyme reaction is slowed. There are deviations in the linearized diagrams. The substrate concentration determined at half-saturation is higher than the K_m value of the free enzyme, as in the external diffusion limitation. At low values ($\Phi_A \leq 1$) the reaction is mostly kinetically controlled and obeys Michaelis-Menten kinetics.

For the determination of the kinetic constants of immobilized enzymes it is important to measure within a broad substrate range, as in non-linear dependences over a narrow range linear regions may appear and may lead to incorrect results. Non-linear curves may be evaluated according to the usual graphical methods, assuming that diffusion limitation will predominate at very low substrate concentrations, and enzyme catalysis at high concentrations. The constants can be obtained from tangents to the extreme regions (see Fig. 2.44). These plots are based on the transformation of Eq. (2.203) for the double-reciprocal plot:

$$\frac{1}{v'''} = \frac{1}{\eta_e V'''} + \frac{K_m}{\eta_e V'''[A]_s}. \tag{2.217}$$

$[A]_s$ is the effective substrate concentration on the surface. Fig. 2.44 shows the types of deviations from the normal linear run caused by diffusion limitation. For very low concentrations at $[A]_s$, η_e approximates the efficiency factor ε for a first order reaction. The apparent Michaelis constant in this region is $K = K_m/\varepsilon$. For high Φ_A values ε becomes $1/\Phi_A$.

2.11.3.3 Inhibition of Immobilized Enzymes

All effects reducing the reaction rate of an immobilized enzyme counteract substrate depletion around the enzyme. Enzyme inhibition and limited diffusion act antagonistically. If both effects exist simultaneously, they diminish each other; in total they are weaker than would be expected from the sum of the indi-

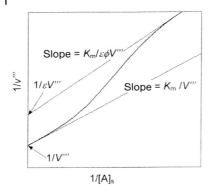

Fig. 2.44 Graphic method for the determination of kinetic constants of immobilized enzymes in the internal diffusion limitation. ε, effectiveness factor (according to Engasser and Horwath 1973).

vidual effects. Thus, under similar conditions, the degree of inhibition appears to be lower for immobilized enzymes compared with native ones. Non-linearity of the curves in linearized plots caused by diffusion limitation is attenuated, but even from non-linearity the type of inhibition remains evident: competitive inhibition only changes the apparent K_m value, but not the maximum velocity, V remains also unaffected by diffusion limitation. Non-competitive inhibition changes both parameters, which is also obvious in diffusion limitation. For a simple non-competitive inhibition ($K_{ic} = K_{iu} \equiv K_i$, see Eq. (2.72) and Section 2.5.3.2) in external limited diffusion, Eq. (2.205) would be extended to:

$$h_A([A]_0 - [A]) = \frac{V[A]}{\left(1 + \dfrac{I}{K_i}\right)(K_m + [A])} \qquad (2.218)$$

Product inhibition is a special case because, with diffusion limitation, product accumulates in the region of the immobilized enzyme and additionally enforces inhibition. Due to this effect diffusion limitation becomes reduced. In total, the immobilized enzyme shows a weaker reaction on changes in product concentration in its vicinity.

All other factors with an effect on enzyme activity also counteract diffusion limitation, e.g. partial inactivation caused by the immobilization procedure of the enzyme. The degree of inactivation is generally underestimated because of a reduction in diffusion limitation, simulating a more efficient immobilization. This gives the impression of apparently improved long-term stability as a consequence of immobilization.

2.11.3.4 pH and Temperature Behavior of Immobilized Enzymes

Immobilized enzymes show an altered dependence on pH and ionic strength, particularly if these parameters are altered by the enzyme reaction itself, like consumption or formation of acids (e.g. proteases) or bases (e.g. urease) as substrates or products of the enzyme reaction. Accumulation of such reaction products by

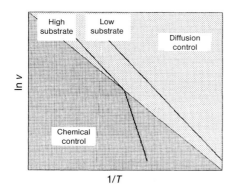

Fig. 2.45 Temperature sensitivity of immobilized enzymes.

diffusion limitation may shift the apparent pH optimum of the enzyme by 1–2 pH values, compared with the free enzyme. Similar shifts of the pH optimum curve occur when the enzyme is fixed to a positively or negatively charged matrix.

In the Arrhenius plot (see Section 2.9.3), immobilized enzymes frequently exhibit inhomogeneities, i.e. transitions between regions of different slopes. This occurs in the lower temperature range when the whole process is chemically controlled, because the enzyme reaction is very slow, and diffusion limitation is not expressed (Fig. 2.45). With increasing turnover rate at higher temperatures substrate depletion occurs, the total reaction exhibiting the characteristics of diffusion limitation, resulting in a lower slope of the Arrhenius plot. At very low substrate concentrations, however, diffusion control covers the whole measurement range, and only a single straight line is obtained.

2.11.4
Transport Processes

Enzyme catalyzed reactions and transport processes through membranes share many similarities. There are, however, also remarkable differences. Enzyme reactions are directed, substrate becomes converted to product and even at complete equilibrium conditions the forward and the reverse reactions differ, as substrate and product are different compounds, while the metabolite before and after the transport process differs only in its relative localization at the *cis* or *trans* site of the membrane. Under equilibrium conditions the situation at both sites must be the same, otherwise an asymmetric component has to be introduced. Correspondingly, there are several types of transport mechanisms through the membrane. Usually apolar low molecular weight compounds penetrate the membrane nearly unhindered by a rapid non-saturable process, which can be described by mere diffusion (see Fig. 2.14). For polar compounds the membrane marks a barrier and special transport systems are required, explicitly the facilitated diffusion and the active transport. Both can be regarded as saturable processes, but while facilitated diffusion is principally symmetric, asymmetric transport, e.g. transport against a concentration gradient, requires energy. Therefore only active transport can be regarded as directed, e.g. there are essential differences in the affinity for the substrate and in the velocity between the *cis* (e.g. inner) and the *trans* (outer) site of the membrane. The treatment of such processes can first be based on enzyme kinetic rules, but must regard the special conditions of the transport systems. It must further be considered that, besides distinct types of transport systems, different transport processes may be present at the same time, e.g. unspecific, non-saturable diffusion through the membrane besides facilitated diffusion or active transport. All these possibilities cannot be treated here, rather the similarities of enzyme kinetics with the transport phenomena should be shown.

In a first approach transport processes can be regarded as enzyme reactions and thus the steady state concept can be applied assuming a linear zero order time progression of the transport of the ligand across the membrane. If this holds, the Michaelis-Menten equation should also be valid and the transport rate should increase with the substrate concentration in a hyperbolic manner.

A ligand [A] will be transported from one (*cis*) to the other (*trans*) site of the membrane. For simplicity it is assumed that the ligand concentration at the *trans* site should be zero, while the *cis* site is exposed to variant ligand concentrations. Principally a relationship according to the Michaelis-Menten equation (Eq. (2.18)) can be applied for the transport velocity from *cis* to *trans* v_{ct}:

$$v_{ct} = \frac{V_{ct}[A]}{K_{ct} + [A]} \tag{2.219}$$

which yields, accordingly, a hyperbolic curve with a maximum velocity V_{ct} at saturation and a dissociation constant K_{ct} corresponding to the ligand concentration at half saturation. Similarly the same data analysis, applying linearized plots, like Lineweaver-Burk, Hanes or Eadie-Hofstee plots, can be performed to determine the constants.

Often it is easier to allow the ligand to equalize between both sites and to add radioactively labelled ligand to the *cis* site in a concentration too low to change essentially the equilibrium distribution. The labelled substrate will pass the membrane with a rate

$$v_{ct}^{eq} = \frac{V_{ct}^{eq}[A]}{K_{ct}^{eq} + [A]} \ . \tag{2.219a}$$

This experiment will be repeated with different equilibrium concentrations of the substrate and will also yield a hyperbolic saturation curve.

Principally, transport occurs in both directions and must be regarded as a reversible process, similar to any chemical and thus to any enzyme catalyzed reaction. Considering this fact, instead of the simple Michaelis-Menten equation the relationship derived for reversible reactions (Eqs. (2.47) and (2.48)) should be applied, where A_1 is the substance to be transported at the *cis* site, and A_2 the substance after the transport at the *trans* site; E is the free transport system:

2.11 Special Enzyme Mechanisms

$$v = \frac{(k_1 k_2 [A_1] - k_{-1} k_{-2} [A_2])[E]_0}{k_1 [A] + k_{-2} [A]_2 + k_{-1} + k_2} \quad . \tag{2.47 a}$$

Regarding only a uni-directed transport from the *cis* to the *trans* site the term $k_{-1} k_{-2} [A_2]$ for the reverse transport may be disregarded and Eq. (2.47a) reduces to:

$$v = \frac{k_1 k_2 [A_1][E]_0}{k_1 [A_1] + k_{-2} [A_2] + k_{-1} + k_2} \quad . \tag{2.220}$$

If conditions are chosen, where the substrate concentration at the *trans* site A_2 is zero, Eq. (2.220) simplifies to the Michaelis-Menten equation (Eq. (2.219)), $k_2 [E]_0$ being the maximum velocity:

$$v = \frac{k_2 [A_1][E]_0}{[A_1] + \dfrac{k_1 + k_2}{k_1}} \quad . \tag{2.219 b}$$

A distinct experimental approach regards the transport system under equilibrium conditions. The transport from the *cis* to the *trans* site can be followed by the addition of radioactively labeled substrate to the *cis* site in such a small amount that it does not really influence the equilibrium conditions. Under these conditions, regarding, $A_1 = A_2 = A$, Eq. (2.220) changes to

$$v = \frac{k_2 [A][E]_0}{[A]\left(\dfrac{k_1 + k_{-2}}{k_1}\right) + \dfrac{k_{-1} + k_2}{k_1}} \tag{2.221}$$

which also obeys the Michaelis-Menten relationship yielding hyperbolic dependences at varying amounts of the equilibrium concentration of A.

In this model it is assumed that the system has one binding site, which accomplishes the transport through the membrane and which is accessible from both sites of the membrane. Alternatively the ligand bound to one site can be transferred to a second site on the opposite face of the membrane. Such intermediate forms are indistinguishable by steady state treatments from the case of one single binding site and the same relationships apply.

Although unidirectional transport is often considered, generally the transport must be regarded from both sites. For each of the sites a Michaelis-Menten equation may be taken, the net flow being the difference between both individual flows:

$$v = \frac{V_1 [A_1]}{K_{m1} + [A_1]} - \frac{V_2 [A_2]}{K_{m2} + [A_2]} \quad . \tag{2.222}$$

Only in the case of asymmetry, with active transport, will the values for V and K_m be different for both sites. If V is the same for both sites:

$$v = V\left(\frac{[A_1]}{K_{m1} + [A_1]} - \frac{[A_2]}{K_{m2} + [A_2]}\right) \quad (2.223)$$

the ratio of the K_m values of both sites determines the overall ratio of the transport. In the case of complete symmetry both V and K_m values will be identical and Eq. (2.223) changes to

$$v = VK_m \frac{[A_1] - [A_2]}{(K_m + [A_1])(K_m + [A_2])} \cdot \quad (2.224)$$

At ligand concentrations much lower than the K_m value the transport velocity will become proportional to the concentration gradient at the membrane:

$$v = \frac{V}{K_m}([A_1] - [A_2]) \cdot \quad (2.225)$$

Corresponding to Fick's law of diffusion (Eq. (1.5)), the transport from the site with the higher substrate concentration will be faster. At substrate concentrations considerably higher than the K_m value

$$v = VK_m\left(\frac{1}{[A_2]} - \frac{1}{[A_1]}\right) \quad (2.226)$$

the conditions become reversed, the substrate with the higher concentration contributes less to the transport velocity because it is already saturated.

2.11.5
Enzyme Reactions at Membrane Interfaces

Interfaces play an important role in cell functions. About half of the enzymes of a cell are connected with or incorporated to the membrane. Structural organization, which is an important feature of membrane bound enzymes, in contrast to soluble enzymes, is a crucial prerequisite for life. Transport processes are one essential function, but direct (more or less intense) interactions of enzymes with the membrane are also important for the functionality of the cell. There are some similarities with the enzymes immobilized to artificial supports already described in Section 2.11.3.

Metabolites, like substrates, distribute between the aqueous, cytosolic space and the membrane according to their polarity, polar metabolites will remain essentially dissolved in the aqueous medium, while non-polar ligands show higher partitioning in the non-polar phase. Membrane bound or associated enzymes convert mainly non-polar substrates e.g. from lipid metabolism (lipases, phospholipases), fatty acids, steroids, eicosanoids or modifed membrane components (phosphorylation, glycosidation, hydrolysis).

The behavior of both the enzyme and the substrate differs essentially from the situation of freely diffusible components in the aqueous surroundings. The

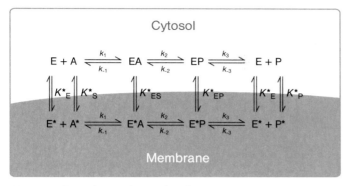

Fig. 2.46 Scheme for interface-mediated enzyme reactions. The upper row shows the components in aqueous surroundings, the lower row components reacting at the interface (with asterisks).

substrate concentration is determined by the density of the substrate molecules in the interface. The fraction of the total enzyme bound to the interface is proportional to the accessible surface area. Enzyme and substrate are in equilibrium between the aqueous solution and the aggregated state at the interface. Fig. 2.46 shows a general scheme for interface-mediated enzyme reactions, where the upper row shows the soluble enzyme, the lower the enzyme reacting at the interface (with asterisks). The scheme gives rise to many mechanisms, dependent on the respective constants, it may be simplified by limitation either to the upper row, which can be described by the normal Michaelis-Menten equation, or the lower row for mere interfacial reactions.

Treatment of enzyme reactions in interfaces is rendered more difficult by the fact that they depend essentially on the experimental conditions, such as stirring or shaking to bring the substrate into the region of the enzyme and to remove the product. Generally, it must be differentiated between matrix enzymes, receiving their substrates from the aqueous solution:

$$E \longrightarrow E^* + A \rightleftharpoons E^*A \longrightarrow E^* + P ,$$

and interfacial enzymes, accessing their substrates exclusively from the interface:

$$E + A \rightleftharpoons E^* + A \rightleftharpoons E^*A \longrightarrow E^* + P .$$

$K_A^* = k_d/k_a$ is the dissociation constant for A from the interface, k_d and k_a being the rate constants for dissociation and association from and to the interface, respectively. Assuming the matrix mechanism, the Michaelis-Menten equation can be used, substituting the flux J for $v/[E]_0$:

$$J = \frac{v}{[E]_0} = \frac{k_{cat}[A]}{K_m + [A]} \,. \tag{2.227}$$

The equation yields the average number of products per enzyme per unit time with the dimension of s^{-1}.

For $[A]_0 < K_A^*$, the concentration of the accessible substrate A will become $[A]_0$:

$$J = \frac{k_{cat}[A]_0}{K_m + [A]_0} \tag{2.227a}$$

while for $[A]_0 > K_A^*$, $[A] = K_A^*$ and will be equal to the *critical micellization concentration* (cmc) (assuming the membranes form micelles of a distinct size).

$$J = \frac{k_{cat} K_A^*}{K_m + K_A^*} = J_{max} \,. \tag{2.227b}$$

With increasing $[A]_0$ the flux increases to the maximum value $[A] \geq K_A^*$. K_m and k_{cat} can be determined according to the normal Michaelis-Menten equation changing the substrate concentration below the cmc.

Steady state kinetics does not distinguish whether the enzyme or substrate are in solution or in the interface for the catalytic turnover. As long as the enzyme accesses the substrate and the substrate partitioning is at equilibrium it makes no difference for steady state kinetics where the substrate or the enzyme is located.

The reaction of an enzyme in the interface can be described by the general Michaelis-Menten equation for reversible enzyme reactions:

$$J = \frac{k_{catA}^* K_{mP}^* X_A^* - k_{catP}^* K_{mA}^* X_P^*}{K_{mA}^* K_{mP}^* + K_{mP}^* X_A^* + K_{mA}^* X_P^*} \,. \tag{2.228}$$

Substrate and product concentrations are given in mole fractions of the respective component in the interface, defined ($X_L^* \sim X_A^* \sim X_P^*$) as

$$X_L^* = \frac{[L^*]}{[M^*]} \tag{2.229}$$

for $[L^*]$ as the respective ligand concentration and $[M^*]$ the concentration of all molecules in the accessible interface. X_L^* becomes 1 for saturation. If the interface contains no other component, $X_A^* + X_P^* = 1$.

Eq. (2.228) can be separated into the contribution of the forward reaction:

$$J_A = \frac{k_{catA}^* K_{mP}^* X_A^*}{K_{mA}^* K_{mP}^* + K_{mP}^* X_A^* + K_{mA}^* X_P^*} \,, \tag{2.230a}$$

and the back reaction

$$J_P = \frac{k^*_{catP} K^*_{mA} X^*_P}{K^*_{mA} K^*_{mP} + K^*_{mP} X^*_A + K^*_{mA} X^*_P} \quad (2.230\,b)$$

The relationship between both reactions is:

$$\frac{J_P}{J_A} = \frac{k^*_{catP} K^*_{mA} X^*_P}{k^*_{catA} K^*_{mP} X^*_A} = \frac{X^*_P}{X^*_A K^*_{eq}} \quad (2.231)$$

for

$$K^*_{eq} = \frac{K^*_{catP} K^*_{mA}}{k^*_{catA} K^*_{mP}} \quad (2.232)$$

Thus the net product flux as the difference between forward and backward reaction $J = J_A - J_P$ is

$$J \equiv J_A \left(1 - \frac{X^*_P}{X^*_A K^*_{eq}}\right) = J_A \left(1 - \frac{X^*_P X^{*eq}_A}{X^*_A X^{*eq}_P}\right). \quad (2.233)$$

In contrast to substrate concentrations in solution, which can be freely modified, the mol fraction X^*_A in the matrix can only be changed by addition of another component, reducing the surface density of A, but the additional component should not modify the features of the system.

The duration of the initial rate at $X^*_S = 1$ increases with the size, but not with the bulk concentration of the vesicles. For the time progression of the reaction the integrated form of the Michaelis-Menten equation (see Section 2.3.2.1) can be applied and also the principles of determination of initial rates for $P = 0$ are valid. However, the range of validity of the initial rate becomes reduced by product remaining bound in the interface. The influence of product inhibition on the initial rate must especially be considered, since the product may rapidly accumulate in the direct proximity of the enzyme.

The rules for distinguishing inhibition types for soluble enzymes are not necessarily valid for interfacial enzymes because the significance of the concentration variables is different. Inhibitory effects, even of specific inhibitors, need not be confined to the binding process in a competitive, non-competitive or uncompetitive manner, but can also intervene with the enzyme–interface interaction, e.g. by weakening the binding and thus favoring the soluble, less active enzyme form. For example, an inhibitor blocking the enzyme–interface interaction would appear as competitive, because it prevents accessibility to the substrate. Such effects also obliterate the distinction between specific and unspecific inhibition. The extent of inhibitions depends not only on the inhibitor and substrate concentrations, but also on the partition of the inhibitor (K^*_I) between solution and matrix.

Box 2.2 Relationship between Enzyme Kinetics and Pharmacokinetics

Enzyme kinetics and pharmacokinetics share many similarities, although there exist far-reaching differences. Without going into detail, some similarities are discussed. Pharmacokinetics describes the time-dependent fate of a drug during its passage through the organism, starting from resorption, followed by its metabolic conversion and elimination. The fate of a distinct drug depends on various aspects, especially its chemical constitution, its distribution and site of action in the organism.

The general rules of binding equilibria and enzyme kinetics hold also for pharmacokinetics. The drug interacts with a specific receptor where it develops its action. The strength of binding, the affinity, is quantified by a dissociation constant, K_d. Accordingly, the same rules for binding are valid, as the general binding equation (Eq. (1.23)). Binding, however, cannot be directly determined, rather, the efficiency of the drug is determined and represented by a *dose–response relationship,* which shows a similar saturation behavior (Fig. 2.47). Since this is a complex process including more steps than just binding, the drug concentration at 50% efficiency is defined, in distinction to a K_d value, as an EC_{50} value (*effective concentration*). An important mechanism for drug action is competition, corresponding to competitive inhibition in enzyme kinetics. The drug competes with the physiological metabolite or with a second drug (antagonist) for binding to the receptor and displaces it at higher concentrations, as, conversely, increasing metabolite concentrations will displace the drug. Competition or competitive antagonism can be detected by shifts of the dose–response curves to higher EC_{50} values without changing the saturation level (Fig. 2.47 A), as saturating amounts of the drug will completely displace the metabolite from its site of action. Non-competitive antagonism is characterized by an indirect interaction rather than displacement and, consequently, the saturation level of the dose–response curve will decrease with increasing amounts of the non-competitive antagonist, the EC_{50}-value may remain unchanged (Fig. 2.47 B). The concentration of the drug required to achieve 50% inhibition in the *dose–response curve* is defined as *inhibitory concentration* IC_{50}, in similarity to the inhibition constant K_i (Fig. 2.47 C).

The first step after application of the drug to the organism is its uptake. This can occur at different sites in the organism, the mouth mucosa, the stomach, the small intestine and the rectum. Resorption depends on various factors:
- Mode of application (oral, e.g. direct or encapsulated; parenteral e.g. subcutaneous or intravenous)
- Accessible surface within the organism
- Chemical constitution of the drug (its water solubility, polarity)

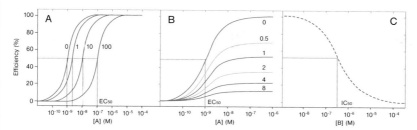

Fig. 2.47 Dose–response curves in semilogarithmic representation for a substance A in the presence of several amounts (indicated as numbers on the respective curves) of (A) a competitive and (B) a non-competitive substance B. Determination of the EC_{50} values is shown. (C) Determination of the IC_{50} value for an antagonistic substance B (after Aktories et al., 2005, with permission from the publisher).

Lipophilic drugs are rapidly reabsorbed through the mucosa of the mouth and appear in the blood circulation circumventing the liver. While resorption in the stomach and the rectum is relatively low, the small intestine, due to its large surface, plays a major role in drug resorption.

Usually it can be taken that there exists a direct relationship between the efficiency and the concentration of a drug at its site of action. This, however, cannot be determined directly. Instead, the concentration of the drug in the blood, the plasma or the urine is taken. Since frequently not the total amount of the drug applied reaches the site of action, the *bioavailability*, i.e. the efficient amount of the drug, relative to the total amount applied, must be determined. While oral applications usually show a lower degree, intravenous application yields nearly 100% bioavailability.

For elimination, drugs are modified by metabolic reactions. Generally, two phases of modification can be discerned:
- *Phase I*: insertion of functional groups by oxidation (essentially mediated by mixed functional monooxygenases, cytochrome P_{450} enzymes), reduction or hydrolysis;
- *Phase II*: conjugation with hydrophilic residues (aminoacids, glutathione, glucuronidation, acetylation, sulfatization by transferases).

The internal time-dependent distribution of a drug in the organism depends on the mode of application. With parenteral application a maximum value is attained immediately, followed by continuous depletion, while with oral application the internal concentration increases initially and depletion occurs after passing a maximum (Fig. 2.48). The integral of the curve, the *area under the curve (AUC)*, is a direct measure of the bioavailability of the respective drug, which is proportional to the amount of the drug applied, M, and divided by the *clearance CL*:

$$AUC = \frac{M}{CL}. \tag{1}$$

The absolute bioavailability B is

$$B = \frac{AUC}{AUC_{i.v.}} \tag{2}$$

AUV_{iv} is obtained by i.v. application.

The relative distribution of the drug within the organism is described by the *distribution volume* V_d, defined as the ratio between the total amount of the applied drug and its concentration c in the plasma:

$$V_d = \frac{M}{c}. \tag{3}$$

It is usually related to the body weight (L kg^{-1}). Considering the accessible water space of the organism, uniform distribution of the drug will yield a value of 0.6 L kg^{-1}, while lower values show low distribution and higher values are an indication of accumulation of the drug in special regions, like the adipose tissues. Therefore, V_d can only be regarded as an apparent distribution volume.

The rate of elimination of the drug, the clearance CL, is usually proportional to the total amount of the drug:

$$CL = \frac{\Delta M}{c \cdot \Delta t} = \frac{M}{AUC} \tag{4}$$

(mL min^{-1}, related to the body weight mL min^{-1} kg^{-1}). It is an indication of the ability of the organism to eliminate the respective drug. The clearance can be obtained from the time dependent progression of the plasma drug concentration (Fig. 2.48). Elimination of drugs usually obeys first order kinetics, i.e. an exponential decrease, which can be linearized by semi-logarithmic representations and a half-life time can be determined as described in Section 2.1.1, which is a measure of the elimination rate. Principally, elimination and, thus, the half-life time, depend both on the clearance efficiency of the organism and the distribution volume:

$$\ln t_{1/2} = \ln 2 \cdot \frac{V_d}{CL}, \tag{5}$$

the larger the volume, the slower the elimination rate.

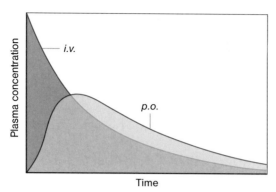

Fig. 2.48 Time progression of the plasma concentration of a drug after intravenous (*i.v.*) and oral (*p.o.*) application. The areas under the curve (*AUC*) are identical for complete bioavailability for both curves (after Aktories et al. 2005, with permission from the publisher).

The knowledge of the half-life time serves also to maintain a constant drug level in the organisms. Just the amount eliminated must be restored:

$$c \cdot CL = B \frac{D_E}{\tau} . \qquad (6)$$

D_E is the single dose applied and τ is the time interval between the doses. To maintain a constant drug level for a longer time τ should be as short as possible and really constant levels can only be established by continuous application of the drug by permanent infusion. Repetitive single doses result in more or less pronounced fluctuations of the drug concentration with maxima and minima around a medium value. For detailed description of such phenomena the reader is referred to relevant textbooks of pharmacology.

2.12
Application of Statistical Methods in Enzyme Kinetics

2.12.1
General Remarks

The analysis of experimental data and the interpretation of curves in different plots are of great importance in enzyme kinetics. Therefore, the application of statistical methods is indispensable. Within the scope of this book they cannot be extensively treated and the reader is referred to the relevant literature on statistics. Here only special problems of application of statistical methods in enzyme kinetics will be discussed.

Studies of enzymes often require special procedures complicating the application of statistical rules. In order to interpret the reliability of the course of curves and to analyse parameters such as equilibrium and Michaelis constants, repeated measurements should be performed to ensure the quality of the data. However, enzymes in dilute solutions for enzyme tests are frequently unstable, so that analysis of extensive test series is often a race against time. The following example will demonstrate this. For a thorough analysis of inhibition or multi-substrate mechanisms test series must be performed with one parameter (substrate) varied, and another (cosubstrate, inhibitor) kept constant. For 10 concentration values of the first parameter per test series and 5 constant concentrations of the second parameter, 50 tests are required. Assuming 5 min per test the total time required is more than 4 h for a single analysis. If three repeated tests are undertaken, it will take more than 12 h. If the enzyme loses half of its activity within 20 h, the first measurement value will differ from the last by 14% in a single analysis, and by as much as 35% in a triple analysis. Thus the deviation caused by activity loss is much larger than the advantage of statistical safeguarding by repeated measurements. By preparing fresh enzyme dilutions within one test series the original activity can hardly be exactly reproduced and further tests would not fit into the series. In such cases it is more favorable to establish the results obtained from single measurements by independent test series.

The diagrams used in enzyme kinetics usually serve to confirm or exclude the laws assumed for the system, e.g., the Michaelis-Menten equation. This is accomplished by analysing how far the measured data follow the curve as predicted by the law. Due to error scattering hardly any of the values will fit exactly to the assumed curves. It must be decided whether observed deviations are due to error scattering or whether the system obeys another rather than the assumed relationship. Normal error scattering should be distributed evenly above and below the assumed function as the average value (*constant absolute error*). *Residual plots* are valuable tools to test the error behavior. Deviation of each value from the assumed function, calculated by regression methods (e.g., a hyperbolic curve according to the Michaelis-Menten equation), is plotted against the independent variable (e.g., substrate concentration) (Fig. 2.49 C, D, F). The points must be distributed evenly around a median line (Fig. 2.49 A, C). A *systematic error* is exhibited if the points drift from the median line in a certain direction (Fig. 2.49 E, F). This may be caused by artificial influences on the measurement or by an alternative mechanism. A *relative error* is exhibited, if the values in the residual plot scatter evenly around the median line, but their extent moves in a certain direction (e.g. the error increases with the intensity of the measurement signal, Fig. 2.49 B, D). This may still confirm the assumed mechanism, but an appropriate weighting has to be considered for regression analysis. Normal regression methods are based on an even error distribution, i.e., a constant absolute error. Residual plots are especially helpful for non-linear curves, where systematic deviations are difficult to detect by eye (Fig. 2.49 E, F). Appropriate statistical methods, e.g., the *W-test* of Shapiro-Wilks or the *student-* or *t-test*

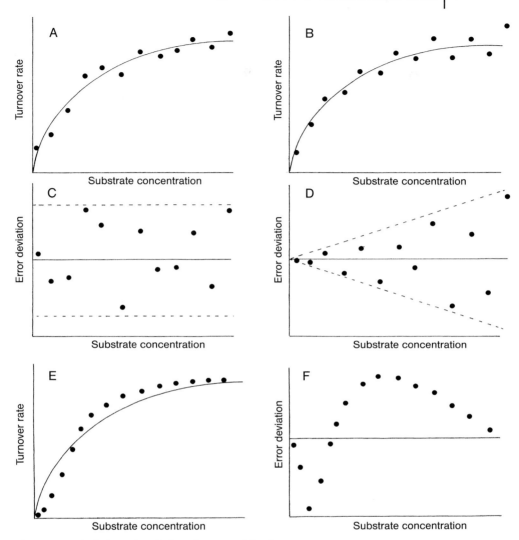

Fig. 2.49 Residual plots (C, D, F) from diagrams of the dependence of the turnover rate on the substrate concentration (A, B, E). Constant absolute error (σ=constant) (A, C), constant relative error (σ/v=constant) (B, D), adaptation of a sigmoidal saturation function to a hyperbolic curve (E, F).

serve to detect runaway data and establish the significance of the measured values. The correlation coefficient indicates the consistency of the data with the assumed curve function.

The application of statistical methods is principally recommended, as it reduces the danger of subjective analysis and interpretation. However, these meth-

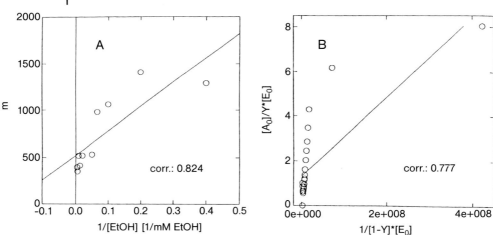

Fig. 2.50 Uncritical application of linear regressions (from seminar protocols). (A) Secondary plot of a bisubstrate reaction (m, slope); (B) Stockell plot (corr., correlation coefficient).

ods also have their limitations and they should be critically evaluated. In enzyme kinetic measurements, mechanisms, e.g. of inhibitions and multi-substrate reactions, are often identified by the patterns of groups of straight lines in linearized plots (common intercepts, parallels). It is rather unlikely that straight lines obtained by regression methods from scattering data of several test series will, without exception, fit clearly into the expected pattern, showing identical slopes (for parallel lines) or meet in exactly the same intercept, and – honestly speaking – such mechanisms will never be established by applying statistical rules. It can of course be stipulated to detect parallels or common intercepts for the least deviation of all data by regression analysis, but then the mechanism is already assumed. Fig. 2.50 shows two examples of uncritical application of linear regression methods from seminar protocols. The aim of scientific studies should be to obtain data of such quality that the result becomes obvious without any statistical analysis.

Before applying statistical methods the following criteria should be considered:
- Is the intended method suitable for the data to be analyzed?
- Can the data be adapted equally well to other relationships?
- Do the data exhibit deviations in certain directions that cannot be explained by normal error distribution?
- Can artificial effects be excluded?

2.12.2
Statistical Terms Used in Enzyme Kinetics

Arithmetic mean: The mean value $\bar{x}$ is the sum of all measurements x_i divided by their number n:

$$\bar{x} = \frac{\sum_{i=1}^{n} x_i}{n} \ . \tag{2.234}$$

Median: Mean value of a test series with data arrayed according to their size, for an odd number of data: $x_{(n+1/2)}$, for an even number the mean value of the two mean measurements: $(x_{(n/2)} + x_{(n/2+1)})/2$.

Mode: The most frequently occurring value of a measurement series.

Variance: Mean square sum of errors:

$$\sigma_x^2 = \frac{\sum_{i=1}^{n}(x_i - \bar{x})^2}{n} \ . \tag{2.235}$$

Standard deviation: Root mean square deviation, RMS.

$$\sigma_x = \sqrt{\frac{\sum_{i=1}^{n}(x_i - \bar{x})^2}{n}} \ . \tag{2.236}$$

Standard deviation of the mean:

$$\sigma_{\bar{x}} = \frac{\sigma_x}{\sqrt{n}} \ . \tag{2.237}$$

Linear regression: Adaptation of data to a straight line after the least square error method. Only the error of the dependent variable y is considered (v in enzyme kinetic measurements). For the independent variable x (e.g., substrate concentration) no error is assumed.

$$y_i = a + bx_i \ . \tag{2.238}$$

The ordinate intercept a is:

$$a = \frac{\sum x_i^2 \sum y_i - \sum x_i \sum x_i y_i}{n \sum x_i^2 - (\sum x_i)^2} \ . \tag{2.239}$$

The slope or the *regression coefficient b*, respectively, is:

$$b = \frac{n\sum x_i y_i - \sum x_i \sum y_i}{n\sum x_i^2 - (\sum x_i)^2} .\qquad(2.240)$$

Standard deviation of the y values:

$$\sigma_y = \sqrt{\frac{1}{n-2}\sum_{i=1}^{n}(y_i - a - bx_i)^2} .\qquad(2.241)$$

Correlation coefficient:

$$r = \frac{\sum(x_i - \bar{x})(y_i - \bar{y})}{\sqrt{\sum(x_i - \bar{x})^2 \sum(y_i - \bar{y})}} .\qquad(2.242)$$

Non-linear adaptation of the Michaelis-Menten equation according to the least square error method (Cornish-Bowden 1995):

$$K_m = \frac{\sum \frac{v_i}{[A]_i}\sum v_i^2 - \sum \frac{v_i^2}{[A]_i}\sum v_i}{\sum\left(\frac{v_i}{[A]_i}\right)^2 \sum v_i - \sum \frac{v_i^2}{[A]_i}\sum \frac{v_i}{[A]_i}} .\qquad(2.243)$$

$$V = \frac{\sum\left(\frac{v_i}{[A]_i}\right)^2 \sum v_i^2 - \sum\left(\frac{v_i^2}{[A]_i}\right)^2}{\sum\left(\frac{v_i}{[A]_i}\right)^2 \sum v_i - \sum \frac{v_i^2}{[A]_i}\sum \frac{v_i}{[A]_i}} .\qquad(2.244)$$

References

General Literature on Steady-State Kinetics

Briggs, G.E., Haldane, J.B.S. (1925) A note on the kinetics of enzyme action, *Biochem. J.* 19, 338–339.

Brown, A.J. (1902) Enzyme action, *J. Chem. Soc.* 81, 373–388.

Henri, V. (1902) Theorie generale de l'action de quelques diastases, *C. R. Acad. Sci.* 135, 916–919.

Michaelis, L., Menten, M.L. (1913) Die Kinetik der Invertinwirkung, *Biochem. Z.* 49, 333–369.

Wharton, C.W. (1983) Some recent advances in enzyme kinetics, *Biochem. Soc. Trans.* 11, 817–825.

Analysis of Enzyme Kinetic Data

Alberty, R.A., Koerber, B.M. (1957) Studies of the enzyme fumarase. VII Series solutions of integrated rate equations for irreversible and reversible Michaelis-Menten mechanism, *J. Am. Chem. Soc.* 79, 6379–6382.

Balcom, J.K., Fitch, W.M. (1945) A method for the kinetic analysis of progress curves using horse serum cholin esterase as a model, *J. Biol. Chem.* 245, 1637–1647.

Boeker, E.A. (1982) Initial rates. A new plot, *Biochem. J.* 203, 117–123.

Cornish-Bowden, A. (1975) The use of the direct linear plot for determining initial velocities, *Biochem. J.* 149, 305–312.

Cornish-Bowden, A., Eisenthal, R. (1978) Estimation of Michaelis constant and maximum velocity from the direct linear plot, *Biochim. Biophys. Acta 523*, 268–272.

Dixon, M. (1965) Graphical determination of equilibrium constants, *Biochem. J. 94*, 760–762.

Eadie, G.S. (1942) The inhibition of cholinesterase by physostigmine and prostigmine, *J. Biol. Chem. 146*, 85–93.

Eisenthal, R., Cornish-Bowden, A. (1974) The direct linear plot. A new graphical procedure for estimating enzyme parameters, *Biochem. J. 139*, 715–720.

Foster, R.J., Niemann, C. (1953) The evaluation of the kinetic constants of enzyme catalyzed reactions, *Proc. Natl. Acad. Sci. USA 39*, 999–1003.

Haldane, J.B.S., Stern, K.G. (1932) *Allgemeine Chemie der Enzyme*. Steinkopff, Dresden & Leipzig.

Hanes, C.S. (1932) Studies on plant amylases. The effect of starch concentration upon the velocity of hydrolysis by the amylase of germinated barley, *Biochem. J. 26*, 1406–1421.

Hofstee, B.H.J. (1952) Specificity of esterases. Identification of two pancreatic aliesterases, *J. Biol. Chem. 199*, 357–364.

Jennings, R.R., Niemann, C. (1954) The evaluation of the kinetics of enzyme-catalzed reactions by procedures based upon integrated rate equation, *J. Am. Chem. Soc. 77*, 5432–5433.

Kilroe-Smith, J.A. (1966) A modified graphical method for determination of equilibrium constants, *Biochem. J. 100*, 334–335.

Lee, H.J., Wilson, I.B. (1971) Enzymic parameters: Measurement of V and K_m, *Biochim. Biophys. Acta 242*, 519–522.

Lineweaver, H., Burk, D. (1934) The determination of enzyme dissociation constants, *J. Am. Chem. Soc. 56*, 658–666.

Markus, M., Hess, B., Ottaway, J.H., Cornish-Bowden, A. (1976) The analysis of kinetic data in biochemistry. A critical evaluation of methods, *FEBS Lett. 63*, 225–230.

Nimmo, I.A., Atkins, G.L. (1978) An evaluation of methods for determining initial velocities of enzyme-catalysed reactions from progress curves, *Biochem. Soc. Trans. 6*, 548–550.

Orsi, B.A., Tipton, K.F. (1979) Kinetic analysis of progress curves, *Methods Enzymol. 63*, 159–183.

Rudolph, F.B., Fromm, H.J. (1979) Plotting methods of enzyme rate data, *Methods Enzymol. 63*, 138–159.

Schwert, G.W. (1969) Use of integrated rate equations in estimating the kinetic constants of enzyme-catalyzed reactions, *J. Biol. Chem. 244*, 1278–1284.

Waley, S.G. (1981) An easy method for the determination of initial rates, *Biochem. J. 193*, 1009–1012.

Walker, A.C., Schmidt, C.L.A. (1944) Studies on histidase, *Arch. Biochem. 5*, 445–467.

Wilkinson, G.N. (1961) Statistical estimations of enzyme kinetics, *Biochem. J. 80*, 324–332.

Enzyme Inhibition

Dixon, M. (1953) The determination of the enzyme inhibition constants, *Biochem. J. 55*, 170–171.

Dixon, M. (1972) The graphical determination of K_m and K_i, *Biochem. J. 129*, 197–202.

Fahrney, D.E., Gold, A.M. (1963) Sulfonyl fluorides as inhibitors of esterases. I. Rates of reaction with aceylcholinesterase, a-chymotrypsin, and trypsin, *J. Am. Chem. Soc. 85*, 997–1000.

Gutowski, J.A., Lienhard, G.E. (1976) Transition state analogs for thiamin pyrophosphate-dependent enzymes, *J. Biol. Chem. 251*, 2863–2688

Kitz, R. & Wilson, I.B. (1962) Esters of methanesulfonic acid as irreversible inhibitors of acetylcholinesterase, *J. Biol. Chem. 237*, 3245–3249.

Schramm, V.L. (2005) Enzymatic transition states and transition state analogs, *Curr. Opin. Struct. Biol. 15*, 604–613.

Wolfenden, R., Radzicka, A. (1991) Transition state analogs, *Curr. Opin. Struct. Biol. 1*, 780–787.

Multi-substrate Reactions

Alberty, R.A. (1959) The rate equation for an enzymatic reaction. *The Enzymes*, 1st edn, Boyer, P.D., Lardy, H., Myrbäck, K. (Eds.), Academic Press, New York, Vol. 1, pp. 143–155.

Cleland, W.W. (1963) The kinetics of enzyme-catalyzed reactions with two or more substrates or products. I. Nomenclature and rate equations, *Biochim. Biophys. Acta 67*, 104–137; II. Inhibition: Nomenclature and theory, *Biochim. Biophys. Acta 67*, 173–187; III. Prediction of initial velocity and inhibition patterns by inspection, *Biochim. Biophys. Acta 67*, 188–196.

Dalziel, K. (1957) Initial steady-state velocities in the evaluation of enzyme-coenzyme-substrate reaction mechanism, *Acta Chem. Scand. 11*, 1706–1723.

Fromm, H.J. (1970) A simplified schematic method for deriving steady-state equations using a modification of the "Theory of Graphs" procedure, *Biochem. Biophys. Res. Commun. 40*, 692–697.

King, E.L. & Altman, C. (1965) A schematic method for deriving the rate laws for enzyme-catalyzed reactions, *J. Phys. Chem. 60*, 1375–1378.

Volkenstein, M.V. & Goldstein, B.N. (1966) A new method for solving the problems of the stationary kinetics of enzymological reactions, *Biochim. Biophys. Acta 115*, 471–477.

pH and Temperature Behavior

Arrhenius, S. (1889) Über die Reaktionsgeschwindigkeit bei der Inversion von Rohrzucker durch Säuren, *Z. Phys. Chem. 4*, 226–248.

Dixon, M., Webb, E.C. (1979) *Enzymes*, 3rd edn. Academic Press, New York.

Eyring, H. (1935) The activated complex in chemical reactions, *J. Chem. Phys. 3*, 107–115.

Laidler, K.J., Peterman, B.F. (1979) Temperature effects in enzyme kinetics, *Methods Enzymol. 63*, 234–257.

Tipton, K.F., Dixon, H.B. (1979) Effects of pH on enzymes, *Methods Enzymol. 63*, 183–234.

Kinetic Cooperativity

Bisswanger, H. (1984) Cooperativity in highly aggregated systems, *J. Biol. Chem. 259*, 2457–2465.

Neet, K.H., Ainslie, G.R. (1980) Hysteretic enzymes, *Methods Enzymol. 64*, 192–226.

Rübsamen, H., Khandker, R., Witzel, H. (1974) Sigmoidal kinetics of monomeric ribonuclease I due to ligand-induced shifts of conformation equilibrium. *Hoppe-Seyler's Z. Physiol. Chem. 355*, 687–708.

Isotope Exchange

Berti, P.J. (1999) Determining transition states from kinetic isotope effects, *Methods Enzymol. 308*, 355–397.

Fromm, H.J. (1975) *Initial Rate Enzyme Kinetics*, Springer-Verlag, Berlin.

Huang, C.Y. (1979) Derivation of initial velocity and isotope exchange rate equations, *Methods Enzymol. 63*, 54–84.

Jencks, W.P. (1969) *Catalysis in Chemistry and Enzymology*, McGraw-Hill, New York, pp. 243–281.

Purich, D.L., Allison, R.D. (1980) Isotope exchange methods for elucidating enzymic catalysis, *Methods Enzymol. 64*, 3–46.

Richards, J.H. (1970) *Kinetic isotope effects in enzymic reactions, The Enzymes*, 3rd edn. Boyer, P. (Ed.), Academic Press, New York, Vol. 2, pp. 321–333.

Ribozymes

Symons, R.H. (1992) Small catalytic RNAs, *Annu. Rev. Biochem. 61*, 641–671.

Kuimelis, R.G., McLaughlin, L.W. (1998) Mechanisms of ribozyme-mediated RNA cleavage, *Chem. Rev. 98*, 1027–1044.

Bevilacqua, P.C., Brown, T.S., Chadalavada, D.M., Diegelman-Parente, A., Yajima, R. (2003) Kinetic analysis of ribozyme cleavage in: *Kinetic Analysis of Macromolecules: A Practical Approach*, Johnson, K. A. (Ed.), Oxford University Press, Oxford, 256 pp.

Polymer Substrates

Chetkarov, M.L., Kolev, D.N. (1984) The Michaelis-Menten equation in the case of linear homopolymer substrates with different degrees of polymerization, *Monatsh. Chem. 115*, 1405–1412.

Hiromi, K. (1970) Interpretation of dependency of rate parameters on the degree of polymerization of substrate in enzyme-catalyzed reactions. Evaluation of subsite affinities of exoenzymes, *Biochem. Biophys. Res. Commun. 40*, 1–6.

Membrane Transport, Interfacial Enzyme Kinetics

Bonting, S.L. & de Pont, J.J.H.H.M. (1981) Membrane transport, in *New Comprehensive Biochemistry*, Vol. 2, Neuberger, A., van Deenen, L.L.M. (Eds.), Elsevier, Amsterdam.

Berg, O.G., Jain, M.K. (2002) *Interfacial Enzyme Kinetics*, John Wiley & Sons, Chichester.

Christensen, H.N. (1975) *Biological Transport*, 2nd edn. W.A. Benjamin, Inc. Reading, MA.

Immobilised Enzymes

Goldstein, L. (1976) Kinetic behavior of immobilized enzyme systems, *Methods Enzymol.* 44, 397–443.

Laidler, K.J., Bunting, P.S. (1980) The kinetics of immobilized enzyme systems, *Methods Enzymol.* 61, 227–248.

Wingard, L.B., Katchalski-Katzir, E., Goldstein, L. (1976) Immobilized enzyme principles, in *Applied Biochemistry and Bioengineering*. Academic Press, New York.

Engasser, J.-M., Horvath, C. (1973) Effect of internal diffusion in heterogeneous enzyme systems: Evaluation of true kinetic parameters and substrate diffusivity, *J. Theor. Biol.* 42, 137–155.

Pharmacokinetics

Aktories, K., Förstermann, U., Hofmann, F., Starke, K. (2005) *Allgemeine und spezielle Pharmakologie und Toxikologie*, 9. Aufl. Urban & Fischer, München, Jena.

Statistics

Cleland, W.W. (1979) Statistical analysis of enzyme kinetic data, *Methods Enzymol.* 63, 103–138.

Cornish-Bowden, A. (1995) *Fundamentals of Enzyme Kinetics*, Portland Press, London.

Cornish-Bowden, A., Endrenyi, L. (1986) Robust regression of enzyme kinetics, *Biochem. J.* 234, 21–29.

Kaiser, R.E., Mühlbauer, J.A. (1983) *Elementare Tests zur Beurteilung von Meßdaten*, 2nd edn. Bibliographisches Institut, Mannheim.

Lehn, J., Wegmann, H. (1985) *Einführung in die Statistik*, B.G. Teubner, Stuttgart.

Taylor, J.R. (1988) *Fehleranalyse*, VCH, Weinheim.

Zar, J.H. (1984) *Biostatistical Analysis*, 2nd edn. Prentice-Hall Inc., Englewood Cliffs, New Jersey, USA.

3
Methods

The variety of methods for the investigation of multiple equilibria and for enzyme kinetic studies is rather broad and many techniques cannot be attributed clearly to a distinct field but are useful for many applications. This is true especially for spectroscopic methods, which serve to study not only binding processes and enzyme reactions, but also conformational changes, which are important for the understanding of regulatory mechanisms like cooperativity and allosteric phenomena, as well as structural aspects. Furthermore, they serve as important detection methods for the study of fast reactions. Therefore much space is given to the different types of spectroscopy. Initially, general aspects are treated before turning to special applications in binding and kinetic studies. Calorimetric and electrochemical techniques are treated similarly.

Special emphasis is given to binding methods, since they are less well known but they supply valuable information about various cellular processes. Since binding measurements are usually not easy to perform, as will be discussed later, various methods, even seldom applied ones, are mentioned in order to give a broad survey and enable selection of the method most appropriate for a distinct problem.

The situation with respect to enzyme kinetic methods is quite different. The methods used are essentially the same as used for enzyme assays, based on the chemical difference between substrate and product. However, enzyme kinetic studies require special procedures and emphasis is laid more on such general aspects than on the various enzyme assays. A special class of kinetic methods comprises techniques to follow fast reactions, especially flow and relaxation methods.

3.1
Methods for Investigation of Multiple Equilibria

Binding measurements are often laborious because they require large amounts of the (mostly valuable) macromolecule, and applying an inappropriate method can become rather wasteful. The general problem is that (reversible) binding causes no intrinsic change, the ligand and the macromolecules remain the same in both the unbound and bound states, in contrast to enzyme reactions where the substrate is converted to a chemically different product and detection

can be based on this difference. Therefore weak differences must be emphasized. The methods applied for binding measurements make use of two features:
- Size difference
- Spectral shifts.

While the first feature is inevitable and can be seen with any binding system, the second one is optional. Obviously, the molecular mass of a complex of two binding components will be the sum of the individual molecular masses. Since mostly a low molecular mass ligand interacts with a macromolecule, the ligand will accept the size of the macromolecule and can be distinguished from the unbound ligand by size determination methods. This looks to be easy, however, the complex will exist only in the (fast) equilibrium with the free components. Size determination methods, like gel chromatography, will separate the components according to their size and, thus, result in a rapid dissociation of the bound ligand from the complex. Therefore, the methods must be modified in order to allow size determination without disturbing the equilibrium. Regarding such conditions these methods are valuable because they are principally applicable for every binding system.

With spectroscopic methods the problem is different. Upon binding of the ligand to the macromolecule a spectral shift can occur, of the macromolecule or the ligand spectrum, or of both, but this depends on the special type of interaction and cannot be anticipated, hence there is a need for extensive preliminary tests. Since binding produces no intrinsic change in the components, only weak interactions can be observed, like changes in the polarity. In some cases charge transfer complexes are formed giving rise to more pronounced spectral changes. Spectral shifts can also be caused by conformational changes of the macromolecule induced by ligand binding. Generally, the shifts are considerably small. Since the intensity of the binding signal is directly proportional to the amount of the macromolecule, it should be applied as concentrated as possible. To demonstrate this consider a macromolecule with a molecular mass of 100 000 binding a ligand with a dissociation constant of 1×10^{-4} M. If both components are applied in the same concentration of 1×10^{-6} M (the actual macromolecule concentration will be 0.1 mg ml^{-1}), then, according to the mass action law:

$$K_d = \frac{[E][A]}{[EA]} = \frac{1 \cdot 10^{-6} \cdot 1 \cdot 10^{-6}}{1 \cdot 10^{-8}} = 1 \cdot 10^{-4} \text{ M},$$

only 1% of the ligand (1×10^{-8} M) will be bound, an amount which will hardly be detectable in the presence of a 100-fold excess of the free ligand. A 100-fold increase in the ligand concentration (1×10^{-4} M) will saturate the macromolecule to 50% (since the ligand concentration becomes equal to the dissociation constant), but the ratio between free and bound ligand reduces to 0.5%. This is even worse to detect.

$$K_d = \frac{[E][A]}{[EA]} = \frac{5 \cdot 10^{-7} \cdot 1 \cdot 10^{-4}}{5 \cdot 10^{-7}} = 1 \cdot 10^{-4} \text{ M}.$$

To obtain bound and free ligand in similar amounts, the macromolecule concentration must also be raised:

$$K_d = \frac{[E][A]}{[EA]} = \frac{1 \cdot 10^{-4} \cdot 1 \cdot 10^{-4}}{1 \cdot 10^{-4}} = 1 \cdot 10^{-4} \text{ M}.$$

To achieve this, the total amount of the macromolecule has to be 2×10^{-4} M, 200 times more than originally applied. The protein concentration is now 20 mg ml^{-1}, this is indeed the upper limit for protein solutions, and also a considerable absolute quantity, which is not realizable for any macromolecule.

3.1.1
Equilibrium Dialysis and General Aspects of Binding Measurements

3.1.1.1 Equilibrium Dialysis

Equilibrium dialysis can be considered as the classical binding method. It is a robust method, which requires only a limited instrumental expenditure, the dialysis device can be produced in a mechanical workshop, the outcome is mostly clear, although the data scatter is relatively high. Compared with other techniques, equilibrium dialysis is a rather reliable technique, which may be seriously considered if there is no special preference for another method. Using it as an example, general features of binding measurements can also be discussed.

The principle of equilibrium dialysis is the same as for conventional dialysis, a macromolecule solution is separated from an outer solution by a semi-permeable membrane, which allows the passage of low molecular weight compounds but not the macromolecule. Usually this method is used to remove high amounts of salt (e.g. after ammonium sulfate precipitation) or to change the buffer. For binding measurements two compartments of the same size are applied, separated by the dialysis membrane. The *inner* compartment (the terms 'inner' and 'outer' are taken from conventional dialysis, where the macromolecule is enclosed in a dialysis bag, floating in an outer solution, while in the equilibrium dialysis device both compartments are quite similar) contains the macromolecule solution, the *outer* compartment the ligand (Fig. 3.1). The ligand will diffuse into the macromolecule compartment until the concentration in both compartments is equalized, thus the ligand concentration in the outer compartment will be halved. If the ligand binds to the macromolecule, the amount of bound ligand will be removed from the solution and correspondingly more of the ligand will diffuse into the inner compartment. The ligand in the inner compartment $[A]_i$ will distribute according to its equilibrium in free and bound ligand,

$$[A]_i = [A]_{bound} + [A] \tag{3.1}$$

and only the free ligand will equalize with the outer compartment. The total ligand concentration in the outer compartment after dialysis $[A]_o$ will be identical to the free ligand concentration in the inner compartment:

$$[A]_o = [A] . \tag{3.2}$$

Consequently, the amount of bound ligand is the difference of the total ligand concentrations between both compartments:

$$[A]_{bound} = [A]_i - [A]_o \tag{3.3}$$

After dialysis the total ligand concentrations $[A]_i$ and $[A]_o$ are determined from aliquots taken from both compartments.

Binding experiments are usually performed with several (10–20) dialysis cells (each dialysis cell containing an outer and inner compartment), the amount of the macromolecule in all inner compartments will be the same, while the outer compartments contain the ligand in varying concentrations, preferentially from one tenth to tenfold of the expected dissociation constant (considering the dilution due to distribution of the ligand between both compartments). Because high sensitivity for detection of the ligand is demanded, and since the volumes of the solutions in dialysis cells will be kept as low as possible to save the macromolecule, the detection methods for the ligand must be of the utmost sensitivity. This can usually only be achieved by applying ligand molecules labelled with a radioisotope.

Equilibrium dialysis has the advantage that both $[A]$ and $[A]_{bound}$ can be directly determined by the experiment. Other methods allow only the determination of one concentration, the other must be derived from the difference from the initially added ligand ($[A]_0 = [A] + [A]_{bound}$), giving rise to artificial influences, like unspecific binding to the walls of the apparatus or the membrane. A disadvantage, in turn, is

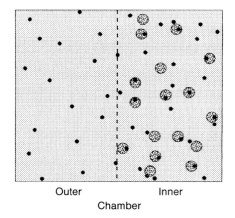

Fig. 3.1 Scheme of equilibrium dialysis. The small ligand molecules pass freely through the semi-permeable membrane and distribute equally between both compartments, while the macromolecule, to which additional ligands bind, cannot leave the inner compartment.

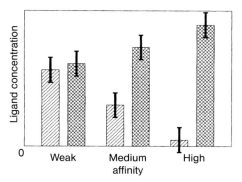

Fig. 3.2 Ratios between the measured values and the error fluctuation in equilibrium dialysis. The left bars of each pair represent the values obtained from the outer compartment (free ligand), the right bars those from the inner compartment (free and bound ligand) for weak, medium and high affinity. The error limits are indicated at the top of the bars.

the fact that [A]$_{bound}$ cannot be obtained directly but only as a difference between the two compartments. Thus, any failure in the determination in either compartment will add to this value. This fact also leads to limitation of the method to medium dissociation constants while weak and very strong binding are difficult to determine ($K_d \approx 10^{-3}$–10^{-7} M). Figure 3.2 illustrates this situation. Weak binding requires high amounts of ligand, but the concentration difference between the outer and inner compartment after dialysis will be small and lost within the error limits. Very strong binding gives high values in the inner compartment but the free ligand amount will be very low and will again be lost within the error fluctuations.

To perform the dialysis a simple dialysis bag can be taken, containing a defined volume of the macromolecule solution and dipping into a tube containing the same volume of the ligand solution. However, such a device requires large amounts of the compounds but will not yield very precise values. Therefore, special devices have been constructed, as shown in Fig. 3.3 (Myer et al. 1962; Englund et al. 1969). The compartments are milled into plastic or Teflon cylinders, the width and depth of the holes determine the test volume, which will be around 50 µl. Two such symmetrical holes are fitted together with a dialysis membrane between, separating the two compartments. Small channels from the outside to the holes allow filling and removal of the solutions after the experiment. The channels are closed by stoppers. To perform a binding experiment about 10–20 of such cells are applied and fixed into a rotating holder. Because of the temperature dependence of the dissociation constant the temperature must be kept constant during the whole experiment, e.g. at 37 °C. Thus all cells must be kept in an incubator or a water bath and should be turned slowly with a driving motor to establish efficient dialysis. Normal dialysis bags (only one layer) manufactured from cellulose with an exclusion limit of 15 000 for proteins, or ultrafiltration membranes with various exclusion limits, can be used as dialysis membranes.

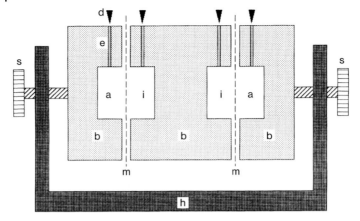

Fig. 3.3 Equilibrium dialysis apparatus with twin dialysis cells. a, outer compartment; i, inner compartment; b, dialysis block; d, stopper; e, filling channels; m, semi-permeable membrane; h, holding device; s, fixing screws. The size of the cells (e.g. 50 µl) is adapted to the required volume (after Englund et al. 1969).

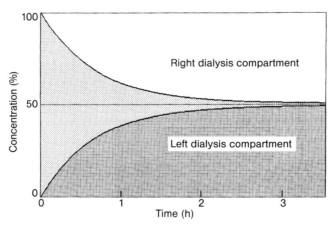

Fig. 3.4 Determination of the minimum dialysis time. At time $t=0$ ligand is present only in one (the right) compartment (100%), while the left compartment contains only the solvent (0% ligand). During dialysis the ligand concentrations in both compartments change, and thus the curves approach each other and meet at the minimum dialysis time.

3.1.1.2 Control Experiments and Sources of Error

Dialysis time. A critical parameter is the duration of the dialysis. Complete equalization of the free ligand between both compartments must be established. On the other hand the macromolecule may not be stable during longer dialysis times and, from this point of view, the experiment should be as short as possible. Because the actual dialysis time depends on several factors (e.g. volume, temperature, size and polarity of the ligand, pore diameter of the dialysis membrane, movement of the cells) the minimum dialysis time should be determined prior to each new experiment. For this the dialysis cells are filled, one compartment with a constant ligand concentration, the other compartment with the same volume of water or a buffer solution. The dialysis is performed with several such dialysis cells, all treated similarly. At time zero and after distinct times aliquots are taken from both compartments and the respective concentration of the ligand is determined. If plotted against time two exponential curves should result. The values taken from the ligand compartment will decrease from 100% of the original ligand concentration at $t=0$ to 50% at complete equilibration. Conversely, the ligand concentration in the buffer compartment will increase from 0 to 50%. The point where both curves just meet at 50% is the minimum dialysis time (Fig. 3.4). To be sure, the actual dialysis experiment should last slightly longer than this; usually dialysis times of about 2 h are required.

Concentration and activity of the macromolecule. The reliability of the results depends on the accuracy of the determination of the concentrations of both the ligand and the macromolecule. The concentration at the beginning of the experiment must be known exactly, which is less a problem with the ligand, but is sometimes a problem with the macromolecule, since determination of exact molarities of proteins is not very easy. Simple protein tests, which are mostly adjusted to a standard protein like bovine serum albumin, are far from accurate, rather, absolute determinations are required. Even if the initial concentration is known, there are many influences which may cause a change in the concentration during the experiment. Less stable macromolecules may become partially or totally inactivated. Sometimes they can precipitate but the greater problem occurs if they remain dissolved, because protein determination will not give any information which part remained active and which became inactive. With enzymes, the activity can be determined before and after the experiment. Also the osmotic pressure, caused by the uneven distribution of the macromolecule between the compartments, can produce a concentration change. This effect cannot be avoided because the flexible membrane will give way to this pressure. Because of this effect concentration determination before and after the experiment is required.

Concentration of the ligand. It must also be established that the total concentration of the ligand remains unchanged during the experiment. This can be a problem with unstable ligands. If the ligand applied is the substrate of the enzyme any conversion to product must be strictly avoided. Enzyme reactions can be prevented by elimination of essential cosubstrates or cofactors. Because in the experiment the enzymes will be present in large amounts, even traces of a cofactor can cause the complete conversion of the substrate during dialysis. If the substrate is able to react with the enzyme alone without any additional factor, inactive substrate analogs can be taken, which may, however, differ in their binding features. A further problem can be unspecific binding or adsorption of the ligand e.g. to the walls of the dialysis cell. If this is not very severe it will not be a profound problem as long as the total amount of the ligand is not included in the calculation, as already discussed above.

Donnan effect. A severe problem, however, can be the uneven distribution of charged ligands between both compartments due to an effect described by W. Gibbs (1876) and G. F. Donnan (1911), the Gibbs-Donnan equilibrium or Donnan effect. Proteins carry surplus charges, which will attract ligands if they are oppositely charged, and will reject them if of the same charge. Thus in the first case the ligand will accumulate in the compartment of the macromolecule and give the impression of binding, while in the second case it will be diminished in this compartment, counteracting a specific binding (Box 3.1).

Box 3.1 Quantification of the Donnan Effect

To quantify the Donnan effect it can be assumed that the chemical potentials of cations and anions (including those of the buffer) must be equal in the inner (μ_{CA}^i) and the outer (μ_{CA}^o) compartment:

$$\mu_{CA}^i = \mu_{CA}^o . \tag{1}$$

The separate chemical potential of anions ($[A_i]$, $[A_o]$) and cations ($[C_i]$, $[C_o]$) in the inner and outer compartment follow the relationship:

$$\mu_{CA}^i = \mu_{CA}^{0i} + RT \ln[A_i][C_i] , \tag{2}$$

$$\mu_{CA}^o = \mu_{CA}^{0o} + RT \ln[A_o][C_o] . \tag{3}$$

The standard potentials μ_{CA}^{0i} and μ_{CA}^{0o} can be regarded as identical. Inserting Eqs. (2) and (3) into Eq. (1) yields:

$$[C_i][A_i] = [C_o][A_o] ; \quad \frac{[C_i]}{[C_o]} = \frac{[A_i]}{[A_o]} . \tag{4}$$

Due to the rule of electroneutrality the ratios of anions and cations in the separate compartments must be equal; z is the amount of surplus charge of the protein and is assumed to be positive:

$$z[E_i] + [C_i] = [A_i] \quad \text{and} \quad [C_o] = [A_o] .$$

Inserting these equations into Eq. (4) gives the following relationships:

$$[C_o]^2 = [C_i]([C_i] + z[E_i]) , \tag{5a}$$

$$[A_o]^2 = [A_i]([A_i] - z[E_i]) . \tag{5b}$$

The difference in cations and anions between the two compartments is therefore:

$$[C_i] - [C_o] = \frac{-z[E_i][C_i]}{[C_i] + [C_o]}, \qquad (6\,a)$$

$$[A_i] - [A_o] = \frac{z[E_i][A_i]}{[A_i] + [A_o]}. \qquad (6\,b)$$

For a negative surplus charge of the protein the sign of both equations will be reversed.

Table 3.1 Dialysis of different amounts of a sodium chloride solution against serum albumin, Na_o^+ is the sodium ion concentration in the outer, and Na_i^+ that in the inner compartment, before and after the experiment, respectively, the indicated values are relative. The negative surplus charges of serum albumin $z[E]_0$ are neutralized with sodium ions, accordingly $Na_i^+ = z[E]_0$ before dialysis

Relative Na$^+$ concentration				Apparent binding (%)
Before dialysis		After dialysis		
Na_i^+	Na_a^+	Na_i^+	Na_o^+	
1.0	0.01	1.000098	0.0099	99.0
1.0	1.0	1.333	0.667	66.7
0.01	1.0	0.508	0.502	1.2

To demonstrate the importance of the Donnan effect, Table 3.1 shows the distribution of sodium ions when sodium chloride is dialyzed against serum albumin carrying a negative surplus charge, which should also be neutralized by sodium ions. Therefore, applying a 100-fold surplus of serum albumin charges compared to sodium chloride in the outer compartment, after dialysis the sodium ions do not equalize between both compartments but the great majority are found together with the protein, giving the impression of a strong binding. If the same cation concentrations are in both compartments before the experiment, thereafter two thirds is found with the protein and only with a 100-fold surplus of the salt in the outer compartment does a nearly equal distribution result after dialysis. Charged ligands would behave in the same way but it is not the ligand but the charge which is efficient. So the Donnan effect can be diminished by applying high salt concentrations, e.g. from a buffer.

3.1.1.3 Continuous Equilibrium Dialysis

Although equilibrium dialysis is a reliable method, a serious disadvantage is the long duration of a few hours, which can be harmful for some macromolecules.

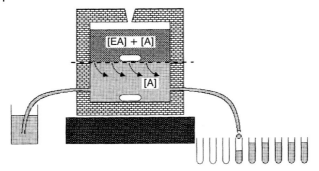

Fig. 3.5 Continuous equilibrium dialysis. The device is separated by a horizontal membrane into an upper and a lower compartment, both stirred with a magnetic stirrer. The upper compartment contains the macromolecule and the ligand, a buffer solution is pumped through the lower compartment and collected in a fraction collector.

A modification of this method, requiring only a few minutes and only one dialysis cell, thus also sparing much material, is continuous equilibrium dialysis. Its principle of dialysis is quite different to that of conventional dialysis, the dialysis equilibrium should not be reached, rather the rate of the ligand penetrating the membrane is determined (the term equilibrium refers not to the dialysis but to the reaction equilibrium). The dialysis cell has the two compartments arranged vertically, one above the other, with the membrane fixed horizontally between them (Fig. 3.5). A magnetic stirrer in each compartment provides homogeneous mixing. The upper compartment is open at its top, it contains a mixture of the macromolecule and the ligand, while the lower compartment is completely filled up to the membrane with buffer solution. The buffer solution is continuously pumped through this compartment via side channels, so that there is a constant flow along the dialysis membrane, removing the ligand that penetrates the membrane. The eluting solution is collected in fractions.

Initially a low amount of radioactively labeled ligand is added to the macromolecule solution in the upper compartment. Part of the ligand will be bound, while the free ligand passes through the membrane. After about 1 min the free ligand will be evenly distributed in the upper compartment and will penetrate through the membrane at a constant rate. The share of the ligand passing through the membrane within a distinct time unit will be constant and due to the constant flow of the buffer solution below the membrane it will be washed away into the fractions. Thus the time-dependent penetration of the ligand resembles a saturation curve, as shown in Fig. 3.6. After a short increase a plateau is reached when a constant flow of the ligand through the membrane is attained. The height of the plateau is directly proportional to the free ligand concentration in the upper compartment. The respective ligand concentrations are determined in the collected fractions by scintillation counting.

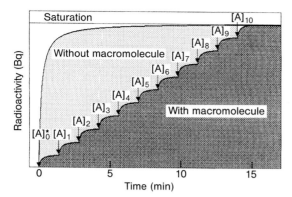

Fig. 3.6 Time course of the continuous equilibrium dialysis. At $t=0$ radioactively labeled ligand is added in a low concentration. In the absence of macromolecule the upper curve evolves. With higher amounts of the macromolecule most of the ligand will bind and only a small step for the free ligand appears. After reaching the plateau of this step defined quantities of unlabeled ligand are added successively after each new plateau. The procedure will be finished when the plateau of the control curve is attained.

In a second step another small amount of ligand is added to the upper compartment, but this time the ligand is unlabeled. This unlabeled ligand competes with the already bound ligand for its binding site on the macromolecule and, therefore, some more labeled ligand becomes displaced and released and therefore, the amount of labeled free ligand increases. This causes a further increase in the flow through the membrane and a higher plateau value. This procedure of addition of small amounts of unlabeled ligand is continued until all originally bound labeled ligand is displaced. This is the case when the macromolecule is completely saturated with the ligand, then the plateau will not increase further upon addition of ligand. The same final plateau will be attained in a separate control experiment where the labeled ligand is added to the upper compartment in the absence of the macromolecule. Then at once the total amount flows through the membrane and a single, high plateau is reached immediately. Just this plateau level must be reached in the binding experiment. The control curve serves also to calibrate the measured radioactivity to the concentration of free ligand passing through the membrane. The individual plateau values, divided by the maximum plateau of the control, provides the ratio of bound ligand. Since it takes about 1–2 min to reach the plateau value, 10 ligand additions are necessary for a complete binding experiment, which can be carried out within about 10 min. This short time is the advantage of the experiment, it is also necessary to avoid depletion of the ligand from the upper compartment. It is assumed that the part of the ligand lost during the performance of the experiment is negligible.

3.1.2
Ultrafiltration

This fast technique to separate macromolecules from low molecular weight components is usually employed for concentration or desalting of macromolecule solutions, but it can also be applied for binding measurements, where the short duration of the experiments is of special advantage, compensating for the relative inaccuracy of these methods. Various devices are commercially available and can be chosen according to the sample volume. The solutions are forced through the ultrafiltration membranes either by high pressure, vacuum, or centrifugal power. Membranes of different exclusion limits are available and consist of various materials, mainly synthetic polymers or nitrocellulose. The materials differ in their affinity for proteins and advantageously protein repelling materials should be used. Consideration must be given to the fact that the pores are not exactly equal in size, so a relatively small exclusion limit should be used. For example 5% of serum albumin ($M_r = 66\,000$) pass through a membrane with an exclusion limit of 30 000. A further problem of ultrafiltration is blocking of the membrane by the particles which become concentrated on it. Stirring during filtration can only partly circumvent this problem. Resolution of protein already attached to the membrane is rarely successful.

For the determination of binding equilibria different ultrafiltration methods can be employed (Fig. 3.7). The unbound ligand passes through the membrane, while the bound ligand is retained. Extensive concentration of the macromolecule should be avoided, as its concentration is considered as constant throughout the experiment. Therefore, only a small portion should be pressed through the membrane to determine the free ligand. Determination of the bound ligand in the solution above the membrane is critical because of the concentration

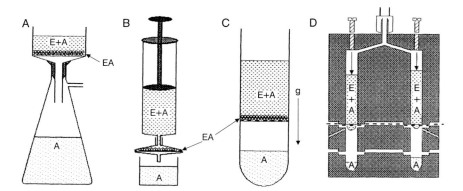

Fig. 3.7 Ultrafiltration devices. (A) Filter holder on a suction flask, (B) filter attachment for a syringe, (C) centrifugation tube with a sealed filter membrane, (D) ultrafiltration apparatus according to Paulus (1969).

effect. This value can be obtained from the difference between total (added) and free ligand, but unspecific binding to the membrane and the walls of the device can cause errors.

The simplest ultrafiltration device is a filter funnel with an ultrafiltration membrane on a suction flask connected to a vacuum (Fig. 3.7 A). As discussed above, only a small part of the macromolecule–ligand solution should be pressed through the filter and the free ligand will be determined in the filtrate. Instead, the solution is often completely pressed through the filter to concentrate the macromolecule–ligand complex on the filter and the radioactivity of the labeled ligand remaining on the filter is counted by scintillation measurement to determine the bound portion. Because the possible impairment of the equilibrium due to concentration is neglected this procedure is rather coarse.

In a similar manner other ultrafiltration devices can be used, like filter kits mounted on a syringe (Fig. 3.7 B) or centrifugation tubes with inserted membranes (Fig. 3.7 C), which can be used in a desktop centrifuge, where several samples with different ligand concentrations can be run simultaneously. Both devices are commercially available in different volumes.

A special apparatus for binding experiments has been developed by H. Paulus (1969). In a Perspex cylinder about 8–10 vertical channels are drilled in a symmetric, concentric arrangement (Fig. 3.7 D). A smaller cylinder with identically arranged channels is attached at the bottom, with a stable ultrafiltration or dialysis membrane fixed between the two cylinders. The channels in the upper cylinder are closed with stoppers or screws from the top, but connected by a central channel at the top with small radial channels. Small channels in the lower cylinder allow the removal of samples of the free ligand directly from the lower side of the membrane. Finally, a third cylinder with wells just at the positions of the vertical channels is attached to the bottom of the second cylinder and serves to collect the filtrate. For a binding experiment the channels in the upper cylinder are filled with about 0.1 ml of the macromolecule solution containing variable ligand concentrations. After closing the channels from the top, pressure of about 275 kPa, from compressed nitrogen, is applied to the central channel at the top; the pressure is conferred equally upon each vertical channel and the solutions are pressed through the dialysis membrane. From the samples taken from the lower membrane site or the vessels the free ligand concentration is determined. This method is reliable and simple and the ultrafiltration apparatus can be manufactured easily by a mechanical workshop.

3.1.3
Gel Filtration

Gel filtration (or molecular sieve chromatography) is a powerful method for the separation of macromolecules like proteins according to their size and can be applied for purification, molecular mass determination and desalting. Porous beads of dextran, agarose, or polyacrylamide are filled into a chromatography column. Upon elution the gel bed can be divided into two different spaces, an outer volume V_o outside the beads, and an inner volume V_i inside the beads. The outer volume is accessible for all molecules of the solution, while molecules can only penetrate into the inner volume if they are smaller than the holes in the beads. The size of the holes can vary with the respective gel type. Most commonly used are dextran gels (Sephadex®, Superdex®) with designations G-10 to G-200, the number being a rough estimate of the exclusion limit, e.g. G-200 for 200 kDa. Because the total bed volume is the sum of the outer and inner volumes: $V_t = V_i + V_o$, molecules which can freely penetrate into the gel beads have available a larger volume compared to macromolecules which are completely excluded and which can only occupy the outer volume. Therefore large molecules migrate faster and elute earlier than small ones. The quality of separation depends essentially on the length of the column, which should be about 1 m (for conventional columns, prepacked FPLC columns are somewhat smaller).

This method can be used to separate the ligand bound to the macromolecule from free ligand. However, simple gel chromatography of a mixture of ligand and macromolecule will result in a complete separation of both components after elution due to fast dissociation of the complex. Therefore procedures must be adopted which allow separation without disturbing the binding equilibrium.

3.1.3.1 Batch Method

This simple method can be used as a first test for binding, rather than for accurate binding analysis. It is carried out in a beaker without the need for a chromatographic column (Fig. 3.8). A gel (preferably Sephadex G-25) is used, which allows free penetration of the ligand, but complete exclusion of the macromolecule. With a small amount of the swollen gel the total volume V_t is first determined by adding a ligand solution of known concentration and volume. The ligand will distribute between the outer and inner volume. After allowing the gel to settle the ligand concentration is determined in a sample taken from the supernatant solution. The same procedure is then carried out with a defined solution of the macromol-

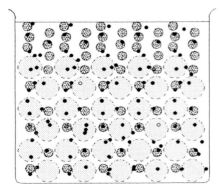

Fig. 3.8 Batch method for the determination of ligand binding. Only the small ligand molecules, but not the macromolecule can penetrate into the pores of the gel particles. Ligand molecules bound to the macromolecules are also restricted to the outer volume.

ecule to determine the outer volume. For the final experiment a ligand–macromolecule solution is applied. The free ligand will distribute in the total volume, but for the bound part only the outer volume is accessible and the ligand concentration in a supernatant sample will be higher than in the absence of the macromolecule. This method can be regarded as a special form of equilibrium dialysis, V_o being equivalent to the macromolecule compartment, V_i to the ligand compartment.

3.1.3.2 The Method of Hummel and Dreyer

This is the most convenient method for binding measurements by gel filtration. It requires a chromatography column filled with a gel which completely excludes the macromolecule, but includes the ligand, like Sephadex G-25. The length of the column must be sufficient for complete separation of macromolecule and ligand. First the column is eluted with a defined ligand solution. The ligand concentration of the eluting solution is measured, e.g. by continuous photometric control, and should be equal to that of the solution applied. If this is the case, a defined, small volume of a macromolecule solution, containing ligand in just the concentration of the original ligand solution, is applied to the column and thereafter elution with the ligand solution is continued. The macromolecule will bind a portion of the ligand, removing this from the free ligand solution. This part migrates faster together with the macromolecule than the circumfluent ligand solution. Upon elution through the column the bound ligand will remain constant, because ligand molecules released during migration will be replaced from the surrounding ligand solution. The portion of bound ligand removed from the free solution will be expressed as a 'valley' in the otherwise constant ligand solution. Because of the slower migration of the ligand on the column, this valley will remain behind the faster migrating macromolecule with its bound ligand, which will be eluted first as a maximum above the constant ligand level (Fig. 3.9 B). The integral of

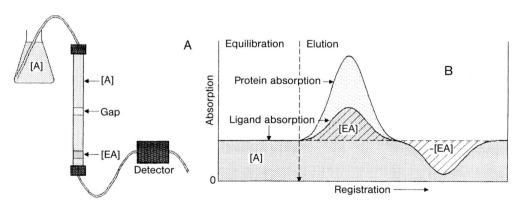

Fig. 3.9 Column chromatographic method for ligand binding according to Hummel and Dreyer (1962). (A) Experimental device, (B) elution profile.

the maximum is proportional to the bound ligand. The valley of missing ligand will elute later and its integral corresponds similarly to the bound ligand. It is advisable to determine the bound ligand from this valley, since the absorption of the macromolecule can add to the ligand maximum. The concentration of the original ligand corresponds to the free ligand in the equilibrium. The advantage of this method is the simultaneous determination of free and bound ligand and radioactive labeling of the ligand is not required, but higher amounts of the macromolecule and ligand are needed.

3.1.3.3 Other Gel Filtration Methods

A simpler and thus less accurate modification of the above method is the *elution of broad zones* (Ackers 1975). In contrast to common gel chromatography, where complete separation of components is intended, the principle of this method is the overlapping of the zones of the different components. The ligand–macromolecule mixture is applied to the gel filtration column in a larger volume and also the column must not be too long, so that in the eluting solution three zones can be differentiated: First the macromolecule that lost the ligand, appears, followed by a fraction of the macromolecule with the ligand still bound, and finally the ligand released during the passage. Its concentration corresponds to the free ligand in the equilibrium [A]. This holds as long as the three zones still overlap and do not become separate from each other. The middle zone is composed of bound and free ligand, but the portion of $[A]_{bound}$ can be better obtained from the difference of $[A]_0$ and $[A]$.

A very accurate method was described by *Brumbaugh and Ackers* (1974). Its application, however, is limited by the considerable technical effort. The gel column, preferentially manufactured from quartz glass to be permeable for UV light, is connected to an absorption photometer in such a manner that the column can be moved through the light pass by a driving motor (Fig. 3.10). The movement will be stepwise, each step replaces the column by the height of the

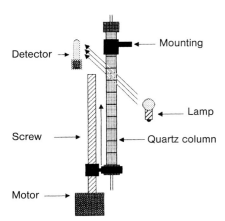

Fig. 3.10 Column chromatographic method for ligand binding according to Brumbaugh and Ackers (1974). The column packed with a molecular sieve gel and equilibrated with a macromolecule–ligand solution is moved stepwise through the light beam of an absorption photometer.

light path. This procedure apparently divides the column into a defined number (column length divided by the height of a single replacement, e.g. 100) of volume elements of similar size, the volume being defined by the cross-section of the column and the height of one replacement. Each volume element may be regarded as a single cell of the batch method (or a dialysis cell) and the outer volume V_o and the inner volume V_i can be determined accordingly by measuring the absorption (of ligand and macromolecule) in each of the volume elements after equilibrating the column with a solution of the macromolecule and the ligand, respectively. The distribution of both components between the inner and outer volume is described by the equation:

$$\zeta_A = V_o + \sigma_A V_i \tag{3.4}$$

where ζ_A is the cross-sectional distribution and σ the distribution coefficient, ranging from 0 to 1, depending on the gel material and the size of the respective molecule. For unhindered penetration through the gel pores with small ligands $\sigma_A = 1$, and Eq. (3.4) becomes $\zeta_A = V_o + V_i$, for complete exclusion of a macromolecule $\sigma_A = 0$ and $\zeta_A = V_o$. Finally, the column is equilibrated with the macromolecule–ligand solution and again the absorption is determined in each volume element. Deviation of these absorption values from those expected from the sum of both control experiments indicate the extent of binding. Due to the high number of volume elements, each being a single cell, a large amount of data is accumulated, conferring high statistical reliability.

3.1.4
Ultracentrifugation

Ultracentrifugation is a gentle method for separation of macromolecules which can be used not only in a preparative manner for purification and concentration, but also for determination of molecular masses. It can further be applied for binding measurements. Because ultracentrifuges are basic equipment of biochemical laboratories, no special device is required. Before describing special methods, some general aspects are mentioned. First we must differentiate between preparative and analytic ultracentrifugation. The latter needs more elaborate instrumentation and is used especially for molecular mass determination, it will not be discussed here. Preparative ultracentrifugation can be subdivided into conventional and gradient centrifugation. For conventional centrifugation fixed angle rotors (Fig. 3.11 A) are applied. They are manufactured from a metal block (mostly titanium) and the holes for the tubes are arranged concentrically, forming a fixed angle with the rotor axis. The sedimenting particles migrate perpendicularly from the rotor axis in the gravity field produced by the centrifugal force. Therefore they will not migrate straight through the tube to its bottom, rather to the tube wall opposite the rotor axis, i.e. free migration in the solution occurs only in part of the tube. The particles impinging on the wall will slip down to the bottom, a movement which cannot be easily described by the common sedimentation rules.

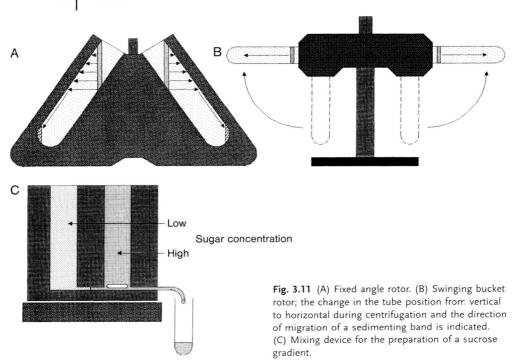

Fig. 3.11 (A) Fixed angle rotor. (B) Swinging bucket rotor; the change in the tube position from vertical to horizontal during centrifugation and the direction of migration of a sedimenting band is indicated. (C) Mixing device for the preparation of a sucrose gradient.

Therefore, this type of centrifugation is suitable only for separation of small from large particles, the latter being collected as a pellet.

Much more information about the size and shape of the sedimenting particles can be obtained by differential centrifugation. For this, free sedimentation in the solution must occur during the whole run. To enable this, swinging bucket rotors are used (Fig. 3.11 B). During centrifugation the tube holders reorient in the direction of sedimentation. Binding methods are described for conventional ultracentrifugation with both fixed angle rotors and swinging bucket rotors.

3.1.4.1 Fixed Angle Ultracentrifugation Methods

Various types of ultracentrifuges and fixed angle rotors exist, differing in, besides the allowed maximum speed, the sample volume, which can vary from several ml to less than 0.2 ml in a special air driven ultracentrifuge (Airfuge®, Beckman), which is especially suited for binding experiments, not only because of the very small volumes, but also due to the short centrifugation times of several minutes compared with the hours of conventional ultracentrifuges.

Upon sedimentation of a macromolecule–ligand mixture the macromolecule will sediment to the bottom of the tube, carrying its bound ligand, while the free ligand will remain at its position. After centrifugation the meniscus region will be depleted of macromolecule and the free ligand can be determined from an aliquot from this zone. Alternatively, the macromolecule together with its bound, radioactively labeled ligand can be sedimented into the pellet and, after complete removal of the supernatant, the radioactivity of the pellet can be determined by transferring the whole tube into a scintillation counter. This is only a rough method neglecting the change in the equilibrium due to concentration of the macromolecule.

Far more accurate is the method described by Chanutin et al. (1942). For this the concentration of the free ligand [A] is assumed to remain constant during the experiment, while the total amount of the ligand $[A]_0$ changes. Accordingly the general binding equation, Eq. (1.23), is rearranged:

$$[A]_{bound} = [A]_0 - [A] = \frac{n[E]_0[A]}{K_d + [A]},$$

$$[A]_0 = \frac{n[E]_0[A]}{K_d + [A]} + [A]. \tag{3.5}$$

$[A]_0$ becomes a linear function of the total macromolecule concentration $[E]_0$, a plot of $[A]_0$ against $[E]_0$ should yield a straight line with intercept [A], the free ligand concentration (Fig. 3.12 B). The experiment is performed by preparing several centrifuge tubes (according to the available tube positions of the rotor) with macromolecule–ligand solutions, varying the ligand concentration at constant macromolecule concentration. After spinning for a distinct time (no pellet should be formed), aliquots are removed from different parts between the meniscus and bottom of each tube and ligand and macromolecule concentrations are determined for each sample. From these data the plot shown in Fig. 3.12 B can be drawn, each tube will yield one straight line. It is, principally, irrelevant from which part of the tubes the samples are removed. If there are enough tubes with different ligand concentrations it may even be sufficient to analyze one sample of the original solution before centrifugation (Fig. 3.12 A, right side), and a meniscus fraction at the end of centrifugation (Fig. 3.12 A, left side). Although this results in only two points for each straight line (Fig. 3.12 B) this is justified because the actual evaluation is derived from the secondary plot of the reciprocal extrapolated values of [A] against the reciprocal slopes m according to the relationship:

$$m = \frac{n[A]}{K_d - [A]} \; ; \quad \frac{1}{m} = \frac{K_d}{n[A]} + \frac{1}{n}. \tag{3.9}$$

The expected straight line extrapolates to the reciprocal values of n and K_d as ordinate and abscissa intercepts, respectively (Fig. 3.12 C).

Steinberg and Schachman (1966) reported a similar method in which the distribution of the components is determined during centrifugation by direct absorption measurement in the analytical ultracentrifuge.

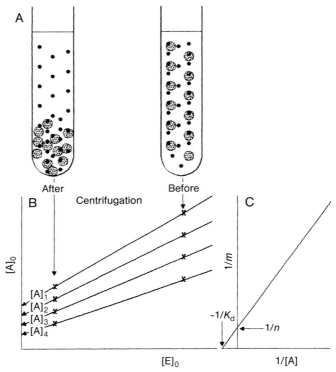

Fig. 3.12 Ultracentrifugation method of Chanutin et al. (1942). (A) Centrifuge tube before (right) and after (left) ultracentrifugation. (B) Plot of the values for $[A]_0$ and $[E]_0$ determined in samples from the tubes. The slopes m and the ordinate intercepts $[A]$ of the straight lines are plotted in a reciprocal manner in (C). The constants derived by extrapolation are indicated.

3.1.4.2 Sucrose Gradient Centrifugation

Swinging bucket rotors (Fig. 3.11 B) have the advantage that the particles migrate in the gravity field of sedimentation freely in solution from the meniscus to the bottom. According to the sedimentation rules, the sedimentation velocity depends on the size, shape and specific volume of the particle, the viscosity of the solvent, the distance from the rotor axis and the rotational speed. Therefore this technique can be applied to determine the molecular mass (and further the rough shape) of macromolecules. However, this method is impeded by diffusion. If the macromolecule solution is applied to the meniscus of the tube in a sharp layer, this will become considerably broadened after spinning for some hours so that exact determination of the sedimentation velocity of the macromolecule is difficult. Methods have been developed to diminish or control this effect, like the sedimentation equilibrium technique in analytical ultracentrifugation, which is just based on the balance between sedimentation and diffusion.

Sucrose gradient ultracentrifugation is another principle to diminish diffusion and to stabilize sharp migrating bands (in principle other media, like glycerol, can also be used for this method). A linear gradient of increasing sucrose concentration from the meniscus to the bottom of the tube is formed, as shown in Fig. 3.11 C. A higher concentration buffered sucrose solution (e.g. 20%) is filled into the front compartment of the gradient mixer and the lower concentration sucrose solution (5%) is placed in the back compartment. The solution in the front compartment is stirred with a magnetic stirrer. The volumes of both solutions must be calculated so that the centrifuge tube will just be filled when both compartments are emptied (actually, the solutions in both compartments should be equal not in volume but in weight). Both compartments are simultaneously emptied by a peristaltic pump, which pours the mixed gradient solution slowly into the centrifuge tube. The sucrose gradient remains stable for several days.

It should be considered that the sucrose gradient centrifugation is not a real density centrifugation, although the density increases continuously within the tube. Actually the centrifugation principle is sedimentation and the particle settles finally at the bottom of the tube (*zonal centrifugation*). Because sedimentation occurs against a medium with increasing density the migrating bands will be sharpened and after standing for a longer time at the end of the run, the band will remain in position without essential broadening. In contrast, in real density ultracentrifugation (*isopycnic centrifugation*), where media like cesium chloride are used, the particles gather at the position of the gradient corresponding to their own density. Such a gradient need not be prepared before centrifugation, only the cesium chloride solution, including the particles to be separated, is put into the tube and is centrifuged at a moderate speed for a longer time (sometimes several days). The gradient is formed during centrifugation and the particles collect at the position corresponding to their own density. Since with this method the principle of centrifugation is actually density and not sedimentation, this method yields no information about the size of the particles (indeed, the broadness of the band is a rough indication of the size of the particles), but is applied to separate particles of differing density, like proteins (which are relatively similar in their density and will gather in one band) from nucleic acids or different types of nucleic acids, like DNA, RNA, single-chain, double-chain.

For an experiment the macromolecule ligand solution is layered on top of the sucrose gradient. The volume of this solution should be very low, not more than one fiftieth of the gradient volume, its density must be lower than that of the sucrose solution. Speed and duration of ultracentrifugation depend on the size of the macromolecule, centrifugation should be stopped before the macromolecule reaches the bottom of the tube. After centrifugation the gradient is withdrawn with a peristaltic pump, either by pricking the tube bottom with a syringe or by inserting a capillary through the gradient to the tube bottom, and fractions are collected. Care must be taken that the fractions are completely equal in volume, it is recommended to count in drops, and the number of the drops is directly proportional to the migration distance from the meniscus to the relative position of the band. The macromolecule band can be detected by absorption or, if it is an enzyme, by its activity. This is also a relatively easy method to estimate the molecular mass of the protein. From the migration distance from the meniscus to the macromolecule position an approximate sedimentation coefficient can be derived by comparing with a known standard protein and, according to an equation given by Martin and Ames (1961), the molecular mass can be calculated. Although this is not completely precise, because the shape of the protein molecules (the standard and the sample) is not considered, the advantage is that no pure preparations are required and even crude extracts can be used, as long as the macromolecule can be specifically identified (e.g. by enzyme assay).

A binding method applying sucrose gradient centrifugation was described by Draper and Hippel (1979). Binding methods discussed so far are based on the assumption of a large difference in size between the macromolecule and the ligand. The differential separation by sucrose gradient also allows the determination of binding equilibria between components of similar size, like subunit–subunit association and interactions between different macromolecules (e.g. protein–protein, DNA–protein). The solution containing the associating components is applied on top of a sucrose gradient, the faster sedimenting component, the macromolecule, must be present in an at least tenfold excess compared with the low molecular weight component (Fig. 3.13). The height of the applied band d can be calculated from its volume and after centrifugation for a distinct time the migration distance l of the band from the meniscus is determined and it is calculated how often the band covered its own height (l/d).

For this method the general binding equation (Eq. (1.23)) will be rearranged, regarding the free macromolecule concentration [E] as the variable in place of [A], $[A]_0$ is assumed to remain constant. As the total amount of [E] is assumed to be large compared with [A], [E] is approximated to $[E]_0$:

$$([A]_{bound})_1 = \frac{[E]_0[A]_0}{K_d + [E]_0} \quad (3.7)$$

$([A]_{bound})_1$ is the portion of bound ligand in the component solution applied to the gradient at the start of the experiment (step 1, Fig. 3.13).

In the second step the macromolecule band migrates just the distance d into the gradient carrying the bound ligand and leaving behind the free ligand. Thus the part of the ligand bound $([A]_{bound})_1$ in the first step will now become the total ligand, which distributes into free and bound ligand $([A]_{bound})_2$ in the second step:

$$([A]_{bound})_2 = \frac{[E]_0([A]_{bound})_1}{K_d + [E]_0} = \left(\frac{[E]_0}{K_d + [E]_0}\right)^2 [A]_0 \quad (3.8\,a)$$

Correspondingly, for the third step of translocation of the macromolecule, now by two distances d:

$$([A]_{bound})_3 = \frac{[E]_0([A]_{bound})_2}{K_d + [E]_0} = \left(\frac{[E]_0}{K_d + [E]_0}\right)^3 [A]_0 \quad (3.8\,b)$$

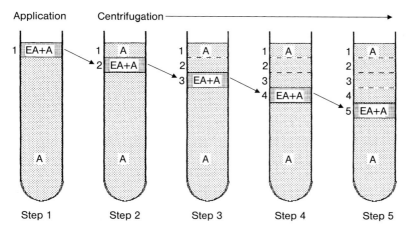

Fig. 3.13 Sucrose gradient centrifugation method for binding measurements according to Draper and Hippel (1979).

and, in general, for step i:

$$([A]_{\text{bound}})_i = \left(\frac{[E]_0}{K_d + [E]_0}\right)^i [A]_0 . \tag{3.9}$$

$$K_d = \frac{[E]_0}{\sqrt[i]{\frac{([A]_{\text{bound}})_i}{[A]_0}}} - [E]_0 . \tag{3.10}$$

At the end of centrifugation the gradient is fractionated and $l/d=i$ is determined. The dissociation constant is obtained from the amount of bound ligand

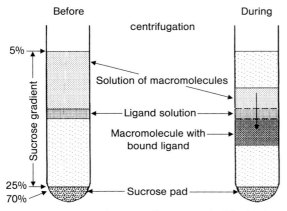

Fig. 3.14 Sucrose gradient centrifugation method for binding measurements according to Yamamoto and Alberts (1974).

in the macromolecule band and the originally applied concentrations of both components.

In an alternative method of Yamamoto and Alberts (1974) the ligand is already inserted as a narrow band at the position of 12% sucrose during the gradient formation. The upper part of the gradient from 5–11.5% sucrose contains the macromolecule (Fig. 3.14). During centrifugation the macromolecule sediments through the narrow ligand band taking with it a portion of bound ligand. After fractionation the amount of bound ligand is analyzed.

3.1.5
Surface Plasmon Resonance

Surface plasmon resonance (SPR) detection is a new technique which employs a special sensor chip. Commercial instruments are available. A thin metal (gold) film (~ 50 nm) is attached with its upper side to a glass plate, while its lower side is linked to an interaction layer (~ 100 nm) consisting of a carboxymethyl-dextran matrix (Fig. 3.15). The chip is incorporated in a cell. One of the interacting components, either the macromolecule or the ligand, is fixed to the dextran matrix. To achieve this, an appropriate reagent solution is applied to activate the carboxyl groups, to which the respective component is then attached, e.g. the macromolecule via its amino groups to form a covalent amide bond. Because the principle of the method is a change in the refractive index upon binding, it is advantageous to fix the smaller component to the matrix and to take the larger component as the free binding component. Since, however, covalent fixation may alter the binding features, for this procedure fixation of the macromolecule is preferred, but a smaller signal must be accepted. To be detected the free component should be larger than 200 Da.

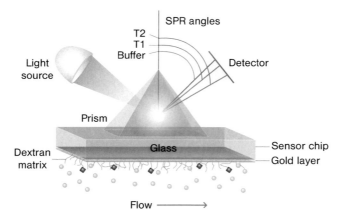

Fig. 3.15 Scheme of the surface plasmon resonance device (after Wilson 2002; with permission from the publisher).

A monochromatic light beam passes through a prism at the upper glass side and impinges on the surface of the flow cell at an angle adapted for total reflection. Oscillation of the conducting electrons (plasmons) at the metal film gives rise to an evanescent field which extends into the sample solution. At a distinct wavelength, resonance occurs and the intensity of the reflected light decreases at a sharply defined angle of incidence (*SPR angle*). This angle depends on the refractive index within the evanescent field around the metal surface and thus on charging the dextran matrix with binding components. The instrument detects the position of the reduced light intensity and evaluates the SPR angle. A basal angle is obtained with buffer flowing around the component fixed to the matrix. Upon adding a low amount of the binding component (ligand) to the buffer the refractive index at the surface and, consequently the SPR angle, will be changed due to binding to the fixed component, the extent of the change being proportional to the amount of ligand bound. The signal is given in resonance units (RU). A signal of 1000 RU, corresponding to an angle change of $0.1°$, is obtained by binding of about 1 ng mm^{-2} protein at the dextran surface. Increasing amounts of the ligand enhance the signal until saturation is attained.

This method allows not only the observation of the binding equilibrium, but also the time dependence of binding. Changing to the buffer solution after saturation has been reached, the time-dependent dissociation of the ligand can be followed. If all ligand is washed out, the basal signal is restored. While the amount of bound ligand is obtained from the SPR signal, the free ligand corresponds to the concentration in the flow buffer.

3.2
Electrochemical Methods

To analyze binding equilibria with the binding methods discussed so far time-independent measurements are executed, while for enzyme kinetic studies methods which allow time-dependent monitoring of the enzyme reaction are advantageous. One such class of methods comprises the electrochemical methods. Two types of enzyme reactions can be investigated with these methods,
- reactions causing pH changes
- reactions releasing or consuming gases.

Before discussing electrochemical methods, in honour of the famous biochemist Otto Warburg, his work on enzymatic reactions releasing or consuming gases should be mentioned. He developed a *manometric apparatus* which served to investigate enzyme reactions of the glycolytic pathway and the citric acid cycle. The reaction occurred in a tightly closed reaction vessel (Fig. 3.16) with a flask for the enzyme solution fused to its side. A central glass ring fused to the bottom of the vessel forms an inner compartment for a stop solution, while the substrate solution is placed in the outer compartment. The vessel is tightly connected to a manometer. The enzyme reaction is started by tilting the apparatus to pour the en-

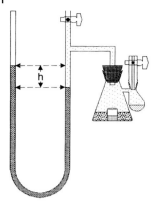

Fig. 3.16 Warburg manometer for measuring the release or consumption of gases during enzyme reactions.

zyme solution into the outer substrate compartment, strictly avoiding pouring out any of the liquid in the inner compartment. To keep the reaction at a defined temperature, the vessel is plunged into a water bath and gently shaken by a driving motor. After a defined time the reaction is terminated by a stronger inclination of the apparatus to pour the stop solution into the outer compartment. The stop solution cause complete expulsion of the gas from the reaction solution and the volume of gas produced or consumed is read from the manometer.

Nowadays gas specific electrodes have completely displaced the Warburg manometer.

3.2.1
The Oxygen Electrode

Oxygen plays an important role in numerous physiological reactions. It is involved in the reactions of oxygenases, hydroxylases and oxidases and it binds to transport proteins like hemoglobin and myoglobin. The oxygen electrode was developed by L. C. Clark in 1953 (Fig. 3.17 A). It facilitates greatly the study of oxygen-dependent reactions. The cathode consists of a platinum wire fixed in a glass tube, the anode is a silver–silver chloride electrode. Both electrodes are immersed in a saturated potassium chloride solution. A constant voltage of 0.5–0.8 V exists between the electrodes. The sample solution is transferred with a microsyringe into a sample compartment, which is connected in an airtight manner to the electrode device, from which it is separated by a Teflon or polyethylene membrane. Stirring with a magnetic stirrer assures a rapid exchange. Dissolved oxygen diffuses through the membrane and is reduced at the cathode:

$$\begin{array}{ll}\text{cathode} & 4\,H^+ + 4\,e^- + O_2 \rightarrow 2\,H_2O \\ \text{anode} & 4\,Ag + 4\,Cl^- \rightarrow 4\,AgCl + 4\,e^- \\ \hline & 4\,H^+ + 4\,Ag + 4\,Cl^- + O_2 \rightarrow 4\,AgCl + 2\,H_2O.\end{array}$$

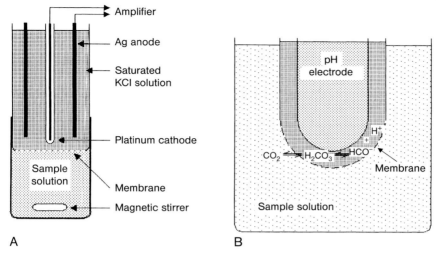

Fig. 3.17 Schemes of an oxygen electrode (A) and a CO_2 electrode (B).

A current is produced which is proportional to the oxygen concentration in the solution. Due to the membrane the oxygen electrode has a delayed response time. In the cathode compartment oxygen accumulates with a half-life time of about 2 min; this impedes the response, especially at transitions from high to low oxygen concentrations. Shorter response times can be achieved with open electrodes without membranes but there is, however, the danger of poisoning by components of the solution. The electrode must be calibrated before measurement, 0% oxygen is achieved by expelling the gas with a nitrogen stream or by addition of sodium dithionite to the buffered solution, air-saturated water is taken for 100%.

Figure 3.18 shows a device for continuous monitoring of oxygen-dependent reactions (Degn et al. 1980). A photometric cuvette, which enables simultaneous monitoring of absorption changes during the reaction, is used as the reaction vessel. A gas stream with defined oxygen content passes over the reaction solution, which is intensively stirred to ensure rapid exchange of the gaseous oxygen with the solution (for proper mixing a hexagonal cuvette is recommended). An oxygen electrode is inserted through a hole in the lower part of the vessel. A second oxygen electrode measures the oxygen content in the upper gaseous phase. The signals of both electrodes and of the photometer are simultaneously monitored by a multi-channel pen recorder or a computer, which also allows direct evaluation of the reaction course. The turnover velocity v_r is measured with linear increase of the oxygen content in the gaseous phase and is plotted in a direct or a linear form. It is assumed that at equilibrium v_r is equal to v_t, the velocity of the oxygen transport from the gaseous phase into the solution; v_t is proportional to the difference between the oxygen pressure in the gaseous phase T_G and in the solution T_S:

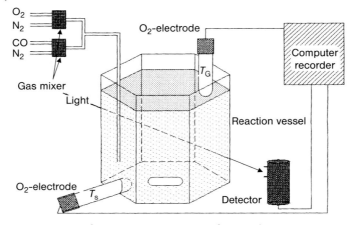

Fig. 3.18 Device for continuous monitoring of oxygen-dependent reactions with separate oxygen electrodes for measurements in solution and in the gaseous phase. As reaction vessel a hexagonal cuvette for absorption measurement is used.

$$v_r = v_t = K(T_G - T_S) \,. \tag{3.11}$$

Under given conditions the factor K depends on temperature, stirring speed, and the ratio between the surface and the volume of the reaction solution. At linear increasing oxygen content in the gaseous phase there no steady state will be reached for the oxygen exchange between the gaseous phase and the solution, i.e. $dT_S/dt \neq 0$:

$$v_r = v_t = K(T_G - T_S) - \frac{dT_S}{dt} \,. \tag{3.12}$$

K is obtained by determination of T_S after varying T_G in an assay without an oxygen-reactive system ($v_r = 0$):

$$\frac{T_S}{dt} = K(T_G - T_S) \,. \tag{3.13}$$

3.2.2
The CO₂ Electrode

The CO_2 electrode is essentially different from the oxygen electrode. A pH glass electrode, covered by a membrane of rubber-coated cellophane or silicon rubber is immersed in the sample solution (Fig. 3.17 B). Dissolved carbon dioxide diffuses into the space between the membrane and the glass electrode and becomes hydrated to carbonic acid. The resulting pH change is proportional by a factor S to the CO_2 content in the solution:

$$\Delta \text{pH} = S\Delta \log p\text{CO}_2 \, . \tag{3.14}$$

This demonstrates that CO_2 determinations are sensitive to pH changes. The pH of the solution must be strictly controlled. The CO_2 electrode is calibrated with a standard hydrogen carbonate solution or with different CO_2 partial pressures.

3.2.3
Potentiometry, Redox Potentials

Oxidation–reduction systems (redox pairs) occurring in many enzyme reactions, and especially in the respiratory chain, can be studied by potentiometric measurements. An electrode (e.g. platinum electrode) in a redox solution becomes charged and shows a potential difference against a reference electrode, which can be measured with a potentiometer. Redox potentials are characteristic values for defined redox systems. They refer to a standard hydrogen electrode, i.e. a platinum electrode aerated by hydrogen gas under atmospheric pressure, immersed in a solution of 1.228 M HCl whose potential is defined as 0. Naturally existing redox pairs are $NAD^+/NADH$, $NADP^+/NADPH$, $FAD/FADH_2$, and cytochrome Fe^{3+}/Fe^{2+}. The redox pair can be oxidized or reduced by applying adequate oxidising or reducing reagents. Determination of the potential difference against the extent of oxidation or reduction yields a potentiometric titration curve.

Redox processes can also be demonstrated with redox indicators. They change their color with the redox state and function as electron donors or acceptors in enzymatic redox reactions. Frequently applied electron acceptors are ferricyanide, 2,6-dichlorophenolindophenol, methylene blue, phenazine methosulfate, and tetrazolium salts, which are often used in histochemical enzyme assays. These dyes serve as indicators for photometric monitoring of the progress of a redox reaction.

3.2.4
The pH-stat

In many enzyme reactions protons are either released or bound, as in dehydrogenases: oxidases, hydrolases, esterases and proteases (by their esterase activity, proteolytic cleavage releases no protons).

reduced substrate + $NAD(P)^+ \longleftrightarrow$ oxidised product + $NAD(P)H + H^+$.

The enzyme reaction could be followed with a pH electrode, however, enzyme reactions are strongly dependent on the pH and the progressive pH change will influence the turnover rate. In order to keep the pH constant enzyme reactions are carried out in buffered solutions. This, however, eliminates the possibility of detecting pH changes caused by the reaction itself. To follow such a reaction, they are performed in non-buffered solutions and the pH is kept constant by adding equivalent amounts of acid or base, their consumption being a direct measure

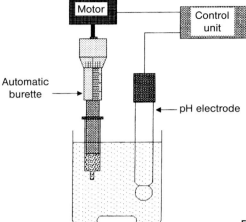

Fig. 3.19 Scheme of a pH-stat.

of the enzyme reaction. This is the principle of the pH-stat (auto-titrator), an instrument consisting of an automatic burette, which maintains a constant pH by release of the respective amounts of acid or base (Fig. 3.19). This automatic burette is driven by a glass electrode with a calomel reference electrode connected to a pH meter and a control unit. A pH change induces an impulse to the burette and titrant is added until the original pH is re-established. The volume of the titrant added is monitored in dependence on time and represents the turnover rate of the enzyme reaction. The concentration of the titrant solution determines the sensitivity of the system, higher dilution causing greater sensitivity, but care must be taken that the total sample volume does not change significantly. During the reaction the solution is constantly stirred. Titration can be achieved by alternating additions and pauses. This, however, affects sensitivity and includes the danger of over-titration. Alternatively, proportional control, supporting quick response of the system is applied. Environmental influences, like CO_2 or electrostatic interactions with other instruments or even with synthetic tissues (clothing of the experimenter) must be carefully avoided. pH-stats are available in different versions, e.g. automatic sampling with intermediate washing and sample preparation, simultaneous maintenance of constant substrate concentration with a second burette, system control and electronic data processing.

In another pH-stat system the pH is not compensated by titration but by an electrolysis stream producing acid or base directly at the electrodes. This has the advantage of a constant reaction volume. The pH is spectrophotometrically controlled using pH indicators (Karcher and Pardue 1971).

pH-stat measurements are often more sensitive than photometric tests. Enzyme reactions can be studied even in highly absorbing homogenates and light scattering suspensions, like cell homogenates, fractionated membranes and immobilized enzymes. On the other hand, handling of the pH-stat is laborious and limited by its slow response period.

3.2.5
Polarography

When two electrodes with a small negative potential difference are immersed in an electro-reducible substance, a small residual current flows between the electrodes (Fig. 3.20 A). Upon continuous increase of the potential a point will finally be reached at which the substance becomes reduced at the cathode. The current increases and this increase continues upon further rise of the potential until the reduction of the substance at the cathode becomes limited due to diffusion. From this point any rise in the potential causes no further increase in the current. A current–voltage curve, a polarogram, as shown in Fig. 3.20 B is obtained. Its inflection point, the half-wave potential $E_{1/2}$, is a characteristic, concentration-independent value for the respective substance. The plateau of the curve, the limiting current, is proportional to the concentration of the substance and can be used for its determination. Electro-oxidizable substances with positive potential will be oxidized at the anode and yield a corresponding, but opposite signal.

Polarography is extremely sensitive, employing highly diluted solutions and small sample volumes. The reduction of the substance is detected by a dropping mercury electrode. Elementary mercury drops from a reservoir through a capillary into the solution (Fig. 3.20 A). The reservoir level is adjusted to release 10–20 drops per minute. The continuous renewal of the surface prevents poisoning of the electrode, e.g. by proteins. The mercury layer formed at the bottom of the vessel serves as the anode. A calomel electrode can be used instead, connected with the sample solution by a salt bridge. The voltage applied is potentially changed and the current produced is measured with a galvanometer. For oxidations, a rotating platinum electrode or a carbon electrode is used as the anode.

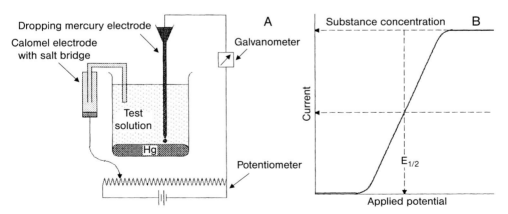

Fig. 3.20 (A) Scheme of a polarographic device with a dropping mercury electrode. (B) Polarogram for the determination of the half-wave potential $E_{1/2}$ and the substance concentration.

When product or substrate of an enzyme reaction generates a polarographic signal, the enzymatic turnover can be recorded at a constant potential. A progress curve is obtained where the time-dependent change in the intensity of the current is a measure of the turnover rate. The method of polarography can be applied for reactions with oxygen (using an oxygen electrode), for thiol compounds (e.g. reactions dependent on coenzyme A), for carbonyl compounds (e.g. pyruvate and NAD^+ or NADH). In contrast to photometric methods the determination can also be performed in turbid and highly absorbing samples.

3.3
Calorimetry

Calorimetry is one of the oldest biological methods. Already in 1780 Lavoisier and Laplace used this method to study respiration in animals. Nevertheless, this method has found little acceptance in biochemistry so far. Most chemical and biological processes are accompanied by release or uptake of heat from the environment. Development of heat is directly related to the reaction process. Because of this calorimetry represents a method with a broad potential of applications and has the advantage of directly studying systems without external influences or modifications, there are also no special requirements for purity. Changes in the range of mJ are detected with microcalorimeters, making micromolar concentrations accessible.

Two calorimetric methods are mainly applied. *Adiabatic calorimeters* have no heat exchange with the environment. The heat quantity $Q=\varepsilon\Delta T$ released or taken up by the system is detected by the relative change in temperature, to which it is related by the calibration constant ε. The calorimeter compartments are shielded from outer environmental influences by an air or vacuum casing (*isoperibolic calorimeters*). However, this does not completely prevent a certain heat exchange, especially for long-lasting experiments. Heat exchange with the environment is prevented by a heatable adiabatic metal shield within the outer casing, that automatically adjusts to the temperature of the inner calorimeter chamber (Fig. 3.21 A). With such a device it is not, however, possible to measure at constant temperature.

Heat conducting calorimeters transfer the heat from the reaction compartment directly to an outer thermal reservoir and the heat flux is controlled by a thermocouple fixed between the two compartments (Fig. 3.21 B). However, these apparatuses are relatively slow and not suitable for fast processes. *Isothermic calorimeters* compensate endo- and exothermal effects by heating or cooling in the measuring cell. The impulses required for compensation are registered in a time-dependent mode. In the *scanning calorimeter* both the reference and the sample cell is kept at the same temperature and the quantity of heat required for compensation in the sample cell is registered.

Manipulations, like initiation of a reaction by adding substrate, or stirring, affect caloric measurements. For compensation, as also for equalization of unspe-

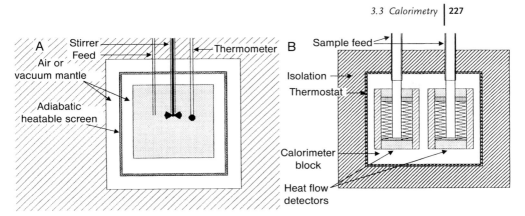

Fig. 3.21 Schemes of an adiabatic isoperibolic calorimeter (A) and a heat-conducting (thermal) calorimeter with twin arrangement (B).

cific heat exchanges with the environment, twin calorimeters equipped with two identical compartments are used. Sample and reference compartments are equally treated, the reaction proceeding only in the sample compartment. Calorimeters can also be equipped with a photometric device to monitor absorption changes. *Flow calorimeters* with mixing or flow cells are especially suited for enzyme reactions. A zero order reaction under steady-state conditions releases a constant quantity of heat per time unit and the inclination from the base line is proportional to the turnover rate.

Ligand binding to macromolecules can also be determined calorimetrically. In separate experiments the dilution heat of the ligand and also of the macromolecule is measured and subtracted from the values measured for binding. Calorimetric titrations of the macromolecule with the ligand yield the dissociation constant K_d and the binding enthalpy ΔH. The calorimetrically determined heat quantity Q shows in the double-reciprocal plot a linear dependence on the ligand concentration for a normal binding process:

$$\frac{1}{Q} = \frac{1}{Q_m} + \frac{K_d}{Q_m[A]} \, . \tag{3.15}$$

Q_m, the heat quantity at saturation, is proportional to the binding enthalpy: $Q_m = \Delta H[A]_{eq}$. $[A]_{eq}$ is the molar amount of bound ligand at complete saturation. In a similar manner aggregation of macromolecules, re-association of proteins from subunits, protonation of amino acid residues, hydrations, conformational changes and denaturation processes can be investigated by calorimetric studies.

3.4
Spectroscopic Methods

Spectroscopic methods, especially absorption spectroscopy, are widely used for enzyme kinetic studies, enzyme tests, ligand binding, conformational changes, investigation of catalytic mechanisms, etc. Easy handling and continuous monitoring of time-dependent processes like enzyme reactions make these methods attractive. Intervention into ongoing reactions, e.g. by additions, is possible at any time. High-quality absorption spectrophotometers can be obtained at moderate prices and used for many applications, like detection and concentration determinations of metabolites, proteins, nucleic acids and lipids. The following gives a survey of the most relevant photometric methods for enzyme studies and their special applications. Common to all these methods is the principle of observing the alterations a light beam suffers upon passage through a sample solution (absorption, ORD, CD) or the light emitted by the sample (fluorescence, Raman effect).

Possible interactions of a photon with a molecule are illustrated in the term scheme shown in Fig. 3.22. An electron exists in a low energy ground state S_0 and a high energy or excited state S_1. The energy difference between the two

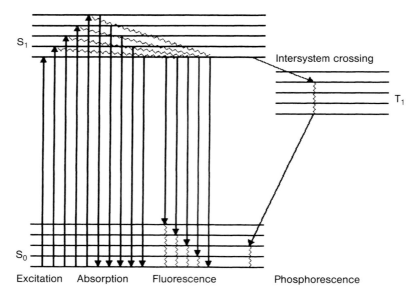

Fig. 3.22 Term scheme of the energy content of an electron. The electron accepts the energy from photon irradiation and goes from the ground state S_0 to the excited state S_1. From there it either returns directly to the ground state (absorption) or remains for a distinct period (in the ns range) at the lower vibrational level of the S_1 state, and returns to the ground state with light emission (fluorescence). A third possibility is intersystem crossing to a less energetic triplet state T_1 emitting light after a longer delay time (phosphorescence). The horizontal lines of the respective electronic states symbolize vibrational levels.

states is 340 kJ mol^{-1}. Each state possesses various vibrational energy levels, differing by about 40 kJ mol^{-1}, and rotational energy levels differing by less than 4 kJ. At normal temperatures the molecules remain preferentially in the lowest vibrational level of S_0, but various rotational levels can be accepted. Absorption of a photon of a distinct frequency induces the molecule to transform into the excited state S_1, where various vibrational and rotational levels can be occupied. According to this an absorption spectrum should consist of many adjacent sharp bands. Environmental influences, especially in solution, and other factors, however, cause broadening of the bands, so that they merge into one or a few broad absorption bands characteristic of the respective molecule.

The molecule cannot remain in the excited state S_1. Mostly the excited state becomes deactivated in a radiationless process and the excitation energy is dispersed as heat to the environment. This is due to collisions with molecules of the same species or with different molecules. Certain compounds, such as dissolved molecular oxygen, are particularly efficient in deactivating the excited states. Such deactivation processes acting from the outside on the excited state are designated *external conversions*. Deactivation can also occur within the molecule by energy redistributions to internal vibrations (*internal conversion*). Finally, the excited singlet state may be transferred by a forbidden radiationless spin exchange into a low energy triplet state T_1 (*intersystem crossing*). Such processes possess a long lifetime ranging from milliseconds to a few seconds and *phosphorescence* light will be emitted. Due to its extremely long lifetime, this state will usually be completely deactivated in solutions by internal or external conversions and may only be observable at low temperatures in the solid phase.

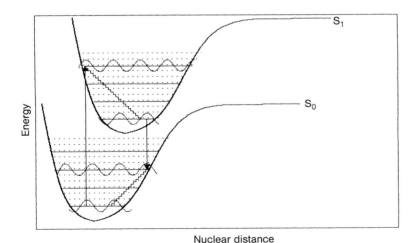

Fig. 3.23 Franck-Condon principle for a two-atom molecule. The transition of the electron occurs with highest probability in a vertical direction in resonance from the ground vibrational state S_0 to a vibrational level of the excited state S_1 oscillating with the same phase.

External and internal conversions compete for the energy of the excited states. If these processes are fast, the excited state will be deactivated and the phenomenon will be observed as *absorption*. If, on the other hand, these processes are comparatively slow and are not able to deactivate the S_1 state immediately, the excited electron changes radiationlessly to the lowest vibrational level (Fig. 3.22). The duration of the respective processes is decisive for the further fate. Excitation of the molecule occurs so rapidly ($\approx 10^{-15}$ s) that the nuclei due to their inertia cannot follow so quickly to the new condition of excited state and retain initially the nucleic distance of the ground state S_0 (*Franck-Condon principle*). The transition occurs vertically over several vibrational levels to the state of highest probability for a nucleic distance closest to the original state (Fig. 3.23). From there the electron moves, within about 10^{-12} s, to the lowest vibrational level where it remains on average for a few ns. Finally, the excited electron reverts to the ground state S_0 emitting light of lower energy, i.e., longer wavelength (*fluorescence*). If the deactivation processes described above are faster than the emission, fluorescence becomes weakened or even completely quenched. In such cases only absorption can be observed.

3.4.1
Absorption Spectroscopy

3.4.1.1 The Lambert-Beer Law

Absorption measurement is based on the Lambert-Beer law describing the attenuation of light intensity I_0 at a given wavelength λ after passage through the solution of an absorbing compound with molar concentration c:

$$I = I_0 e^{-\varepsilon d c} . \tag{3.16}$$

d is the length of the light path in the solution and ε the molar absorption coefficient (l mol^{-1} cm^{-1}). I/I_0 is the *transmittance* or permeability and is usually indicated in percent. Unhindered passage of light, $I = I_0$ corresponds to a transmittance of 1 or 100% at the respective wavelength, while 0% transmittance is total impermeability i.e. a closed optical path. These limiting values serve for the calibration of photometers for a given wavelength. According to Eq. (3.16), light intensity decreases exponentially with the concentration of the absorbing substance. This can be converted into a linear dependence by applying the negative logarithm of the transmittance:

$$A = -\log(I/I_0) = \varepsilon d c . \tag{3.17}$$

A is the measure of absorption. The previously used terms 'extinction' or 'optical density' are no longer valid (actually the quotient I_0/I represents optical density or opacity).

3.4.1.2 Spectral Properties of Enzymes and Ligands

Principally all biological compounds exhibit absorption and are, therefore, accessible to absorption (UV/Vis) spectroscopy, although the absorption maxima of many substances, e.g., carbonyl groups or peptide bonds, lie in the far-UV region. This is a hardly accessible region and overlapping spectra from many contributions make interpretation difficult. For many studies, however, like enzyme kinetics, ligand binding, or conformational changes, not the absolute absorption, only spectral changes are important. For example, an enzyme reaction cannot be followed photometrically, even if the product shows an intense absorption band, if the substrate has just the same spectral features. The difference between both absorptions is decisive, but the relationship to the absolute absorption must also be considered. In the presence of relatively low total absorptions even a small difference can be detected, but the same difference will be difficult to detect at very high absorptions, since under such conditions the instrumental sensitivity is low, resulting in a high scattering. Actually, in most cases the spectral differences between an enzyme substrate and its product are not very large and only small shifts can be observed. A good example is the cofactor $NAD(P)^+$. It shows an intense absorption band at 260 nm which decreases upon reduction to $NADP(H) + H^+$. This decrease may be taken as a signal for dehydrogenase tests. However, in addition, a new absorption band emerges at 340 nm. This is not only an easily accessible region in the visible region but there is also the great advantage that the oxidized form does not absorb in this region and any absorption observed is directly indicative of the appearance of the reduced form. This is the principal of the *optical test* for dehydrogenases. Because of the convenience of this test, it is not limited only to dehydrogenases but is also applied for the analysis of numerous enzyme reactions which can be coupled (sometimes even via one or two intermediate reactions) to a dehydrogenase reaction. If, for example, the product of an enzyme reaction, which cannot directly be followed photometrically, can act as a substrate in a dehydrogenase reaction, both reactions can be performed in the same enzyme assay and the reduction of NAD (or oxidation of NADH) can be taken as a signal for the conversion of the substrate of the first reaction, which provides the substrate for the dehydrogenase. The reaction conditions of such *coupled assays* are rather complex to ensure that the test reaction and not the indicator reaction becomes rate limiting, e.g. by a surplus of the dehydrogenase. Coupled assays are helpful for the determination of enzyme activity, but are not recommended for enzyme kinetic studies, as it cannot be excluded that the indicator reaction will become rate limiting and will affect the test reaction when certain parameters such as cofactor concentration or pH are changed.

Most substrate/product pairs in enzyme reactions do not show such marked spectral differences as NAD^+/NADH. Often there are only hardly detectable spectral shifts or differences in intensity, as in many isomerizations. Ligand binding and conformational changes in proteins also exhibit minimal spectral alterations that may be detected only with highly sensitive photometers, e.g., dual beam and dual wavelength photometers (see below).

While for enzyme kinetic measurements the absorption properties of substrates and products are of relevance, binding and conformational studies concentrate more on the spectral properties of proteins and, if present, their coenzymes. As proteins are composed of the 20 proteinogenic amino acids, their spectral features determine the shape of the protein spectra, and as larger proteins contain usually all 20 amino acids, spectra from different proteins are essentially similar to one another. In the far-UV region (190 nm) where the contribution of the peptide binding is found, the amino and carboxyl groups predominate. Inorganic ions like Cl^- (181 nm) and OH^- (187 nm) also absorb in this region, as does oxygen in gaseous or dissolved form. For measurements in this region the optical system of the photometer has to be gassed with nitrogen. Dissolved oxygen must be removed from solutions in vacuum. With the conventional photometric equipment the far-UV region is hardly accessible due to the weak lamp intensity, but since there is usually no change in the backbone region of native protein this far-UV range is of minor interest.

More interesting is the spectral range between 190 and 210 nm. Elements of the secondary structure of proteins contribute to the absorption. The bands of the random coil and the β-sheet are relatively similar to one another, but more intense than the α-helix band, which shows a characteristic shoulder between 200 and 210 nm (Fig. 3.24). From these absorptions some information about the secondary structure of the respective protein and about conformational changes can be gained. A large contribution to this region with pronounced maxima derives from the three aromatic amino acids (Fig. 3.25). The spectral maximum of phenylalanine is at the shortest wavelength and highest intensity (< 190 nm with a shoulder at 206 nm). Tyrosine has the lowest maximum at the longest wavelength

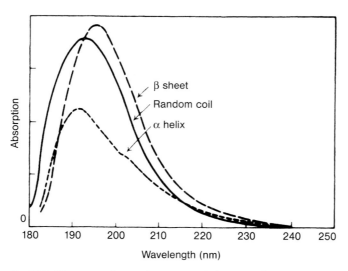

Fig. 3.24 UV spectra of secondary structural elements of proteins (Rosenheck and Doty 1961).

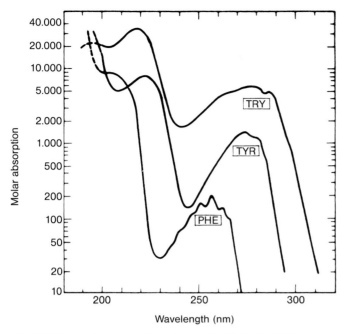

Fig. 3.25 UV spectra of aromatic amino acids (Wettlaufer 1962).

(220 nm), while the tryptophan spectrum is located between them. Histidine shows a band, methionine a shoulder at 210 nm. Cysteine absorbs between 200 and 210 nm, its intensity increases upon deprotonation of the thiol group, and an additional band is formed between 230 and 240 nm. The aliphatic amino acids exhibit absorptions in the region below 200 nm, mainly originating from the carboxyl group. Thus various contributions gather in the UV region between 190 and 220 nm, resulting in a characteristic strong absorption band observed in typical protein spectra, but it is more difficult to recognize specific effects.

A second absorption band of proteins is in the nearer UV range, between 260 and 300 nm. This is less intense, but consists solely of contributions from the three aromatic amino acids (Fig. 3.25). Phenylalanine has the weakest absorption of the three aromatic amino acids, its maximum is eight times lower than that of tyrosine and 35 times lower that of tryptophan (Table 3.2) and becomes visible only in the absence of both other amino acids. Its absorption is located at shorter wavelengths with a maximum at 257 nm. Characteristic for the phenylalanine spectrum is a pronounced fine structure of several small bands which, in the presence of tyrosine and tryptophan, can be observed on their short-wave flank, especially in the solid phase at low temperatures. The tyrosine absorption at 274 nm is usually overlapped by the strong tryptophan absorption at 280 nm. Native proteins, containing all aromatic amino acids, show essentially the characteristics of the tryptophan spectrum with a maximum at 280 nm. The intensity of this band is taken as

Table 3.2 Spectral features of amino acids and coenzymes in aqueous solution. ε: molar absorption coefficient, Φ_F: quantum yield (partially after Fasman 1989)

Compound	Absorption maxima		Emission maxima	
	λ_{max} (nm)	ε (l·mol⁻¹cm⁻¹)	λ_{max} (nm)	Φ_F (%)
Phenylalanine	206	9 300	282	4
	242	86		
	257	197		
	267	91		
Tyrosine	224	8 800	303	21
	274.6	1 420		
Tryptophan	219	35 000	350	20
	280	5 600		
Histidine	211	5 860		
Cysteine	250	360		
Cystine	248	350		
Adenine	259	14 900	321	0.026
Coenzyme A	260	16 000		
FAD	260	46 200	536	2.5
	375	9 300		
	438	14 600		
	445	11 300		
FMN	450	12 200	536	24
NAD, NADP	260	18 000		
NADH, NADPH	260	14 100	470	1.9
	340	6 220		
Pyridoxal phosphate	295	6 700	392	
	388	6 500		
Thiamin diphosphate	235	10 100		
	267	9 200		
p-Aminobenzoic acid	305		360	
ANS	350		515	0.4
ANS in ethanol			470	37
TNS	317		500	0.08
TNS in ethanol			429	52
DANSYL amide	320		580	5.5
DANSYL protein-bound			468	84
Fluoresceine	495		518	92
Rhodamine B	575		630	80

a measure of protein concentration. Ionization of the phenolic hydroxy group of tyrosine in an alkaline environment causes an increase in the absorption intensity and a shift of the maximum by about 20 to 295 nm. This effect can be used for quantitative determination of tyrosine by alkaline titration.

It should be possible to deduce the absorption spectrum of a distinct protein by superimposing the individual spectra of all its amino acids. Such calculated spectra, however, deviate considerably from those obtained by direct spectral analysis, especially, the intensity of the bands is weaker than expected from the sum of the individual amino acid contributions. This is because amino acid residues are shielded within the folded proteins. Partial unfolding of the protein molecule with urea, guanidine or other detergents exposes these residues and the intensity of the spectral bands increases. Successive denaturation of a protein yields information about its structure. Conformational changes of proteins, accompanied by hiding or exposing of aromatic amino acid residues, may cause spectral alterations. Changes in the polarity in the vicinity of distinct amino acid residues lead to spectral shifts. A decrease in polarity causes a red shift in the absorption maxima of aromatic amino acids, due to the difference in the energy levels of the ground and excited states. Spectral alterations are also observed with charge-transfer or metal complexes.

The determination of ligand binding of proteins by spectroscopic titration with difference spectroscopy is based on such effects. Interaction of a ligand with the protein either causes directly an alteration of the protein or the ligand spectrum, or the change is mediated by an induced conformational change.

3.4.1.3 Structure of Spectrophotometers

Julius Elster and Hans Geitel, two physicists from Göttingen, developed in 1891 the first vacuum photo cell. This can be regarded as the hour of birth of the photometers, because the quantitative determination of light intensity was the limiting factor for photometry. Nevertheless, the development of a convenient photometer for research proceeded relatively slowly and the first broadly applied photometer with high precision was the filter photometer developed by Theodor Bücher in collaboration with the German company Eppendorf in 1950 and the Zeiss spectrophotometer PMQ 2 at the same time. As the photometer emerged as a standard instrument for biochemical laboratories a rapid development followed. Modern instruments with their benefits of self-adjustment and self-tests spare the experimenter some work and thinking, however, it is easy to forget the functional principles, which are important to ensure proper application and to avoid erroneous measurements and misinterpretation of data. A short introduction to the general principles and some special applications of absorption photometers is now given. The main parts of photometers: the light source, a monochromator for spectral dispersion of the light and a photocell or a photomultiplier to measure the light intensity, and equipment for the display and recording of the measured signals are discussed (Fig. 3.26).

Light source. An ideal light source for spectral photometers should cover the total spectral region with constant light intensity, a continuum. Unfortunately, such light sources do not exist, all available lamps show distinct spectral characteristics with an intensity maximum within a limited wavelength range. It is not possible to completely cover both the UV and the visible spectral regions with a single lamp. Commercial spectrophotometers are therefore equipped with a hydrogen or a deuterium lamp for the UV region from 190–340 nm and

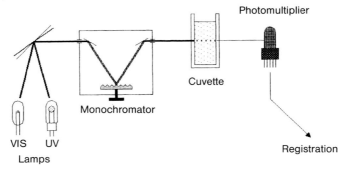

Fig. 3.26 Construction scheme of a spectrophotometer.

a tungsten or a halogen lamp for the visible region from 320–900 nm. A pivoting mirror directs the light of the respective lamp through the sample. Simple instruments possess only the visible lamp and cannot used for UV measurements. For measurements at constant wavelength *filter photometers* with a mercury lamp emitting distinct lines (line spectrum) are generally sufficient. They are of high stability and suitable for various enzyme tests and protein determinations. The line of the respective wavelength is collimated by interference or liquid filters so that no monochromator is required.

Monochromator. Most photometers perform spectral light resolution with the aid of diffraction gratings. Compared to the glass or quartz prisms used previously, they have the disadvantage of a larger degree of light dispersion and thus yield monochromatic light of minor spectral purity. The advantage of gratings lies in the registration of spectra. The monochromatic light is collimated through the exit slit at constant rotation of the grating and directed into the sample solution. For gratings, the degree of rotation is proportional to the wavelength, the progression of the driving motor serves as a direct signal for the wavelength scale of the spectrum. Prisms show no linear light dispersion, thus complicating the registration of continuous spectra. The spectral purity of gratings, i.e., their quality, depends on the number and shape of the grooves. A high burr between the furrows causes strong light dispersion. This can be prevented by flattening the burr, the grating is blazed to a given wavelength and incidence angle. Dispersion-free holographic concave gratings produced by laser-holographic methods are frequently used. To obtain light of the highest optical purity, two monochromators are connected in series, one behind the other. The width of the entrance and exit slits of the monochromator determines the sharpness of the spectral cut-out. The measure is the *bandwidth* of the incident light, i.e., the spectral width at half intensity. The gain in optical purity at narrow spatial collimation, however, is accompanied by a loss in light intensity and thus in sensitivity. The slit width must correlate with the quality of the grating; narrowing the slit cannot compensate bad spectral resolution from a simple grating.

Cuvettes. To eliminate reflections and dispersion, rectangular cuvettes with ground surfaces are used. They usually have an inner thickness of 1 cm. For the visible spectral range (340–800 nm) glass or plastic cuvettes are used, which are impermeable for UV light. For this range cuvettes prepared from quartz glass have to be applied.

Photomultipliers. After passing the sample cuvette the intensity of the light beam is measured in a photocell or a photomultiplier. Photocathodes are used which release electrons from alkaline material. They have an optimal response only within a given spectral range. It must generally be considered that all optical materials, the light source, diffraction gratings, mirrors and collective lenses, exhibit special wavelength characteristics. These effects superimpose the spectrum of an absorbing sample and causes significant distortion. To obtain authentic spectra, photometers possess corrective functions. The photomultiplier signal is adjusted to a constant level, dependent on wavelength, by resistor strips. With computer-controlled equipment corrections are executed with software programs. Photomultipliers always retain a certain level of dark current, even in the total absence of light, causing a permanent background noise that affects the sensitivity of the unit.

Diode array photometers are based on a multi-channel photodetector instead of a photomultiplier. A series of photodiodes are linearly arrayed. The whole polychromatic light of the lamp is directed through the cuvette. Spectral resolution occurs in the grating monochromator behind the cuvette and the complete spectrum is projected onto the multi-channel photo detector and recorded as a whole. With this method spectra can be recorded within milliseconds. This technique is applied in rapid-scanning stopped-flow units that are able to measure several hundred spectra per second.

Recording. For recording the current induced in the photomultiplier is transmitted into a proportional voltage. The transmission signal is transformed into an absorption signal according to the Lambert-Beer law by a transmission–absorption transformer and monitored by a recorder. Even the smallest absorption differences are detectable by amplification, limited, however, by the noise of the unit. The quality of instruments is determined by the signal-to-noise ratio, indicating the weakest possible measuring signal still detectable above the background noise. For kinetic measurements, time-dependent recordings are applied and, for spectral studies, wavelength-dependent measurements. Computer-controlled instruments display the primary data on a screen; these can be saved and documented by printer. Computer control permits direct processing and calculation of data. Sections may be cut out, slopes calculated by regression methods (e.g., the determination of reaction velocities) and data sets (spectra) subtracted from one another.

3.4.1.4 Double Beam Spectrophotometer

Ordinary photometers are based on the *single beam technique*; there is only one cuvette in the light beam. Reference or blank values are measured separately and are deducted from the sample values either by calculation or suppression of the blank to zero. Sometimes, besides the genuine enzyme reaction, spontaneous reactions are observed, e.g., due to instability of a substance or oxidative processes caused by

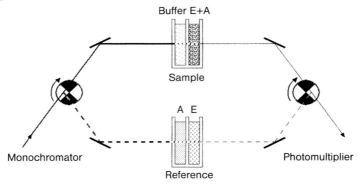

Fig. 3.27 Scheme of a double beam spectrophotometer with tandem cuvettes for difference spectroscopy.

dissolved oxygen. There is no constant blank to be easily subtracted from the sample. For such cases the *double beam technique* represents a significant improvement (Fig. 3.27). Passing the monochromator, the beam meets a rapidly rotating diaphragm with a reflecting mirror. This mirror allows, alternately, the light to pass or to be totally reflected. The two alternating beams are directed on mirror-symmetric pathways through a sample and a reference cuvette, respectively. After penetrating the cuvettes, both beams are conducted over a second diaphragm, identical to the first one and rotating at the same frequency, so that they rejoin and reach the photomultiplier over exactly the same pathway. In connection with the rotating diaphragm the photomultiplier discriminates between the two beams and the difference between sample and reference intensity is displayed as a measuring signal. In another configuration the beam is split by a beam splitter into two beams, each of half intensity, that are sent through the two cuvettes on parallel pathways. The rotating diaphragm is not necessary, but the beams must be detected separately, each with its own photomultiplier.

If both sample and reference cuvette contain the same absorbing substance in identical concentration, than both light beams detect the same absorption spectrum, the sample beam with a positive, the reference beam with a negative sign, and both measuring effects compensate each other, the photometer will record a base line. This allows the detection of very small differences between two samples, even if both show high absorptions. However, at high absorption even very small unintended deviations, like concentration differences, badly adjusted light beams or differences in the thickness of the cuvettes will produce considerable effects and give the appearance of spectral differences between both cuvettes.

3.4.1.5 **Difference Spectroscopy**

The double beam method is suitable for the study of spectral shifts caused by interactions of different components and it is especially appropriate for the study of structural features of proteins. The principle of this technique of difference spectroscopy will be explained with the example of macromolecule–ligand binding. First it must be established that binding indeed induces a spectral shift. In principle, any spectral change can be investigated, either of the protein or the ligand spectrum, or of both, but due to its macromolecular structure shifts in the protein spectrum are mostly clearer. Spectral shifts are caused either by direct interactions between ligand and groups on the binding site of the macromolecule or by ligand-induced conformational changes. The effect must be specific, not due to indirect effects like changes in pH or ion strength upon addition of the ligand solution. To detect spectral shifts, the absolute spectra of both the protein and the ligand have to be compensated. The double beam arrangement will compensate the absorption of one component, e.g., the protein, if the same solution is in the sample and the reference cuvette and remains constant during the experiment. The ligand, however, as the variable component, is added only to the protein in the sample cuvette to show the binding effect in comparison to the unbound protein in the reference. However, the ligand absorption will add to the sample signal and overlap any binding effect. Therefore the ligand absorption must also be compensated for. This is achieved applying *tandem cuvettes*. They are divided into two even compartments by a transparent window (Fig. 3.27). For binding measurements, the protein solution in identical concentration is filled into only one compartment of the sample and the reference cuvette. The two remaining compartments are filled with buffer solution. At this step a base line must be measured, proving that both cuvettes contain the same solutions. Deviations indicate differences in concentration and must be corrected. To study the binding process the ligand is added successively in small aliquots to the protein solution of the sample cuvette and, in parallel, to the buffer compartment of the reference cuvette to compensate for ligand absorption. As repetitive ligand additions dilute the protein solution in the sample cuvette, the protein solution in the reference cuvette must similarly be diluted by additions of the same volumes of buffer. Because of the minimal spectral shifts to be detected in highly absorbing solutions, this technique is very sensitive to any concentration differences, which can appear during the titration process if the additions are not completely equal. Ligand additions should therefore be small in volume and no solution should be spilled from the compartments when stirring or mixing the solutions. A small magnetic stirrer at the bottom of the cuvette is recommended. To evaluate the binding process ligand is added to the protein in many small aliquots until saturation is reached (*spectroscopic titration*). Initially, the spectral shift grows proportionally to the addition of ligand and approaches a plateau value at saturation of the binding sites. The evaluation of spectroscopic titrations is described in Section 1.3.2.2.

Spectral changes can appear as modifications of intensity or shifts of absorptions bands, or of both together. If caused only by a spectral shift without intensity changes, difference spectra show the shape of the first derivative of the absorption spectrum. The smallest changes will be observed at the maxima of the absorption spectrum, while shifts at the flanks result in maxima or minima of the difference spectrum. Aromatic amino acids exhibit a bathochromic shift to-

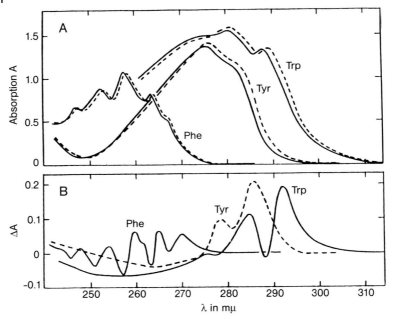

Fig. 3.28 Spectral shifts and difference spectra of the aromatic amino acids phenylalanine, tyrosine, and tryptophan at transition from the aqueous phase into 20% dimethylsulfoxide. (A) Absorption spectra (shift in dashed lines), (B) difference spectra (Herskovits 1969).

wards longer wavelengths on transition from a polar to a less polar environment, due to energy differences between the ground and excited states in media of different polarity (Fig. 3.28). This occurs when such residues are shielded by ligand binding or they move from the surface to the inside of the protein. Changes of this kind may be demonstrated by *solvent-dependent difference spectra*. The polarity of the solvent water in the sample cuvette becomes gradually reduced compared to that in the reference cuvette by adding less polar solvents, e.g. glycerine, glucose, ethylene glycol or polyethylene glycol. Residues at the protein surface come into contact with the increasingly non-polar environment, while the environment inside does not change, so that information about distinct residues may be obtained. A more detailed analysis can be obtained by applying solvents of varying molecular size. The low molecular weight ethylene glycol may penetrate into regions of the protein molecule not accessible to the high molecular weight polyethylene glycol. Differences between both solvents of comparable polarity are indications of narrow folds and pockets, e.g. active or regulatory centers, at the protein surface. Complete unfolding of the protein structure and gradual exposure of individual residues from the inside of the molecule can be followed by addition of urea, detergents or chaotropic substances. Enthalpic and entropic processes of protein folding can be studied by *temperature-dependent difference spectra,* where the temperature in the sample cuvette is gradually raised in comparison to the constant temperature of the reference cuvette.

Protonation processes of distinct residues (e.g. tyrosyl and thiol residues) are investigated by *pH-dependent difference spectra.* Spectral changes can also be caused by alterations of charges in the vicinity of chromophores, splitting and formation of ionic bonds and hydrogen bridges, and conformational changes caused by such effects. Reversible aggregation processes of subunits of the protein may be studied with *concentration-dependent difference spectra.* Decrease or increase in the concentration of the macromolecule in the sample cuvette shifts the aggregation equilibrium relative to the unchanged one in the reference cuvette. To compensate for the concentration differences between both cuvettes their pass length is correspondingly adapted, a tenfold dilution of the sample solution is compensated by a tenfold thickness of the sample cuvette.

When the protein displays no significant spectral changes, the effect can be enhanced by introducing chromophore groups at defined positions, e.g. thiol or amino groups. Binding of ligands to chymotrypsin can be followed by the prior addition of proflavin, a chromophore which binds to the protein and becomes subsequently replaced by addition of another specific ligand.

3.4.1.6 The Dual Wavelength Spectrophotometer

In 1954 B. Chance developed the dual wavelength photometer for measurements in turbid and strongly scattering solutions, e.g. membrane-bound cytochromes of the respiratory chain in mitochondria suspensions. A monochromator with two separately adjustable gratings creates two light beams of different wavelengths: a sample and a reference beam (Fig. 3.29). With a rotating mirror disc both beams pass alternately through the cuvette on the same path. The photomultiplier signals of both beams are separately detected and recorded as a difference signal.

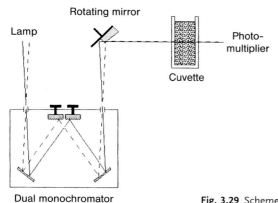

Fig. 3.29 Scheme of a dual wavelength photometer.

The loss of light intensity by scattering is largely independent of wavelength. When measuring in suspension, the wavelength of the sample beam is adapted to the absorption band of the compound to be studied, the reference wavelength is set outside the absorption band. The reference beam registers the scattering share, which is then deducted from the absorption of the sample beam. The great advantage of this technique is that fluctuations within the cuvette, that always occur in suspensions, are compensated by the identical light path of both beams, as the one cuvette is sample and reference at the same time.

This method can also be applied for the detection of very small spectral changes in binding studies or enzyme kinetic measurements. The sample beam is adjusted to the wavelength of the largest absorption change, frequently the flank of the absorption band, the reference beam is fixed to an isosbestic point of constant absorption, often located near the absorption maximum. The latter corrects fluctuations in the cuvette. This technique allows the measurement of effects with maximum absorption changes of 0.005. Commercial instruments usually combine double beam and dual wavelength optics.

3.4.1.7 Photochemical Action Spectra

Otto Warburg discovered that hemoglobin inactivated by CO can be regenerated by light exposure. Castor and Chance (1955) developed a specific apparatus for measuring such photochemical action spectra, which is suitable for the study of the reactive state of pigments like cytochromes, phytochromes and chlorophylls (Fig. 3.30). A cell suspension or a cell extract is exposed to a fixed $CO:O_2$ ratio. By irradiation with a lamp the interaction of the pigment with a component of the system is altered, e.g., CO is released and oxygen can bind. An oxygen electrode detects the change in oxygen concentration in the solution. Application of liquid color lasers enhances the emission intensity at narrow spectral bandwidth.

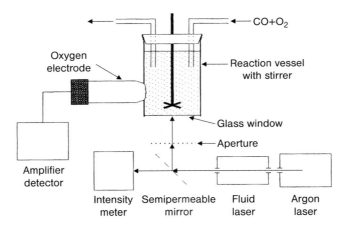

Fig. 3.30 Scheme of a device to measure photochemical action spectra.

3.4.2
Bioluminescence

Bioluminescence is an extremely sensitive technique for the determination of ATP, FMN and NAD(P)H. It is widely used in enzyme analysis. Enzyme reactions which are directly dependent on these compounds can be analyzed and they can also be linked to other enzyme reactions. Two bioluminescence systems are commercially available. ATP reacts with the cofactor luciferin of luciferase from glowworm (*Photinus pyralis*) by emitting light of 562 nm. The luciferase from the bacterium *Photobacterium fischeri* is FMN-dependent and reacts with NADH and NADPH. As the detection of the substrate to be studied (e.g. ATP) is itself based on an enzyme reaction, these bioluminescence measurements are always coupled tests with the disadvantages already mentioned for the determination of initial velocities for enzyme kinetic studies. In the initial phase of the reaction (e.g. on adding ATP), the intensity of the emitted light reaches a maximum value after about 3 s, thereafter it declines due to product inhibition. The maximum light intensity is proportional to the ATP concentration. Greater precision is achieved by integration of the curve over a defined time period.

The instrumental requirements for luminescence measurements are comparatively small. Luminometers consist of a sensitive photomultiplier, an amplifier and a digital meter. Test combinations with the respective components are commercially available. Fluorescence spectrometers can also be used for luminescence measurements.

3.4.3
Fluorescence

3.4.3.1 Quantum Yield

Fluorescence, as already described in Section 3.4, is based on the phenomenon, that an excited electron emits light upon returning from the S_1 to the ground state S_0. Compared with the absorbing (exciting) light, fluorescence is always to longer wavelengths and radiation occurs in all directions. The light must be of high intensity to excite as many molecules as possible, but only a minor share of the light will actually be absorbed while the greater part of the incident light penetrates the cuvette. Furthermore, only some of the excited molecules contribute to fluorescence. Therefore the light emitted is only a very small fraction of the incident light. The extent of the light emitted by an excited molecule depends on the nature and rate of processes competing for the energy of the excited state and is defined as quantum yield Φ_F, the ratio of emitted (q_e) to absorbed (q_a) photons:

$$\Phi_F = \frac{q_e}{q_a} = \frac{k_e}{k_e + k_{ic} + k_{is} + \Sigma k_{ec}} \ ; \tag{3.18}$$

k_e, k_{ic}, and k_{is} are the rate constants for emission, internal conversion and radiationless intersystem crossing (forbidden spin exchange to a low energy triplet state T_1), respectively; Σk_{ec} is the sum of external conversions. Thus a compound will only emit fluorescence if the competing processes are slower than the lifetime of the excited state. While Σk_{ec} includes contributions from external factors in the vicinity of the chromophore and can be influenced (and thus minimized) from the outside, k_{ic} and k_{is} are determined by the configuration of the molecule and must be regarded as constants.

The quantum yield is, like the absorption coefficient, a substance constant. It may adopt values between 0 and 1, respectively 0–100%. At a quantum yield of 1 (100%) the total light absorbed by the chromophore is emitted as fluorescent radiation, at 0% there is no fluorescence. To determine the quantum yield, the fluorescence intensity of the respective compound is measured and the absorbed photons are analyzed under identical conditions, either in a highly scattering solution or, in place of a cuvette, by a totally reflecting magnesium oxide screen. For compounds with similar spectral properties their quantum yields behave in nearly the same way as their maximum fluorescence intensities, so that the quantum yield of an unknown sample may be determined by comparison with a standard compound. Quantum yield is strongly temperature sensitive, it decreases at higher temperatures because of the increase in deactivating collisions.

3.4.3.2 Structure of Spectrofluorimeters

The structure of spectrofluorimeters (Fig. 3.31) resembles that of spectrophotometers, allowing for the peculiarities of fluorescence detection. High light intensity is needed for excitation, therefore a stronger light source is required. Since excitation occurs usually from the UV to the medium visible range, only one lamp, mostly a xenon high-pressure arc lamp is applied. It possesses a maximum intensity between 300 and 500 nm, while below and above this region the intensity decreases considerably, in the far-UV region it is very weak. A disadvantage of the lamp type is also that fluctuations of the arc can occur during the measurement. To compensate for such fluctuations, a ratio system is often

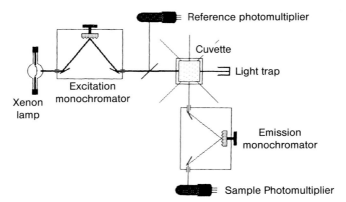

Fig. 3.31 Scheme of a spectrofluorimeter.

employed, with a beam-splitter directing part of the light beam before entering the cuvette into a reference photomultiplier. The instrument indicates as a signal the ratio in intensities of the sample and reference beams.

As in an absorption photometer, the light is spectral dispersed by a grating monochromator and the respective wavelength directed into the cuvette. Since the principle of fluorescence measurement is to detect the weak emitted light, the strong exciting light beam directly passing the cuvette must be quantitatively eliminated. To achieve this, the cuvette housing must be completely black inside and a light trap is placed where the light beam impinges. Since fluorescent light is emitted in all directions it can be detected at any angle from the exciting light beam. To avoid reflections, however, a right angle is preferred, mostly to the sides, but sometimes also to the bottom of the cuvette. Therefore all the windows of fluorescence cuvettes are polished. Apart from that they resemble absorption cuvettes in their shape, normally with a 1 cm inner path length. Round tubes sometimes applied for simple tests are less recommended. For UV measurements quartz cuvettes must be used.

The emitted light passes a second monochromator before reaching the photomultiplier. Due to the weak fluorescence radiation highly sensitive photomultipliers are used. This is a major difference between fluorescence and absorption measurements. The emitted light is directly related to the amount of the chromophore and very low amounts can be detected, because the photomultiplier measures between complete darkness in the absence of a chromophore and very low light intensities in the presence of the chromophore. In absorption spectroscopy absence of the absorbing substance implies maximum light intensity for the photomultiplier and increasing amounts of the absorbing substance reduce the intensity. Thus the photomultipliers of absorption photometers are always confronted with high light intensities and low amounts of the substance have to be detected as a faint decrease in the high light intensity. Hence, the fluorescence method is more than about 100 times more sensitive than the absorption method. On the other hand, the sensitive fluorescence photomultipliers can be damaged by high intensities e.g. strong scattering of the sample, incoming daylight or room lighting.

As a further difference the intensity of the emitted light is directly proportional to the amount of the chromophore while, in absorption photometry, the absorbing substance is exponentially related to the transmitted light. The linearity of fluorescence with increasing chromophore concentration is a good control for fluorescence measurements, especially to disclose perturbations (see below).

As spectrofluorimeters possess two separate monochromators, two different types of spectra can be recorded. The *excitation* monochromator, placed in front of the cuvette, measures excitation spectra of the chromophore, which are essentially absorption spectra, since the chromophore will emit light to the same extent as it absorbs light. To measure an excitation spectrum, the second monochromator must be fixed at a wavelength of maximum emission, this wavelength must be above the range of the excitation spectrum as emission is always of lower energy than absorption. In spite of the expected similarity of excitation and absorption spectra, large differences are often observed. This is due to artificial reasons, since the optical features, especially the lamp characteristic, superimposes the spectra, deforming their shape. Special instruments possess correction functions to obtain authentic spectra, but for many applications this is not required since the virtual absorption spectra can be scanned in an absorption spectrophotometer.

The *fluorescence* or *emission* monochromator, placed behind the cuvette, yields emission spectra at wavelengths higher than the fixed excitation wavelength. They exhibit the fluorescence characteristics of the chromophore. For the excitation wavelength any absorbing wavelength of the chromophore can be chosen but, usually, the largest absorption maximum will be used in order to obtain the highest fluorescence gain. Sometimes, however, other wavelengths are recommended, e.g. if there are disturbing influences, e.g. from scattering or Raman bands (see below). By changing the excitation wavelength the emission intensity will correspondingly be altered while the shape of the spectrum is retained. Changes in the shape of the spectra are indicative of disturbances, like impurities, the presence of different chromophores, Raman or scatter peaks.

Different versions of spectrofluorimeters are commercially available. Since even the basic set-up described here is relatively expensive, simpler instruments are offered e.g. applying a filter instead of an excitation monochromator or recording the total fluorescence. This can be sufficient for routine assays, but not for detailed studies.

Phosphorescence can also be measured with spectrofluorimeters. A rotating diaphragm allows one to alternate between the excitation and the emission beam. At the moment when the excitation beam becomes shielded only the longer-lasting luminescence can be seen, while the short-term fluorescence disappears immediately. However, in solution phosphorescence is often quenched and becomes visible only in the solid (frozen) state.

3.4.3.3 Perturbations of Fluorescence Measurements

As a very sensitive method fluorescence is subject to numerous disturbances, usually termed *fluorescence quenching*. Some of these processes are subsumed under the designation *external conversions*, the rapid deactivation of the excited state, mainly due to collisions. Molecular oxygen dissolved in solutions is especially effective and must be removed by degassing with inert nitrogen. Impurities of the substances applied or in the solvent are frequently responsible for fluorescence quenching and, therefore, only components of the highest purity should be used. The chromophores can also deactivate themselves, especially at higher concentrations (*concentration quenching*). Occasionally two molecules in the excited state enter into resonance and form dimers (*excimers*). Concentration quenching causes a deviation from the linear increase in fluorescence intensity with the amount of chromophore and, therefore, fluorescence measurements are confined to highly diluted solutions.

Reabsorption (*inner filter effect*) is observed when light emitted by the chromophore is absorbed by another substance in the solution whose absorption band overlaps with the chromophore emission. The extent of this effect depends on the concentration of the absorbing substance and on the distance the emitted light has to cover in the solution. It depends on the cuvette dimensions and can be discerned by applying cuvettes of differing size.

Light scattering is an inevitable perturbation in fluorescence measurements. When excitation and emission wavelengths become similar, a symmetric, often very intense, band is observed, simulating an emission maximum. Several events contribute to this band: Rayleigh scattering (see below), Tyndall scattering in macromolecular (proteins) and colloidal solutions, impurities (dust, air bubbles) in the cuvette or fingerprints on the cuvette walls, reflections and incomplete elimination of the exciting light in the cuvette housing. Light scattering can easily be identified by changing the excitation wavelength, the scatter peak will move in the same sense. To minimize light scattering, the highest purity of samples and solutions, and cleanliness of cuvettes have to be observed.

Especially at low fluorescence intensities the *Raman effect* can become very disturbing. Photons interact with vibrational and rotational levels of the molecules, conferring or accepting a distinct amount of vibrational energy v', and emitting photons whose frequency v_R is either above or below the excitation frequency v_0: $v_R = v_0 \pm v'$. At distinct distances from the exciting wavelength, longer-waved *Stokes* and shorter-waved *anti-Stokes* lines are observed. Compared with the mostly relatively broad fluorescence bands, Raman lines are sharp and symmetrical and upon changing the exciting wavelength they move in the same direction. They also appear in the solvent in the absence of the chromophore. The Rayleigh bands found at the position of the excitation wavelength are caused by elastic scattering of photons by the (solvent) molecules without exchange of energy.

3.4.3.4 Fluorescent Compounds (Fluorophores)

From Eq. (3.18) it can be deduced that without profound knowledge of the molecular constitution of a substance it is impossible to predict whether and with what intensity it will emit fluorescent light. There exist some rules for potentially fluorescent candidates but most of these compounds show no detectable emission, so fluorescent chromophores (fluorophores) must be regarded as exceptions and their fluorescent features identified experimentally. Fluorescence may be found in compounds with conjugated double bonds, in aromatic and heterocyclic ring systems, and also in lanthanides (europium, gadolinium, and terbium in the trivalent oxidation state). Naturally bound metal ions can sometimes be replaced by such fluorescent ions. To complicate the situation, emission further depends on the state of the compound, e.g. the ionization state or the polarity of the environment. The indole ring system of tryptophan exhibits strong fluorescence with maximum intensity between pH 9 and 11 where the carboxyl group is deprotonated and the amino group is uncharged (Fig. 3.32). In the hybrid ion form between pH 4 and 8 the emission is weaker. In the acid

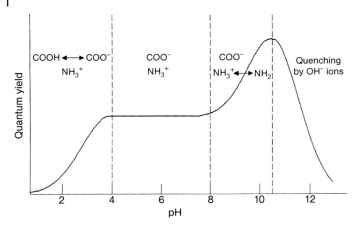

Fig. 3.32 Dependence of the quantum yield of tryptophan on the pH value of the solvent (after Brand and Witholt 1967).

region below pH 2, as well as in the alkaline region above pH 12, the indole fluorescence becomes completely quenched, by the protonated carboxyl group, and by interactions with hydroxy ions, respectively. NAD^+ and $NADP^+$ emit only in their reduced forms as NAPH or NADPH (see Table 3.2). The very sensitive fluorimetric dehydrogenase test makes use of this property.

A frequently observed phenomenon is a hypsochromic shift (blue shift) of the emission maxima of fluorophores, often linked with an increase in intensity, upon the transition from a polar to a non-polar environment. A prominent example is 1-anilino-naphthalene-8-sulfonate (ANS), a strongly lipophilic compound (Fig. 3.33). In water it exhibits only weak fluorescence, which rises considerably upon transition into increasingly non-polar solvents, e.g. alcohols of growing chain lengths, simultaneously shifting the emission maximum by 20–30 nm towards shorter wavelengths (Fig. 3.34). Blue shifts are to be observed when the dipole moment of the chromophore is higher in the excited state than in the ground state. Interactions occur with the polar molecules of the solvent. On return to the ground state a smaller amount of energy is released than in non-polar solvents, due to the loss of solvation energy. Because of its lipophilic character, ANS binds with high affinity to non-polar regions of proteins, frequently to distinct binding centers. This is accompanied by a high increase in fluorescence intensity. Upon binding of ANS to the porphyrin binding pocket of apomyoglobin the quantum yield rises from 0.4 to 98%. Other chromophores, like toluidine naphthalene-6-sulfonate (Fig. 3.33) show similar effects. This is the case also with aromatic amino acids. The emission maximum of tryptophan dissolved in water, 350 nm, shifts to 330–340 nm when the amino acid becomes buried within the apolar inner core of a protein.

This sensitive interaction of fluorophores with their environment makes fluorescence spectroscopy a valuable tool for enzyme studies. Structural changes in

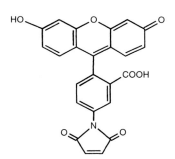

Fig. 3.33 Structures of some important fluorescent dyes.

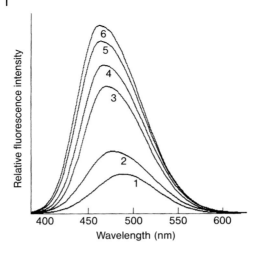

Fig. 3.34 Blue shift and intensity increase of the fluorescence spectrum of anilinonaphthalene sulfonate at decreasing polarity of the solvent: (1) ethylene glycol, (2) methanol, (3) ethanol, (4) n-propanol, (5) n-butanol, (6) n-octanol (after Stryer 1968).

the protein molecule, interactions with other components, ligand binding, protein aggregation, membrane association, etc., may be measured with high sensitivity. The fact that only a few compounds emit fluorescent light is more an advantage of the method, the observed effects can easily be attributed to a distinct chromophore as in absorption spectroscopy. For performing protein or enzyme studies the first choice is to make use of the naturally occurring fluorophores, which do not require modification of the macromolecule.

In proteins, only the three aromatic amino acids exhibit fluorescence emissions. Phenylalanine shows the weakest emissions, while tryptophan and tyrosine are comparable in their quantum yield (Table 3.2). This, however, applies only to free amino acids. In the native protein, tryptophan has the highest fluorescence intensity, due to partial quenching of the tyrosine fluorescence. In the excited state the pK value of the phenolic hydroxy group decreases, the resulting phenolate has a very low quantum yield. Tryptophan can always be identified in proteins by its fluorescence; the lack of tryptophan fluorescence is a clear indication of the absence of this amino acid. The weaker tyrosine fluorescence becomes visible only in the absence of tryptophan, at best it will be detectable at the short wavelength flank of the tryptophan band. Phenylalanine is detectable only in the absence of other aromatic amino acids. So practically each protein possesses its own chromophore whose spectral characteristics will supply valuable information for the study of conformations and bindings.

A special case is the green fluorescent protein (GFP) found in the jelly-fish *Aequorea victoria*. Three amino acids (Ser, Tyr, Gly) rearrange to form an intensely green fluorescent mesomeric structure. (Fig. 3.35). The protein is widely used as a genetic label.

Other naturally occurring fluorescent compounds are NAD(P)H, flavins, pyridoxal phosphate, steroids (cholesterol, bile acids, and steroid hormones), anthra-

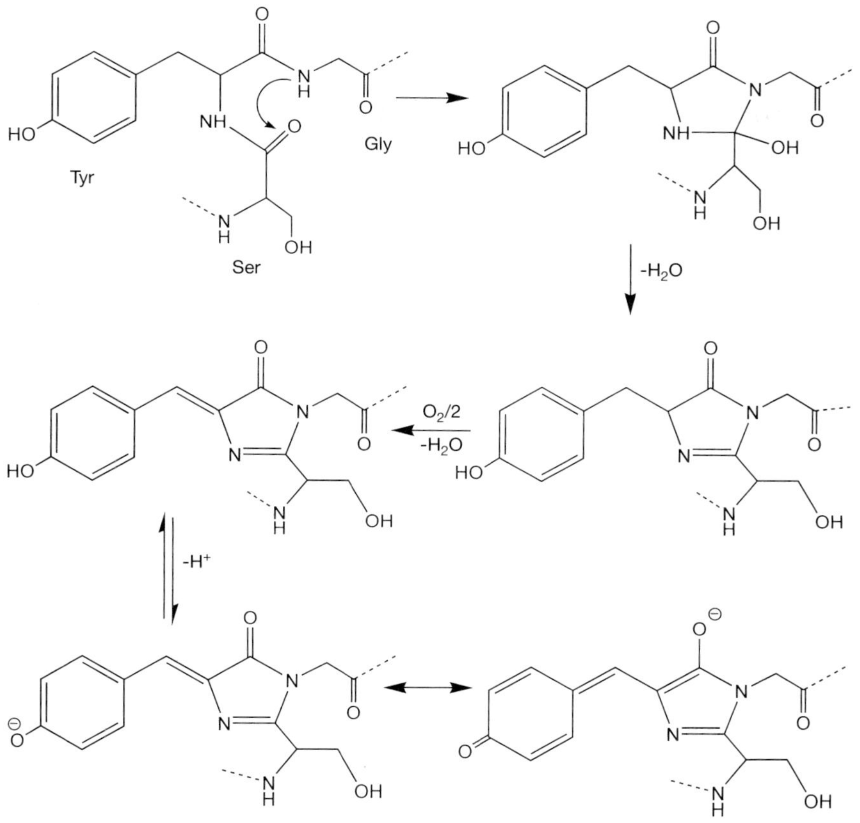

Fig. 3.35 Formation of the chromophore of the green fluorescent protein from the primary sequence.

nilate and the components of folic acid, pteridine and *p*-amino benzoate (Table 3.2). Purines and pyrimidines and their nucleotides (e.g. adenosine phosphates and the nucleic acids) show very low quantum yields. Some of these compounds can be converted to fluorophores by modification, e.g. the adenosine compounds AMP, ADP, ATP, NAD, NADP, and coenzyme A by inserting an N-6-etheno group (ε-group) at the pyrimidine ring (Fig. 3.33). They partially retain their biological activity. A fluorescent analog of thiamine and its derivates is thiochrome.

When the natural fluorescence of proteins and their cofactors is not sufficient, artificial fluorophores can be introduced. There are two types of such compounds. First, the already mentioned strongly apolar chromophores ANS and TNS bind non-covalently to hydrophobic regions, like binding or active sites of enzymes. Secondly, fluorophores can be linked with group-specific reactive residues, which bind covalently to the protein. Iodoacetamide, maleimide, aziridine or disulfide react with thiol groups, isothiocyanate, succinimide, sulfonyl chloride, and sulfonyl fluoride attach to amino groups, hydrazine and amine residues to aldehyde and ketone groups. Azides are photoreactive groups which become activated and covalently bound to the target molecule upon irradiation with light. These reactive groups are connected to various fluorophores like fluorescein, rhodamine, eosin, NBD chloride (7-chloro-4-nitrobenzofurazan), and the 5-dimethylaminonaphthalene-1-sulfonyl- (DNS-, DANSYL-) chromophore (Fig. 3.33). A recently introduced, powerful group of fluorophores are the Alexa Fluor chromophores, which are available in different modifications, differing in their spectral features (Haugland 2002).

3.4.3.5 Radiationless Energy Transfer

A peculiarity of fluorescence is the phenomenon of radiationless transfer of excitation energy from one singlet state to a second one (FRET, fluorescence resonance energy transfer). A certain fluorophore (*donor*) is excited while a second, non-excited fluorophore (*acceptor*) emits light. This effect occurs when the emission spectrum of the donor overlaps with the absorption spectrum of the acceptor (Fig. 3.36) and the distance between both fluorophores is less than 8 nm. According to the theory of Theodor Förster (1948) the rate of the energy transfer k_T is:

$$k_T = 8.71 \cdot 10^{-23} \cdot k_e r^{-6} J \cdot n^{-4} \kappa^2 \ (s^{-1}) \tag{3.19}$$

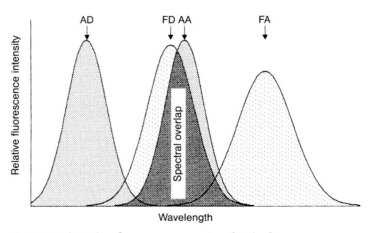

Fig. 3.36 Radiationless fluorescence energy transfer. The fluorescence spectrum of a donor fluorophore (FD) almost overlaps the excitation spectrum of an acceptor fluorophore (AA). Upon excitation of the donor (AD, excitation spectrum of the donor), the fluorescence spectrum of the non-excited acceptor (FA) is observed.

and its efficiency E:

$$E = \frac{r^{-6}}{r^{-6} + R_0^{-6}} \tag{3.20}$$

is reversibly related to the sixth power of the distance r of both fluorophores from each other. R_0 is the distance at 50% transfer efficiency:

$$R_0 = 978.5 \cdot (J \cdot n^{-4} \kappa^2 \Phi_D)^{-6} \text{ (nm)} . \tag{3.21}$$

J is the integral of the spectral overlap of donor fluorescence and acceptor absorption, n is the refractive index of the solvent, Φ_D is the donor quantum yield, and k_e the rate constant of the emission, κ is an orientation factor of the dipole moments of both fluorophores towards each other:

$$\kappa = \cos \gamma - 3 \cos a \cos \beta ; \tag{3.22}$$

a and β are the inclinations of the dipoles to the joint axis and γ is the angle formed by the dipoles with themselves. The factor κ can adopt values between 0 (vertical orientation) and 4 (parallel orientation), and requires detailed knowledge of the molecular arrangement of both fluorophores. This can be hardly known without detailed structural data, but frequently relatively free mobility of the fluorophore on the macromolecule surface can be assumed. In such cases a medium value of 2/3 may be taken for κ.

The distances between two fluorophores fixed on the same macromolecule in aqueous solution may be measured with the Förster equation, otherwise this can only be done using X-ray or NMR structure analysis. A pair of fluorophores with suitable overlaps must be successfully inserted at defined locations of a macromolecule, e.g. at active or regulatory centers. Then their distance from each other can be determined. Either analogs of the original ligands or fluorophores fixed covalently to defined amino acids may be employed. Applying this technique it was possible to demonstrate that the three catalytic centers of the pyruvate dehydrogenase complex are more than 4 nm distant from each other. This finding ruled out the previous assumption of a 1.4 nm swinging lysyl–lipoyl arm which alone should be able to reach all centers from its attachment point. It turns out that the swinging arm is fixed to a mobile peptide domain. Employing the same method the assembly of the same enzyme complex from its subunits could be followed. For the bifunctional allosteric enzyme aspartokinase:homoserine dehydrogenase the distance of the two catalytic centers was determined to be 2.9 nm. A change in conformation induced by binding of the allosteric inhibitor threonine causes the two centers to move apart by 0.7 nm. Further, the distance between the two tRNA binding sites on a ribosome could also be measured. These examples demonstrate the wide variety of applications for this technique. Ligand binding can be measured by this technique, the ligand taking one (donor or acceptor) part, a fluorophore fixed in the neighborhood of the binding site the other part.

Alternatively, binding of the ligand can be recognized by its effect on the interaction of a pair of fluorophores placed in the vicinity of the binding site.

An energy transfer effect is identified by fluorescence quenching of the excited donor or, even better, by the emission of the non-excited acceptor. Energy transfer may also occur with a non-fluorescent acceptor when its absorption spectrum overlaps with the donor emission. In this case only fluorescence quenching of the donor is observed. The radiationless energy transfer must be distinguished from the simple reabsorption of fluorescence light described in Section 3.4.3.3, which is significantly weaker and only occurs at higher concentrations.

3.4.3.6 Fluorescence Polarization

Fluorescence polarization is a valuable extension of fluorescence spectroscopy as it allows one to gain more insight into the structure of the macromolecule and, although measurements are performed in normal times, fast processes can be studied. Fluorescence polarization can be studied with normal spectrofluorimeters, however, a distinct modification is required. The light irradiated from the lamp is linearly polarized by a polarization filter before entering the cuvette. Two light beams rectangular to the incident light beam, one leaving the cuvette to the left, the other to the right, pass analyzer filters and their intensities are detected by two separate photomultipliers (Fig. 3.37).

$$P = \frac{F_{II} - F_\perp}{F_{II} + F_\perp} . \tag{3.23}$$

If linearly polarized light enters a solution of a fluorophore, only those molecules whose dipoles are arranged in the polarization plane will be excited, while those with perpendicular orientation are not. Only those molecules properly

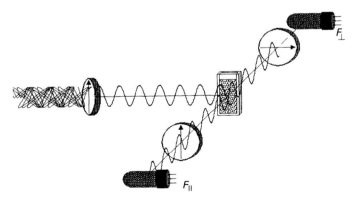

Fig. 3.37 Scheme of a fluorescence polarization device. A vertical polarization filter in front of the cuvette produces linearly polarized light. The fluorescent light emitted from the cuvette is analyzed for the share of polarized light parallel F_{II} and vertical $F_\perp$, to the polarization plane of the incoming irradiation.

oriented will be excited and in turn will emit light (*photoselection*). If the excited molecules remain in their position, the emitted light will be similarly polarized and light can only pass the analyzer filter adjusted in parallel to the linearly polarized light $F_\parallel$ but not the filter perpendicular to this direction ($F_\perp$, Fig. 3.37). However, due to molecular motion the original orientation will change during the lifetime of the excited state, leading to a partial depolarization of fluorescent light, the fluorescence intensity between the two light paths will converge. It is obvious that the extent of convergence depends on the rate of rotation of the molecules and thus on their size, small molecules will cause greater depolarization than larger molecules. According to the Perrin equation the decrease in the polarization P depends on the molar volume of the fluorophore V (ml mol^{-1}), the absolute temperature T, the solvent viscosity η and the lifetime of the excited state τ_0:

$$\frac{1}{P} = \frac{1}{P_0} + \left(\frac{1}{P_0} - \frac{1}{3}\right) \frac{R\tau_0}{V} \cdot \frac{T}{\eta} . \tag{3.24}$$

P_0 is the polarization of immobile molecules ($T=0$ K) and R the gas constant. A linear dependence of $1/P$ on T/η results when the polarization of the emitted light is measured by variation of temperature or solvent viscosity (or both). The straight line intersects the ordinate at $1/P_0$, so that the slope only contains τ_0 and V as unknowns. This method allows one to determine the volume of a fluorophore if τ_0 is known (or vice versa, the lifetime of the excited state if V is known). With the molecule mass known, information on the shape of the molecule can be obtained. It must be realized that the mobile element determines the volume. If a small fluorophore is attached to a large protein molecule, the latter dictates mobility, including the hydrate shell and shape (spherical, elliptical, etc.). If the fluorophore is freely mobile on the surface of the molecule, slow movements of the protein and rapid movement of the fluorophore are observed simultaneously. In the Perrin plot two overlapping straight lines appear. Such an experiment can also provide the information required for the orientation factor κ in the Förster equation (Eq. (3.21)).

3.4.3.7 Pulse Fluorimetry

Pulse fluorimetry, a further development of fluorescence polarization, is basically a fast measuring method (see Section 3.5). The fluorophore is excited by a flash of light shorter than the lifetime τ_0 of the excited state. After the flash of light the decline of fluorescence is followed over a given time period (Fig. 3.38A). Fluorescence intensity S_0 at the moment of excitation ($t=0$) decreases exponentially over time t:

$$S_t = S_0 e^{-t/\tau_0} . \tag{3.25}$$

In a semi-logarithmic plot of $\ln S_t$ against t, τ_0 is obtained from the slope of the resulting straight line. Upon excitation with linearly polarized light, the time-de-

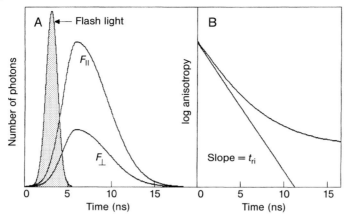

Fig. 3.38 Pulse fluorimetry. (A) Time course of fluorescence intensities in parallel ($F_{\parallel}$) and perpendicular ($F_{\perp}$) to the polarization plane of irradiation after excitation with a short flash of light. (B) Semi-logarithmic plot of the time-dependent change in anisotropy upon overlay of two motions with differing velocities.

pendent course of fluorescence intensity may be registered separately, i.e., in parallel ($F_{\parallel,t}$) and at right angles ($F_{\perp,t}$) to the polarization plane. The anisotropy A_t is the measured:

$$A_t = A_0 e^{-t/\tau_{ri}} = \frac{F_{\parallel,t} - F_{\perp,t}}{F_{\parallel,t} + 2F_{\perp,t}} \;. \tag{3.26}$$

The denominator is composed of the fluorescence intensities of the three spatial directions (x parallel, y and z with identical intensities at right angles to the excitation). The denominator corresponds to the total fluorescence S_t. A_0 is the anisotropy at the moment of excitation, τ_{ri} is the rotation relaxation time of the fluorophore. Its reciprocal value is a linear combination of the rotation diffusion constant D_i ($i = 1,2,3$) of the three main rotational axes of a particle:

$$D_i = \frac{k_B T \gamma_i}{6 \eta V} \;. \tag{3.27}$$

V is the hydrodynamic volume of the particle, k_B the Boltzmann constant, η the solvent viscosity, T the absolute temperature and γ_i a function of the axial ratio of an ellipsoid. τ_{ri} results from the slope of a semi-logarithmic plot of A_t against t. From the number and magnitude of the relaxation times conclusions can be drawn about the size of the molecule (Fig. 3.38 B). With a spherical shape ($D_1 = D_2 = D_3$) a linear dependence is obtained with a uniform rotation relaxation time:

$$\tau_{\text{ri}} = \frac{1}{6D} = \frac{\eta V}{k_B T} \quad . \tag{3.28}$$

Two different relaxation times result at free mobility of a fluorophore bound to a macromolecule.

A rapid sequence of very short flashes (≈ 1 ns) from a flash lamp, a laser or a Kerr cell excites the sample solution. A photon counter performs a time-dependent registration of the emitted photons. The Kerr cell contains a highly dipolar liquid (nitrobenzene, acetonitrile, benzonitrile). The molecules are aligned by an external field so that an incoming linearly polarized light beam is turned by 90°. A polarization filter is then positioned so that the beam cannot pass. By switching off the external field for a short period of a few ns the molecules in the Kerr cell lose their alignment, the light beam regains its original polarization plane and passes the filter.

3.4.4
Circular Dichroism and Optical Rotation Dispersion

Asymmetrical structures possess the property of optical rotation and deflect the plane of polarized light. Except for glycine, all amino acid residues in proteins possess at least one optically active center. These asymmetric carbon atoms remain unchanged during enzyme reactions, binding processes or conformational changes and are not suitable to investigate such processes. However, the protein molecule itself with its secondary structure elements, the α-helix and β-sheet, exhibits asymmetrical properties that may be influenced by catalytic and regulatory processes.

Optical activity may be measured in two different ways. *Optical rotation dispersion* (ORD) registers the difference in refraction indices, *circular dichroism* (CD) records absorbance differences between right- and left-handed circularly polarized light. Circularly polarized light can be regarded as the sum of two linearly

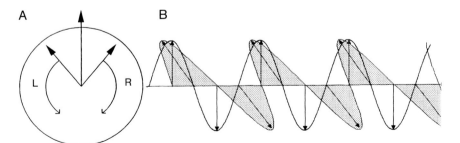

Fig. 3.39 Plane-polarized light (A) develops from superposition of equal parts of right (R)- and left (L)-handed circularly polarized light. (B) Circularly polarized light develops from two phase-shifted perpendicularly oriented parts of linearly polarized light (after van Holde 1985).

polarized components with a phase difference of 90°, while linearly polarized light is composed of two opposite circularly polarized beams (Fig. 3.39).

The ORD method measures the dependence of the refraction angle of linearly polarized light from the wavelength:

$$\Phi = \frac{180 \cdot d \cdot (n_L - n_R)}{\lambda} . \tag{3.29}$$

Φ (degrees) is the optical rotation, d is the path length of the beam in the cuvette, λ the wavelength, n_L and n_R the refractive indices for left- and right-handed circular light. In protein and nucleic acid solutions with a chromophore concentration of $\approx 10^{-4}$ M the polarized light rotates at 0.01–0.1 deg cm^{-1}. Precise instruments still register refractions of 10^{-4} deg.

For the CD method:

$$\Theta = \frac{2.303 \cdot 180 \cdot (A_L - A_R)}{4\pi} . \tag{3.30}$$

$A_L - A_R$ is the absorption difference of right- and left-handed polarized light. Ellipticity $\Theta(°)$ is defined as an arc tangent of the ratio of the short to the long axis. The difference is about 0.03–0.3% of the total absorption. To compare different samples the molar rotation $[\Theta]$:

$$[\Phi] = \frac{100 \cdot \Phi}{cd} = \frac{[a]M_r}{100} \tag{3.31 a}$$

or the molar ellipticity $[\Theta]$:

$$[\Theta] = \frac{100 \cdot \Theta}{cd} = \frac{[\Psi]M_r}{100} = 3.300 \cdot (\varepsilon_L - \varepsilon_R) \tag{3.31 b}$$

is indicated. c is the sample concentration, $[a]$ the specific rotation, $[\psi]$ the specific ellipticity, M_r the relative molecular mass of the sample. ε_L and ε_R are the absorption coefficients for left- and right-handed circular light.

The CD spectrum of a compound appears in the region of its absorption spectrum, beyond the absorption $[\Theta]=0$. Its shape resembles that of the absorption spectrum, however, depending on the structure of the molecule, either with a positive or negative Cotton effect and differing intensity (Fig. 3.40). ORD spectra appear as the first derivatives of CD spectra, passing (from shorter wavelengths) a minimum, followed by a maximum for a positive Cotton effect and the reverse sequence for a negative Cotton effect. The turning point in both cases correlates with the maximum or the minimum, respectively, of the CD spectrum. At higher and lower wavelengths the ORD spectra extend far beyond the absorption region. Previously, when instruments for the far-UV region were not yet available, this feature was utilized to obtain information from the optical rotation at an accessible wavelength λ, on the optical rotation at the wavelength λ_0 of the turning point with the aid of the *Drude equation*:

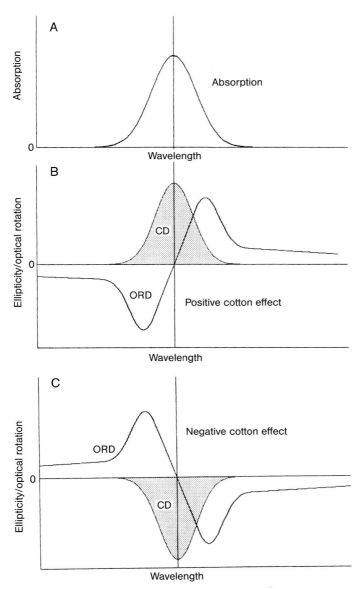

Fig. 3.40 Schematic comparison of absorption spectra (A), ORD and CD spectra (B, C) of an asymmetrical compound with positive (B) and negative (C) Cotton effect.

$$[\Phi]_\lambda = \frac{k}{\lambda^2 - \lambda_0^2} \tag{3.32}$$

k is a constant.

ORD and CD spectroscopy are closely related and can be interconverted by applying the *Krönig-Kramers* transformation:

$$[\Theta(\lambda)] = -\frac{2\lambda}{\pi} \int_0^\infty \frac{[\Phi(\lambda')]\lambda'}{\lambda^2 - \lambda'^2} d\lambda' . \tag{3.33}$$

Hence, only one of the methods has to be employed, while the other method reveals no additional information. In spite of the greater technical expenditure CD spectroscopy is usually preferred. The spectra obtained by this method are simpler, an advantage especially in cases of atypical behavior or of overlapping effects which are more difficult to interpret from the far-reaching ORD spectra.

CD spectra are widely used for the characterization of secondary elements of protein structure. As shown in Fig. 3.41, α-helix and β-sheet on the one hand and coil structure on the other show opposite CD spectra. The α-helix is characterized by *exciton splitting* into a double band between 210 and 220 nm caused by the formation of excimers. This allows predictions on the relative distribution of these structural elements in proteins. Conformational changes of such structural elements, induced by substrate or effector binding, may also be observed.

Disulfide bridges and the side chains of aromatic amino acids (Fig. 3.42) also exhibit optical activities within protein structures. Comparable to the absorption

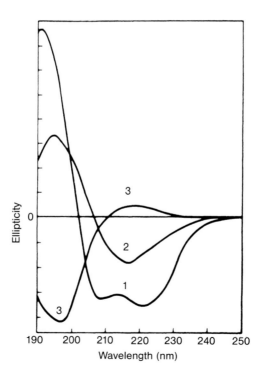

Fig. 3.41 CD spectra of the secondary structure elements of proteins. 1) α-helix, 2) β-sheet, 3) random coi (after Greenfield and Fasman 1969).

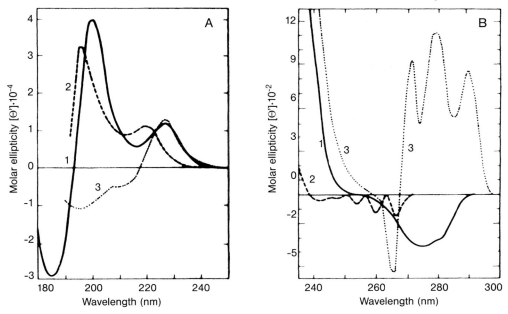

Fig. 3.42 CD spectra of aromatic amino acid derivatives in the far- (A) and near- (B) UV region. (1) N-acetyl-L-tyrosine-amide, (2) phenylalanineamide, (3) N-acetyl-L-tryptophan-amide (after Shikari 1969).

spectra, these amino acid residues show the most outstanding effects in the far-UV region between 200 and 240 nm, but these spectra are superimposed by those of the secondary structures. At 210 nm histidine also exhibits a CD spectrum. Characteristic for these amino acids is the near-UV region between 250 and 300 nm, where especially tryptophan shows a prominent CD spectrum. Tyrosine has only a weaker negative band, while phenylalanine demonstrates a less intense but characteristic structure. CD spectroscopy is the best suited method for analysis of aromatic amino acids in proteins. Prosthetic groups and coenzymes, e.g. porphyrins and NADH, also yield CD spectra.

For the study of regulatory or kinetic mechanisms, spectral changes as a consequence of the respective process are of greater interest than the spectra themselves. Apart from the analysis of conformational changes, the CD method has the advantage that even optically inactive ligands may produce CD signals by their interactions with the asymmetrical structures of the protein molecule, so that such specific signals represent a direct measure of ligand binding.

The structure of an ORD apparatus resembles that of an absorption photometer (Fig. 3.43 A). Light is spectrally dispersed by a monochromator, passes the sample cell and is detected on a photomultiplier. Specific for the ORD unit is a polarization crystal positioned in front of the sample cell, producing plane-polarized light. Contrary to the rectangular absorption and fluorescence cuvettes, cyl-

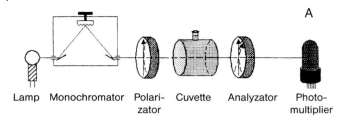
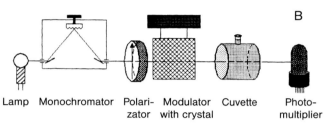

Fig. 3.43 Construction scheme of an ORD (A) and a CD spectrometer (B).

indrical cells are used. The light passing the cell penetrates an analyzator filter which is applied to measure the deflection angle. In a CD instrument, monochromatic light is also plane-polarized by a polarizer (Fig. 3.43 B). It then penetrates a birefringent quartz plate set at an angle of 45° to the plane of polarization, splitting the light into two components of equal intensity but different refractive indices, oriented at right angles to each other. The thickness of the plate is chosen to delay the slower beam by a quarter-wave on exit (electro-optic modulator, lambda quarter plate). Applying an electric field the refraction indices of both beams and the orientation of the rotation can be reversed.

3.4.5
Infrared and Raman Spectroscopy

Infrared (IR) and Raman spectroscopy detect transitions between vibrational levels of molecules. On account of their different frequencies, information may be gathered about the nature of the vibration (torsion, translation, valence, or deformation vibration), about the participating atoms as binding partners (C–O, C–N, and C–H), and about the nature of bonds (single, double, or triple bonds). This allows conclusions to be drawn on the molecular structure. For macromolecules like proteins, however, highly complex results are obtained. Principally, both IR and Raman spectroscopy supply the same information. They are similarly related to one another as absorption and fluorescence spectroscopy. IR spectroscopy analyzes absorptions at the frequency of molecular vibration, while Raman spectroscopy observes the dispersion of photons modified by the amount of vibration energy. For vibration spectra, generally the wavenumber $v = 2\pi/\lambda$ (cm^{-1}), also termed "Kayser", is applied.

3.4.5.1 IR Spectroscopy

The most pronounced IR signals are obtained from asymmetrical or polarized groups. Liquid substances can be measured directly, solid substances must be embedded in materials like nujol or pressed in potassium bromide. Water and dissolved salts, e.g. buffer substances, cause strong absorptions and therefore limit the application of the IR method for enzymes. It is necessary to operate outside the absorbance range of water. By using heavy water (D_2O) the spectrum shifts in such a way that the spectral region initially overlaid by water may become accessible. A significant improvement allowing a wider application for studies of proteins and enzymes in aqueous solution was realized by combining IR with Fourier transformation (FT-IR spectroscopy), thus reducing the originally very long measuring time of 10–20 min to a few seconds. As a light source for IR spectroscopy a Nernst stick, consisting of a mixture of zirconium and yttrium oxide or a sintered silicon carbide stick (globar) is used. The light is directed through a monochromator and into the cuvette. For detection, heating by the incoming IR irradiation is measured with a thermoelement. A bolometer determines the temperature-dependent changes in the resistance of platinum wires or semiconductors. The pneumatic Golay cell determines the motion of a flexible diaphragm induced by heating a gas due to the adsorption of the light on a blackened metal plate.

3.4.5.2 Raman Spectroscopy

In Raman spectroscopy, molecules are excited by photons (UV or visible light) with higher energy than necessary for vibrations. They either transmit a distinct amount of energy as vibrational energy to the molecule, losing their own energy (*Stokes lines*), or they take this amount of energy from the molecule and enhance their own energy (*anti-Stokes lines*). The excitation band (*Rayleigh scattering*), caused by elastic scattering of the photons at the molecules, has by far the highest intensity, the longer-wave Stokes lines possess higher intensity than the short-wave anti-Stokes lines. Usually excitation wavelengths which do not overlap with the absorption band of the compound are chosen. In contrast to this, for the resonance-Raman method excitation and absorption wavelengths are comparably. This enhances the vibration coupled to the electronic transitions, producing very specific vibration bands. Raman spectroscopy is affected by the fluorescence phenomenon in the same way as fluorescence spectroscopy is interfered by the Raman effect. The frequency of the Raman bands is indicated as the difference between the excitation frequency and the respective Raman band. The method detects even weak interactions such as hydrogen bonds and is, therefore, well suited for the study of catalytic mechanisms. In aqueous solution Raman spectroscopy becomes accessible by excitation with intense long-wave laser light, such as the Nd:YAG laser (1064 nm), an yttrium–aluminum garnet doped with neodymium. As in fluorescence spectroscopy, scattering radiation of the sample is measured at right angles to the excitation. Pulse lasers allow one to follow very rapid processes.

3.4.5.3 Applications

IR and Raman spectroscopy are applied in structural studies of proteins, for the analysis of secondary structures (a-helix, β-sheets), conformational changes, ligand binding, prosthetic groups, and metals. These methods were used for example for the study of interactions of oxygen and carbon monoxide with hemoglobin. Groups containing sulfur show strong Raman scattering because of the high polarization potential of sulfur. Especially, the formation and cleavage of disulfide bridges may be studied. With IR spectroscopy the R and T states of hemoglobin could be differentiated by ligand-induced alterations of thiol spectra. Metal proteins exhibit intense effects, especially in resonance Raman spectroscopy, e.g. porphyrin heme proteins (cytochrome, hemoglobin), proteins with Fe–S complexes (components of the respiratory chain, ferredoxin), and copper proteins. Various enzyme reactions and enzyme catalytic mechanisms have been studied with FT-IR and Raman spectroscopy, e.g. the cleavage of fructose biphosphate by aldolase into glyceraldehyde phosphate and dihydroxyacetone phosphate, and the appearance of the acyl-intermediate in the proteolytic mechanisms of chymotrypsin and papain. These methods, however, are not well suited for observing enzyme reactions in enzyme tests and enzyme kinetic studies.

3.4.6
Electron Paramagnetic Resonance Spectroscopy

IR and Raman spectroscopy are methods only peripheral to the subject of this book. This holds even more for electron paramagnetic resonance spectroscopy (EPR). Nevertheless, this technique will be briefly discussed, because of analogies with other spectroscopic (notably fluorescence) methods, especially the labeling technique and the high specificity for distinct groups. Broad application of this method, however, is limited by the considerably high instrumental requirement. For details and related methods, especially NMR spectroscopy, textbooks of physical chemistry are recommended.

Molecules with paramagnetic properties, i.e., with unpaired electrons, can be studied by EPR spectroscopy. Unpaired electrons have a spin quantum number $S = 1/2$, correlating to a magnetic moment of $m_s \pm 1/2$. In a magnetic field they either move parallel ($m_s = +1/2$) or anti-parallel ($m_s = -1/2$) to the field axis z. An oscillating magnetic field at right angles to the field axis induces transitions between the two spin states, when the field frequency lies in the region of the Larmor frequency of the spinning electron. The energy content E_m of an electron spin in the magnetic field H is:

$$E_m = g \mu_B m_s H \; ; \tag{3.34}$$

μ_B is the Bohr magneton (9.273×10^{-24} J T^{-1}), H the field strength, g is called the g factor (see below). The difference in the energy contents of two electrons a and β with magnetic moment $m_s = +1/2$, respectively $m_s = -1/2$ is:

$$\Delta E = E_\alpha - E_\beta = g\mu_B H \, . \tag{3.35}$$

Resonance stands for

$$h\nu = g\mu_B H \, . \tag{3.36}$$

h is Planck's constant and ν the microwave frequency. In EPR spectroscopy molecules with unpaired electrons that enter into resonance with a monochromatic radiation are studied.

Local permanent fields H_{loc} in the environment of the unpaired electron, especially the magnetic moment of the nucleus, overlap with the external magnetic field H and lead to a *hyperfine splitting*, a special property of EPR spectra. If an electron is located around a nucleus with a nuclear spin I, then H_{loc} may adopt along the external field axis $2I+1$ values, according to the $2I+1$ values of the magnetic moment of the nucleus m_I:

$$H_{loc} = H + a \cdot m_I \, ; \tag{3.37}$$

a is the hyperfine coupling constant. For a hydrogen atom ($I=1/2$), half of the radicals of the sample are in the state $m_I=+1/2$. They enter into resonance, when the external field meets the conditions:

$$h\nu = g\mu_B(H + 1/2a) \tag{3.38}$$

respectively

$$H = (h\nu/g\mu_B) - 1/2a \, . \tag{3.38a}$$

The other half with $m_I=-1/2$ meets the resonance conditions at:

$$H = (h\nu/g\mu_B) + 1/2a \, . \tag{3.38b}$$

Instead of one line, the spectrum now exhibits two lines, each with half of the original intensity, separated by the coupling constant a and centered around a field defined by the g factor.

If the radical contains a nitrogen atom ($I=1$), it will split into three lines ($m_I=-1, 0, 1$), as the nitrogen nucleus has three possible spin orientations, each of which will be taken by one third of the radicals (Fig. 3.44A).

Most commercial instruments use a magnetic field of 0.3 T (1 T = 10^4 G), corresponding to resonance with an electromagnetic field with a frequency of 10 GHz and a wavelength of 3 cm ("X-band", microwave region). The magnetic field is varied, and the EPR spectrum is recorded by measuring the absorption of the microwave radiation from a microwave generator (klystron) by a microwave detector. It is represented as the first derivative of the absorption spectrum.

EPR spectra are characterized by the shape of their lines, the g factor and the coupling constant a. The latter results from the distance of the spectral lines

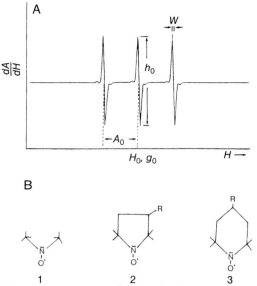

Fig. 3.44 EPR spectrum of a nitroxyl radical (A). A_0 is the anisotropic coupling constant, g_0 the isotropic g factor, W the line width, h_0 the amplitude, H the magnetic field strength (after Graupe, 1982). (B) Structures of stable nitroxyl radicals: (1) di-*tert*-butyl-nitroxyl, (2) tetramethylpyrolidine-nitroxyl (proxyl), (3) tetramethylpiperidine-nitroxyl (tempo).

and is a measure of the interaction of the electron with the local field of the nucleus. The g factor is obtained from the center of the spectrum (Fig. 3.44 A). It is a measure of the total moment of spin and orbital momentum interacting with the external magnetic field. The g factor for an electron is $g_e = 2.0023$. The deviation of the g factor determined by the EPR spectroscopy from this value, $g = (1-\sigma)g_e$, depends on the ability of the magnetic field to influence the local electron stream of the radical. This interaction is dependent on the orientation of the magnetic field vector to the molecular axes. In anisotropic alignments (in a single crystal) the shape of the spectrum depends on the angle of the crystal to the magnetic field. In an isotropic distribution in solution a uniform spectrum is obtained. By restricting mobility, e.g. by binding a radical to a macromolecule, the spectral lines will be broadened and become asymmetrical. This property allows the direct observation of the binding of paramagnetic ligands or study of the influence of bound radicals.

The transition metals Cr^{3+}, Mn^{2+}, Co^{2+}, Cu^{2+} and the lanthanides are naturally occurring paramagnetic compounds. A stable paramagnetic intermediate arising during the catalytic mechanism of an enzyme could be demonstrated for the pyruvate-formiate lyase. Stable radicals can be synthesized and, similar to fluorophores (Section 3.4.3.4), fixed covalently to the protein by side-chain reactive groups, e.g. maleimide or isothiocyanate (*spin-label*). Mostly nitroxyl radicals shielded by tertiary butyl groups are used in the form of stable tetramethylpiperidine- or tetramethylpyrolidine-1-oxyls (Fig. 3.44 B).

3.5
Measurement of Fast Reactions

A variety of methods for measuring fast reactions have been developed, one of the fastest being pulse fluorimetry, already described in Section 3.4.3.7, with a time resolution in the nanosecond range. Generally the various methods can be reduced to a common scheme consisting of three main modules. Characteristic for the special method is the module which serves to initiate the fast reaction. The construction principle of this module determines the time range and the type of the reaction to be analyzed by the respective method (Fig. 3.45). The principle of *flow methods* is based on rapid mixing of reactants. These methods are suited for the analysis of fast enzymatic reactions and provide valuable information for the understanding of catalytic mechanisms. With a time resolution in the range of milliseconds they are comparatively slow within the scale of fast reactions. With *relaxation methods* short-term disturbances of equilibria are induced and, consequently, only reactions in equilibrium and no turnover can be observed, e.g. ligand binding, isomerizations, or spontaneous and induced conformational changes in macromolecules. These methods are about 1000-fold faster than flow methods. Even more rapid are irradiation methods, which require the sensitivity of the system to light impulses. For the characterization of a defined process it is often necessary to combine different methods and experimental approaches in a suitable manner.

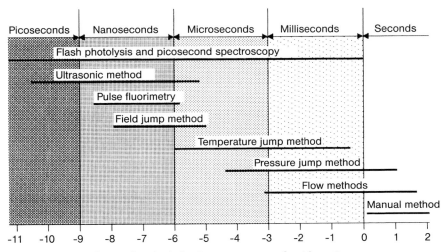

Fig. 3.45 Time resolution of methods for the measurement of rapid reactions.

The second module is for detection of the rapid change in the reaction compartment. Though this depends also on the special method and the required time resolution, similar techniques can be applied in different methods. Optical methods are especially suited for enzyme reactions, absorption photometry being a relatively simple and universal method is most widely used, but its sensitivity is limited and also fluorescence, CD, EPR and NMR spectrometry are combined with fast methods. Further, light scattering, electrical conductivity or determination of oxygen are applied. As most of these methods have already been discussed, only the relevant adaptation to the rapid method will be described here.

The very short signals of these techniques require rapid registration/recording as the third essential module. Previously, storage oscillographs were used which show directly the reaction traces on the screen, but they became widely displaced by computers with rapid data storage and subsequent evaluation and processing of the data.

3.5.1
Flow Methods

3.5.1.1 The Continuous-Flow Method

The continuous-flow method was the first technique for the observation of fast reactions. In 1905 Raschig constructed a simple device to observe a reaction in the gaseous phase for up to 25 ms. In 1923 Hartridge and Roughton developed a continuous flow apparatus for the study of carbon monoxide binding to hemoglobin. Even modern instruments adopt the general principle of this early device, based upon rapid mixing of two separate reaction solutions (Fig. 3.46 A). Two components, e.g. an enzyme and a substrate solution, are filled into reaction syringes. These syringes are emptied with constant velocity by a driving motor, and the solutions simultaneously enter a mixing chamber. Upon mixing the reaction starts and the reactant solution flows continuously through an observation tube where the progress of the reaction is monitored, either optically (by absorption or fluorescence photometry), or with a thermo-element. In an optical device the tube serves as a cuvette and must be constructed accordingly (quartz glass for UV measurements).

The distance between the mixing chamber at the start of the reaction and a distinct observation point in the observation tube is directly proportional to the time coordinate of the progressing reaction. Thus the age of the reaction can be described:

$$\text{Age (s)} = \frac{\text{Volume observation} - \text{mixing point (cm}^3)}{\text{Flow rate (ml/s)}} = \frac{\text{Flow distance (cm)}}{\text{Flow (cm/s)}}.$$

As long as a continuous flow is maintained, the reactant mixture always has the same age at a fixed observation point, e.g. 1 cm behind the mixing chamber the reaction will be 1 ms old at a flow rate of 10 m s^{-1}. Each point of the reaction

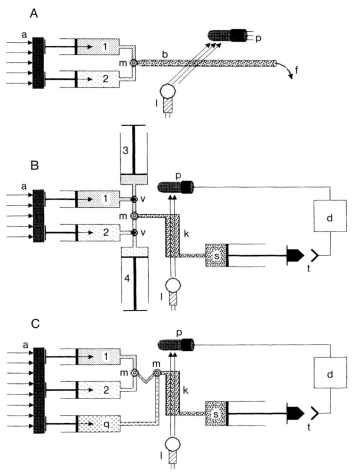

Fig. 3.46 Schematic representations of flow instruments: (A) Continuous-flow apparatus, (B) stopped-flow apparatus, (C) multi-mixing apparatus. (1, 2) Reaction syringes, (3, 4) reserve syringes, (a) drive, (b) observation tube, (d) detector, (f) efflux, (k) cuvette, (l) lamp, (m) mixing chamber, (p) photomultiplier, (q) additional reaction syringe (for a third reactant or a quenching liquid), (s) stop syringe, (t) trigger, (v) three-way valve.

coordinate can, therefore, be measured independently of time, i.e. short time periods may be observed over longer periods of time and the signals can be made reliable by averaging of times. This is an obvious advantage in comparison to other fast methods, where the signal must be detected in real time, and originally even no fast registration is needed. However, this has to be paid for with a large wastage of substance. Observing a single point for 1 s requires 200 ml flow through a tube of 5 mm diameter. To characterize the time progress of the reaction, several points at different distances from the mixing chamber must be observed and, accordingly, several liters of the reactant solutions would be required. For hemoglobin studied by Hartridge and Roughton this was a minor problem, but for the study of enzyme reactions, such volumes are usually not available. The development of storage oscillographs (and finally of rapid data storage by computer) and the higher sensitivity of modern photometers led to the real breakthrough in fast methods, and also the continuous flow apparatus. The measuring time and a reduction in the tube diameter to 1 mm permitted the reduction of the reaction volume for one measurement to 10 ml.

A second advantage of the continuous-flow method compared with the other flow methods is the shorter dead time, i.e. the time between initiation of the reaction and the onset of detection. The first observation point in the continuous-flow method is located directly behind the mixing chamber, at the entrance to the observation tube, so that the unobservable path between these points is very short. Dead times of 0.2 ms can be reached by perfect instruments.

The dead time determines the time resolution of the apparatus and is the most critical factor for instruments designed to detect fast reactions. Therefore, it is generally the goal to shorten the dead time and make the instruments faster. However, any method has its limitation. Flow methods will become faster by increasing the flow rate. This can be achieved by faster emptying of the reaction syringes, increasing the velocity of the driving motor. However, the sine qua non of these methods is complete homogeneous mixing, otherwise no reliable measurements can be obtained. Homogeneous mixing within fractions of milliseconds requires perfect construction of the mixing chamber. From various nozzles the reactant solutions are dispersed into the mixing chamber in counter current flow, causing strong turbulence (Fig. 3.47). Too high flow rates produce friction effects and heating of the reactant solution in the narrow nozzles. This influences the reaction process and causes streaks in the reaction solution.

The *pulse-flow method* developed by Britton Chance is a modification to reduce the reaction volume. Instead of a continuous flow, the solutions are pressed into the observation tube with short impulses. A special application of the continuous-flow method is its combination with scanning EPR spectroscopy (Yamazaki 1960).

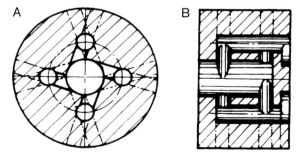

Fig. 3.47 Cross-sections through a mixing chamber for flow instruments. (A) Front view, (B) side view (after Gibson and Milnes 1964; with permission from the publisher).

3.5.1.2 The Stopped-flow Method

Albeit the various advantages of the continuous-flow method its broad application is still limited by the requirement for large quantities of the reactants thus it was the stopped-flow principle that led to the breakthrough for the investigation of fast enzyme reactions. The first stopped-flow apparatus was developed by Britton Chance in 1943. The same arrangement with two reaction syringes and a mixing chamber is used as for the continuous-flow method. In contrast to this, however, the reaction syringes are emptied into the mixing chamber with one rapid shot and not continuously emptied. An observation cell is placed behind the mixing chamber, followed by a stop syringe (Fig. 3.46 B). Usually the observation cell is a photometric cuvette, arranged perpendicular to the flow direction to allow free passage of the optical beam for absorption measurements. For fluorimetric measurements the emitted beam is detected perpendicularly to the excitation in the third space dimension. The reactant solution must pass the cuvette and enter the stop syringe, which triggers the registration. Because the solution must cover this long distance before onset of registration, the dead time is considerably high. Precise custom-made instruments achieve dead times below 1 ms, but commercial instruments hardly reach dead times lower than 5 ms. A shorter dead time can be achieved by reducing the path length of the cuvette (e.g. to 0.2 mm), but at the expense of sensitivity. As already discussed for the continuous-flow method, homogeneous mixing is essential. This can be tested by filling a dye solution in one syringe and water in the second reaction syringe. Drifts of the signal observed after a shot indicate incomplete mixing. Since temperature differences in the flow system cause streaks, all components of the apparatus coming into contact with the solutions must be carefully maintained at the same temperature with a thermostat.

The impulse for emptying the drive syringes is produced either hydraulically or by a gas pressure-driven piston, which simultaneously pushes both reaction syringes via a driving block. In another system, the reactant solutions are directly subjected to a high gas pressure and the flow is released by opening an electromagnetic valve. The two reaction syringes normally have the same volume so that both reactants are dissolved in a ratio of 1:1 after mixing. By using syringes with different volumes, an unequal mixing ratio may be achieved if, for example, the enzyme solution should not essentially be diluted (*generative flow apparatus*; Chance 1974). Reservoir syringes placed at the sides allow refilling of the reaction syringes and repeated measurements. As any air bubbles in the system will be dispersed into a fine opaque fog, it is not recommended to empty and refill the apparatus after each shot. After each experiment the solution in the cuvette will be replaced by the solutions of the following shot and pressed into the stop syringe. This causes the plunger to move back and trigger the recording (*end-stop system*). The stop syringe must be emptied after each shot. In another device the driving block, and thus the flow, is stopped by a resistor located near the reaction syringes (*front-stop system*). This allows a more simple construction because the observation cell is not exposed to high pressures. To avoid release of gas bubbles in the observation chamber due to relaxation of the

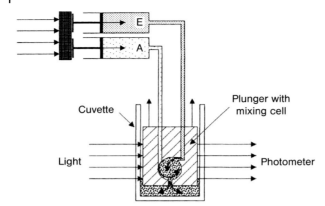

Fig. 3.48 Stopped-flow apparatus after Strittmatter (1964).

solution, a limited counter-pressure is produced by a constriction between the cell and the terminal reservoir.

The data collected with the optical system are either displayed in an oscillograph and photographed or stored in a digital data memory. They may then be projected onto a screen, recorded on a printer, and processed in a connected computer (data calculation, adaptation of curves, etc.).

P. Strittmatter (1964) developed an easy-to build but effective stopped-flow apparatus (Fig. 3.48). The mixing chamber is a plunger of synthetic material, fitted exactly into a photometric cuvette, which in turn is placed in a photometer. The reactant solutions from the reaction syringes are pressed by a motor drive through flexible tubes into the plunger. A mixing chamber is incorporated in the plunger and, after passing, the reaction solution leaves the plunger through a hole at its bottom. The reactant mixture pressed into the cuvette pushes the plunger upwards and opens the light path for the photometric measurement. Dead times of only a few milliseconds can be achieved with this device. Normal laboratory photometers can be employed, however, they must be adapted for registration of rapid signals. If microcuvettes are used, only some tenths of a milliliter of the sample solution are required.

UV–Vis absorption spectrophotometers are the most frequently used detection systems for stopped-flow instruments. Based on the small changes in absorbance to be registered within a short period of time, the instrument must be much more sensitive than normal spectrophotometers. The lamp current must be stabilized and higher light intensity, e.g. by using larger slits, may reduce the background noise, but at the expense of the spectral resolution. A *double beam device*, as shown in Fig. 3.49, further enhances the sensitivity of the instrument. Before entering the mixing chamber the reactant solutions pass through reference cuvettes, each half of the thickness of the sample cuvette. The difference signal between the sample after mixing and the reference before mixing is recorded. Especially for minimal absorption differences and for measuring in turbid solutions the *dual wavelength principle* is applied, where the reference

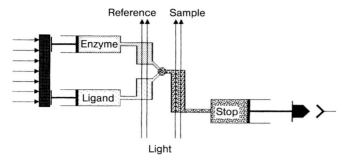

Fig. 3.49 Double beam-stopped-flow apparatus.

beam follows the same path through the sample cell, but at a different wavelength than the sample beam. As with the dual wavelength photometer (see Section 3.4.1.6), the reference beam is adapted to the wavelength of the isosbestic point or that of the turbidity beyond the absorption maximum of the reactant. This device can also be employed to observe the time-related absorption changes in the sample cell at two different wavelengths simultaneously. *Rapid-scan stopped-flow* units allow the time-resolved observation of spectra. Light from the light source is sent directly through the sample cell and becomes spectrally dispersed after passing the cell. A multi-channel photo detector (diode array detector) scans the complete spectrum within a few milliseconds. More than 100 spectra can be registered within one second (Hollaway and White 1975).

For *fluorescence measurements*, excitation is effected with deuterium or halogen lamps, also used for absorption measurements, or with a xenon arc lamp. The emitted light is collected perpendicularly after passage through an emission filter or a monochromator. Fluorescence polarization requires two photomultipliers perpendicular to the linearly polarized excitation beam and polarization filters oriented perpendicular position to each other for the two emission beams (Section 3.4.3.6). For the measurement of conformational changes in enzymes stopped-flow units equipped with *circular dichroism* optics with a mercury or xenon arc lamp as light source are used (Bayley and Anson 1975). *NMR stopped-flow units* are suitable for the observation of conformational changes, but they require higher reactant concentrations and possess a poorer time resolution (more than 10 ms) (Grimaldi and Sykes 1975). For the study of association–dissociation processes in macromolecules, the *light scattering stopped-flow apparatus* is employed, using light from a laser or other sources including X-ray radiation from a synchrotron (Flaming and Parkhurst 1977; Moody et al. 1980). Light-sensitive reactions may be initiated by short irradiation with a flashlight or laser. Combining this method with a stopped-flow apparatus, reactants may be rapidly mixed and the reaction activated by flash, e.g. the reaction of oxygen with cytochrome oxidase blocked by CO. CO is released after irradiation and vacates the binding sites for oxygen (see Section 3.5.3). Stopped-flow units are also combined with the *temperature-jump method* (Section 3.5.2.1). Temperature changes induced by enzyme reactions are measured with a *calorimeter stopped-flow apparatus*, a thin thermo-element or a thermistor serving as sensor. The response time is near 50 ms. Reaction-dependent changes in the concentration of protons are measured with a *pH stopped-flow apparatus* with a glass electrode.

A modification of the stopped-flow method is the *multi-mixing system* (Fig. 3.46 C). The two reactants pass a defined distance after mixing, while already reacting unobserved. Another reactant from a third reaction syringe is fed to this reaction mixture via a second mixing chamber. The (photometric) observation only starts at this point. By changing the flow rate or the path length between the two mixing chambers, the reaction times of the first two reactants can be altered. A simplification of this method is the *quenched-flow technique* where the third reaction syringe contains a quench solution which immediately stops the reaction of the first two reactants, e.g. by pH changes or denaturation of the enzyme with perchloric acid or trichloroacetic acid. It may also contain an indicator. The turnover is subsequently analyzed chemically. The apparatus requires no device for fast registration (Fersht and Jakes 1975). Bray (1961) used isopentane cooled in liquid nitrogen as quenching liquid. The reactant solution is shot into it and freezes immediately. This *rapid-freezing* method is especially suited for EPR measurement with free radicals or paramagnetic metals.

3.5.1.3 Measurement of Enzyme Reactions by Flow Methods

The stopped-flow technique is the most appropriate method for the observation of fast enzymatic turnovers. Reactions may be analyzed that cannot be recorded with normal photometric methods, e.g. the real initial velocity at low substrate concentrations that is already running out before onset of measuring by manual mixing. An even more important application concerns the investigation of rapid processes preceding the normal enzyme reaction, the *pre-steady-state reactions*. They start with the mostly diffusion-controlled binding of substrate to the enzyme. This process is principally faster than the time resolution of the method, but as a second-order reaction it can become observable by a proper choice of reactant concentrations. This process is often followed by isomerization of the enzyme into an active conformation. The substrate molecule interacts with the catalytic site of the enzyme to form a transition state from which it will be converted into the product. To observe these processes, a detectable signal must be available. The most obvious signal is substrate degradation or product formation, as observed already in conventional enzyme kinetics. However, the binding process of the substrate cannot be observed separately. Following product formation, *burst-kinetics* will be observed (Fig. 3.50). The burst π appears as a rapid increase which results from the interaction of substrate with the free enzyme immediately after mixing and the initiation of the first reaction cycle. In this initial phase the substrate is not hindered in its binding to the enzyme by already-bound substrate or product molecules, as will be the case in the subsequent steady-state phase, which proceeds correspondingly slower in a linear manner.

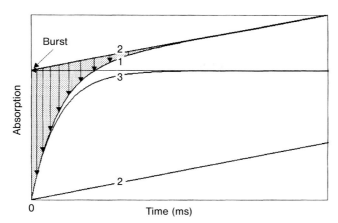

Fig. 3.50 Progression of a burst reaction with (1) and without (3) subsequent steady-state phase; (2) is the extrapolated steady-state phase. The vertical arrows show the time-related absorbance change of the pre-steady-state reaction.

The amplitude of the burst indicates the amount of product molecules formed in the first reaction cycle. An example is the cleavage of a peptide by trypsin reaction, which can be formulated according to the scheme (Bender et al. 1965):

$$E + A \underset{k_{-1}}{\overset{k_1}{\rightleftarrows}} EA \xrightarrow{k_2} EA' \xrightarrow{k_2} E + P$$

EA is the enzyme complex with the still uncleaved peptide, which becomes cleaved to form EA' and one part of the peptide dissociates directly (not indicated), the second one (P) in the following step. By extrapolating the linear steady-state phase to the ordinate at the time $t=0$, the burst π is obtained, from which the number of active centers participating in this process can be derived according to the equation

$$\pi = [E]_0 \left(\frac{\frac{k_2}{k_2+k_3}}{1 + \frac{K_m}{[A]_0}} \right)^2 \tag{3.39}$$

which can be plotted in a reciprocal form:

$$\frac{1}{\sqrt{\pi}} = \frac{1 + \frac{k_3}{k_2}}{\sqrt{[E]_0}} + \frac{K_m \left(1 + \frac{k_3}{k_2}\right)}{[A]_0 \sqrt{[E]_0}} \tag{3.40}$$

of $1/\sqrt{\pi}$ against $1/[A]_0$. The ordinate intercept corresponding to infinite substrate concentration yields $(1+k_3/k_2)/\sqrt{[E]_0}$, for $k_2 \gg k_3$ $1/\sqrt{[E]_0}$ is obtained directly. This is a valuable method for the determination of the real number of active sites involved in the catalytic turnover and to discern them from inactive enzyme molecules in the same solution, which otherwise are difficult to differentiate. For a detailed evaluation of the pre-steady-state process, e.g. the analysis of the reaction order or in the case of overlapping processes, the linear steady-state phase must be deduced from the burst reaction.

Compared with ordinary steady-state measurements, relatively high enzyme concentrations are needed in order to obtain detectable effects within this short pre-steady-state period. To observe processes occurring directly at the enzyme, e.g. binding processes or conformational changes, the protein absorbance or fluorescence can be measured and superposition of the signal by the steady-state reaction can be avoided. However, spectral changes in protein absorbance are mostly very small. Larger effects can be obtained if there are chromophoric groups bound, such as flavines or haem. Introduction of a chromophore as a label can also support the study of pre-steady-state processes. With the stopped-flow method, interactions of the enzyme with cofactors, metal ions, inhibitors, allosteric effectors, etc. can be studied. Reactions releasing or binding protons, as in dehydrogenases, can be observed in non-buffered or weakly buffered solutions with suitable pH indicators. The pK_a value of the indicator should be as near as possible to that of the reactant mixture.

A special application is the *pH-jump-stopped-flow method*. The enzyme is filled into a reaction syringe in weakly buffered solution while the second syringe contains a buffer set at another pH value. Upon mixing, the enzyme experiences a sudden jump into the other pH environment and changes of the enzyme molecule itself as well as interactions with ligands, under the condition of altered protonation of distinct groups, are to be observed. Jumps in ionic strength and in reactant concentrations are produced similarly, or solvents are added under change of polarity.

3.5.1.4 Determination of the Dead Time

The *dead time* determines the time resolution of fast kinetic equipment and is an important criterion for its quality. For the flow methods the dead time is defined as the time between the start of a reaction in the mixing chamber and the onset of registration by the detection system. Dead times can be determined experimentally with dye reactions. A chromophore is mixed with a reactive substance, which changes its color. This second order reaction will be performed under pseudo-first order conditions, maintaining one component in a large surplus. Then the reaction proceeds exponentially and can be linearized in a semi-logarithmic plot. The straight line obtained is extrapolated to a blank value containing the dye in the corresponding concentration before the start of the reaction. The difference between this point and the start of registration on the time axis correlates to the dead time (Fig. 3.51).

Reduction of 2,6-dichlorophenolindophenol by ascorbic acid or alkaline hydrolysis of 2,4-dinitrophenyl acetate to 2,4-dinitrophenol serve as indicator reactions. The chemiluminescent reaction of luminol (3-aminophthalhydracine)

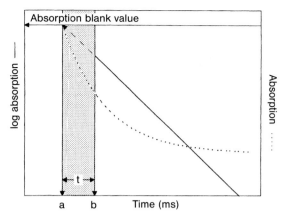

Fig. 3.51 Determination of the dead time t of a stopped-flow apparatus. An exponential pseudo-first order reaction (......) is linearized by semi-logarithmic plotting (———) and extrapolated (– – –) to the absorption of the blank value. (a) Start of the reaction in the mixing chamber, (b) start of registration.

with hydrogen peroxide and potassium hexacyanoferrate in alkaline solution, which does not require a light source, is also applicable for absorption and fluorescence optics. The high increase in fluorescence of anilinonaphthalene-sulfonate (ANS) when binding to serum albumin makes it especially suitable for fluorimetric detectors.

3.5.2
Relaxation Methods

The term *relaxation*, first used by J.C. Maxwell for the return of a molecular system to its thermal equilibrium, determines a group of methods for the observation of the processes of restoring the equilibrium after a short perturbation. According to the van't Hoff reaction isochore, the thermodynamic equilibrium constant K is dependent on the absolute temperature T

$$\left(\frac{\partial \ln K}{\partial T}\right)_P = \frac{\Delta H^0}{RT^2} \ . \tag{3.41}$$

A similar relationship holds for the dependence of the equilibrium constant on the pressure P:

$$\left(\frac{\partial \ln K}{\partial P}\right)_T = \frac{\Delta V^0}{RT} \tag{3.42}$$

and on the electric field strength E:

$$\left(\frac{\partial \ln K}{\partial E}\right)_{T,P} = \frac{\Delta M}{RT} \ . \tag{3.43}$$

ΔH^0 is the standard reaction enthalpy at a temperature T, ΔV^0 is the change in volume per formula turnover of the reaction under standard conditions, R is the gas constant. The difference in partial molar polarizations between products and substrates ΔM represents the change in charges of the reactants induced by the reaction. For each of these dependences a separate technique was developed (originally by Manfred Eigen and his group in Göttingen, Germany). Compared with flow methods the time-resolution could be raised by several magnitudes. The principle of these techniques is that a reaction system in equilibrium becomes perturbed by short changes (*jumps*) in temperature, pressure or electric field strength. Due to inertia of the system it is not able to adapt immediately to the altered conditions established by the perturbation and the delayed adaptation is monitored. Enzymatic turnovers cannot be measured directly, rather the enzyme reaction must be in the equilibrium state (between forward and back reaction) and the relative shift of the equilibrium after the jump is followed. Ligand binding, conformational changes, association equilibria, etc. can be observed. The shifts in equilibrium are comparatively small and the signals are weak and require great precision in the detection.

3.5.2.1 The Temperature Jump Method

Of all relaxation methods, the temperature jump method (*T-jump*) has the widest range of application, especially in enzymatic studies. The reaction mix in the observation chamber is heated by a sudden temperature shock and observed with optical or polarographic techniques (Fig. 3.52). The temperature jump is triggered by discharging a high voltage condenser over a spark gap. The electric surge discharges over gold or platinum electrodes through the solution in the observation cell and is then grounded. The friction of ions migrating in the electric field induces a rapid temperature increase (Joule heating).

The degree of the temperature jump can be adjusted according to the relationship:

$$\Delta T = \frac{CU^2}{8.36 c_p \rho V} \tag{3.44}$$

by the voltage U (normally between 10000 and 100000 V) and the capacitance C (0.01–0.1 µF). It is mostly between 5 and 10 °C. V is the reaction volume, c_p the specific heat and ρ the density of the solution. ΔT depends on the resistance which is highest in the solution. To limit the temperature increase, especially in enzyme solutions, the resistance R must be kept as low as possible in the observation cell. According to $R \sim d/QL$, this may be achieved by a narrow distance between the electrodes d, a broad electrode diameter Q and a high conductivity L. To enhance the conductivity, the solution should contain a high concentration

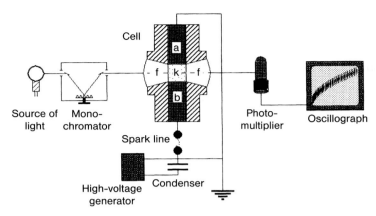

Fig. 3.52 Scheme of a temperature jump instrument.
(a, b) Electrodes, (f) observation window, (k) cuvette.

of electrolytes (0.1–0.2 M KNO_3, which is relatively inert towards the electrodes). The duration of the temperature jump depends on the resistance and capacity and may be reduced by lowering these factors. However, times below 1 μs cannot be reached, as a shock wave is created that interferes with the measurement and releases gas bubbles in aqueous solutions. Due to mass inertia, the temperature rise is faster than thermal expansion. This induces an increase in pressure of 5 MPa for a temperature difference of 10 °C in aqueous solution, which relaxes in a shock wave. Although a pulse duration of only 50 ns could be achieved by the application of a 5 m coaxial cable instead of a condenser (Hoffmann 1971), the pressure effect limits the method to a range of 1 μs. This effect becomes least when measurements are carried out at 4 °C.

Instead of Joule heating, the temperature jump can be created by irradiation with microwaves or laser light, especially for measurements at low ionic strength or in non-aqueous solutions. Absorption of microwaves of a frequency of 10^{10} s^{-1} from a microwave generator in aqueous solution generates a temperature increase of 1 °C in 1 μs. This relatively minor effect, as well as the high price of the microwave generator, prohibits wider use of this method. Pulses from a neodymium laser, whose wavelength is enhanced by liquid nitrogen from 1060 to 1410 nm, can produce a rise in temperature of several °C within 25 ns as water absorbs strongly in this region. This method is thus more rapid than the shock wave appearing in the region of microseconds. However, heating of the measurement solution with a light path of more than 1 mm is inhomogeneous, as light absorbance in the aqueous phase rises exponentially with the length of the light path. Therefore, only small volumes may be analyzed.

A T-jump apparatus employs basically the same detection techniques as used for stopped-flow instruments. Because of the relatively weak signals the demands on the sensitivity of the equipment are equally high. Absorption spectroscopy is the method most frequently applied, fluorescence and fluorescence

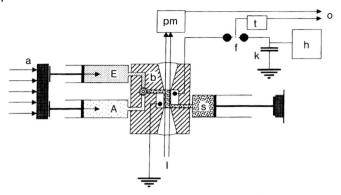

Fig. 3.53 Scheme of a stopped-flow temperature jump device. (A, E) Reaction syringes, (a) drive, (b) cuvette block, (f) spark line, (l) lamp, (h) high-voltage generator, (k) condenser, (pm) photomultiplier, (o) oscillograph, detector, (s) stop syringe, (t) trigger.

polarization measurements are also used and, to a small degree, optical rotation which, however, requires a longer light path. Equipment for the observation of light scattering and electrical conductivity has also been developed. The latter method is not applicable for T-jumps based on Joule heating.

Combination of the T-jump method with a stopped-flow apparatus extends the areas of application. The fact that *T-jump* may be used only for reactions in equilibrium requires the reactants to be present in comparable quantities to make a shift in equilibrium accessible, i.e. reactions with an equilibrium clearly preferring one side will hardly be detected. This holds for many quasi-irreversible enzyme reactions. With the *stopped-flow temperature jump method* the reactants are rapidly mixed and immediately exposed to the temperature jump. Time resolution of the *T-jump* method is faster by three orders of magnitude than the stopped-flow method, so that an enzyme reaction running in the millisecond range does not affect the *T-jump* measurement. Figure 3.53 shows the scheme of such an apparatus. A combined observation cell with electrodes is built into a regular stopped-flow apparatus linked to the equipment of a *T-jump* apparatus.

Compared to the dead time for the stopped flow apparatus, the characteristic factor for time resolution for the *T-jump* equipment is the time required to reach 90% of the maximum temperature level. It can be determined by protonation reactions with a pH indicator.

T-jump instruments are employed for the study of enzyme reactions, ligand binding, conformational changes in enzymes, spontaneous or ligand-induced transitions of allosteric enzymes, catalytic mechanisms or aggregation processes. The observed signal may often be composed of several individual processes with different time constants, e.g. binding of a ligand followed by an isomerization reaction of the enzyme. Reactions giving only weak signals can be linked with other reactions producing a strong signal. In this manner, the temperature-dependent pH shift of a fast reacting buffer system can be coupled with a distinct proton-dependent process, e.g. a dehydrogenase reaction, converting the *T-jump* instrument into a pH jump apparatus.

3.5.2.2 The Pressure Jump Method

Various pressure jump systems (*P*-jump) for the generation of short-term pressure differences in solution have been described. Figure 3.54 shows a device developed by Strehlow and Becker (1959). The sample is inserted into the cell of an autoclave, closed by a flexible Teflon membrane and surrounded by a pressure-conducting medium (water, paraffin). This medium is covered with an elastic polyethylene membrane and the remaining space in the autoclave is filled with compressed gas to a pressure of 5.5 MPa. An upper hole in the autoclave is tightly covered with a conductive membrane consisting of a pressure-resistant copper beryllium bronze. A magnetic trigger releases the locking device of a metal needle fixed above the membrane. This needle punches the membrane and simultaneously starts recording. The membrane bursts under high pressure and the pressure in the solution relaxes. The time resolution of this method (20–100 µs) is significantly lower than that of the *T-jump* method. It is limited by the duration of bursting of the membrane and by the velocity of propagation of the shock wave front in the solution. An improved model of this device (Knoche 1974) omits the gas chamber. The paraffin solution is put directly under increasing pressure until the membrane locking the autoclave bursts.

In another construction principle the autoclave is formed by a long tube, which is divided into two cells by an aluminum disk. Both cells contain water, one at normal, the other at high pressure. The sample chamber is located below

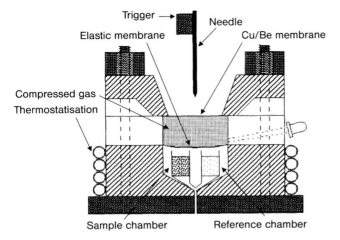

Fig. 3.54 Construction scheme of a pressure jump apparatus.

the low-pressure compartment and separated from it by a rubber membrane. The pressure in the high-pressure compartment is raised until the separating disk breaks and the pressure is transmitted to the sample solution in the form of a shock wave. The value of the high pressure is determined by the tear strength of the separating disk. Pressures of up to 100 MPa are achieved (Knoche 1974).

Fast repetitive pressure changes are produced with the apparatus of Macgregor et al. (1985). The observation cell is locked by an arrangement of piezoelectric crystals. The crystals expand and contract cyclically, induced by a high-voltage switch-on/switch-off cycle. The sample solution is under a stationary pressure of about 10 MPa which increases with each repetitive pressure jump by 500 kPa.

The ultrasonic method is also based on the principle of periodic pressure changes induced by sound waves in solution. This method is only briefly mentioned, as it is of minor importance for the study of enzymes. The perturbation is not a sudden jump but a rapidly oscillating sine wave. When the frequency of this wave approaches the range of the rate constant (respectively the reciprocal relaxation time, see Section 3.5.4) of the reactant solution, the system oscillates in resonance with the sound wave. The concentration change in the system in order to regain its equilibrium follows with reduced amplitude and delays the sound wave front. From this the rate constant of the reaction can be obtained. In the dispersion method the sound velocity, and in the absorption method the absorption coefficient, as a function of frequency are determined. The ultrasonic frequency is varied from 0.05 to 100 MHz. Reactions in the nanosecond range can be observed in aqueous solutions. However, large volumes in high concentrations (10 mM) are required.

The signals of the P-jump method are detected by optical or conductivity measurement. The transparent parts of the observation cell are equipped with pressure-resistant sapphire windows.

The fields of application of the pressure jump method for biological systems, especially in enzyme studies, are limited. As can be seen from Eq. (3.42), the pressure-dependent change in the equilibrium constant is proportional to the volume change of the system, i.e. only reactions exhibiting a considerable volume change are accessible. Biological macromolecules are highly pressure-resistant, despite their complex structure. Organisms from deep sea regions normally resist pressures of up to 100 MPa without suffering any damage. For the dependence of the rate constant on pressure the following relation applies, in analogy to Eq. (3.42):

$$\left(\frac{\partial \ln k}{\partial P}\right)_T = \frac{\Delta V^{\ddagger}}{RT} \,. \tag{3.45}$$

$\Delta V^{\ddagger}$, the activation volume, is the difference between the volumes of the transition and initial states. Increasing pressure shifts the equilibrium according to the Le Chatelier-Braun principle in the direction of volume reduction. Relatively large changes in volume, and thus good pressure dependence, are exhibited by protonation reactions and ionic and hydrophobic interactions, while cleaving and formation of hydrogen bridges cause only insignificant volume changes. Proteins, due to their quasi-crystalline structure, are highly resistant against pressure. The denaturation of a protein progresses in many individual steps, partially accompanied by either volume increases or decreases, so that the single steps compensate each other essentially over the total process. The resulting change in volume is comparatively insignificant, so that pressure-dependent denaturation may hardly be achieved. Generally, proteins have a negative reaction volume, they are destabilized by pressure, while DNA is stabilized due to its positive reaction volume.

3.5.2.3 The Electric Field Method

The electric field method (E-jump) is even less applicable to biological systems than the P-jump method and is mentioned only for the sake of completeness. The equipment is similar to the T-jump apparatus. The sample solution is subjected to a high-voltage pulse instead of an electric surge. At a tension of 100 000 V water would boil within seconds. The electric field is, therefore, built up over a rectangular impulse of only a few nanoseconds to microseconds and discharged again. Instead of a condenser, a coaxial cable, 0.3–1 km in length, is charged by a high-voltage generator. A spark gap charges a condenser. This has the shape of two electrodes attached to the observation cell, which becomes rapidly discharged. In this manner a sample is exposed briefly to a high-voltage electric field. Protonation reactions can be studied with this method. Relaxation times of a haemoglobin–oxygen mixture have been successfully determined, and helix–coil transitions in proteins could be observed.

3.5.3
Flash Photolysis, Pico- and Femto-second Spectroscopy

Flash photolysis was originally developed by Norrish and Porter (1949) for reactions in the gaseous phase and later transferred to reactions in solution. It serves to study unstable compounds with half-life times of less than 1 s. This method was applied to observe processes in photosynthesis and vision, porphyrin–metal complexes, binding of O_2 and CO to hemoglobin and myoglobin. The release of ATP from a photosensitive ATP derivative by UV irradiation with a laser pulse is a good example of the observation of an enzyme reaction by flash photolysis. A component of the reaction system is activated by a short, very intense flash of light and the reaction initiated by this process is observed. The molecules must be photosensitive and are transformed into free radicals or a triplet state. This limits the range of application to systems already possessing such compounds or systems being suited to the introduction of such compounds.

Flashes are generated by discharge flash lamps filled with argon, krypton or xenon. Time resolutions of a few milliseconds can be obtained. The picosecond

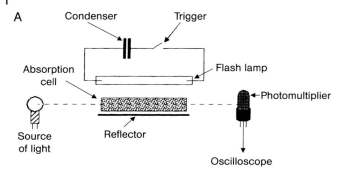

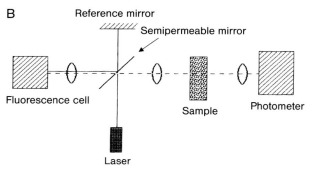

Fig. 3.55 Construction scheme of a flash photolysis apparatus with a flash lamp (A) and a laser (B).

range becomes accessible with pulse lasers. Figure 3.55 A shows the scheme of an apparatus where flashes are generated by discharging a condenser. The lamp is aligned in parallel to a quartz cylinder containing the reactant solution. For optimal irradiation the unit is surrounded by a cylinder covered on the inside with a reflector screen of magnesium oxide. The progress of the reaction within the cylinder is measured by a photometric device attached lengthwise to the cylinder. The photometric lamp is also a flash lamp, emitting a flash somewhat delayed compared to that of the photolysis flash (*flash spectroscopy*). The whole wavelength range is registered simultaneously. For the observation of the time-dependent progress of the reaction at a certain wavelength (*kinetic spectrophotometry*) a continuous light source is employed.

Introduction of lasers made the nano- and pico-second range accessible for flash photolysis. With pulsed solid-phase lasers with rubies (694 nm), Nd^{3+} in glass or yttrium–aluminum garnet (YAG, 1063 nm) pulses of 10–30 ns duration are obtained. Mode locking of the laser yields pulses in the range 30–100 fs. By modulating the laser light wavelengths of 347 (ruby), 532, 355 and 266 nm (Nd:YAG) are accessible. Dye lasers pumped up with flash lamps are also used. They have a significantly lower light intensity but a larger wavelength range and are less expensive. Their pulse duration lies mostly above 100 ns. To obtain the necessary excitation energy for the molecules, light impulses have to be enhanced. In the UV region, excimer lasers are employed, utilizing the formation of dimers between an inert gas and a halogen in the excited state, e.g. ArCl (308 nm) and XeCl (248 nm). They have a pulse length of 20 ns. For flash spectroscopy, the light pulse is split into two beams, one for excitation of the sample, the other as a spectroscopic light pulse. For a time delay it is deflected by a reflection mirror and directed into a fluorescent solution, that in turn emits a continuous light within a defined spectral range (Fig. 3.55 B). A spectral continuum is also obtained by focussing the light pulse into a water or water-alcohol cell. For kinetic spectrophotometry in the nanosecond range, the light intensity of the photometric lamp must be very high to eliminate interferences by scattering light and the photolysis flash. To avoid fatigue of the photomultiplier, measuring is done in very short pulses with a flash lamp or a pulsed xenon arc lamp.

A modification of flash photolysis is *pulse radiolysis*, employing electron pulses of 1–100 ns from a microwave linear electron accelerator (LINAC) instead of light impulses. Different from photoexcitation, where a defined compound in solution is specifically excited and the solvent is permeable to radiation, the electron energy in diluted solutions is almost completely released to the solvent. Electron irradiation induces the formation of radicals, ionization and excitation of molecules into permitted (singlet–singlet) and forbidden (singlet–triplet) transitions. In water, H_3O^+, OH^-, $H^\bullet$, H_2, and H_2O_2 are formed, and especially hydroxyl radicals ($OH^\bullet$) with oxidizing and hydrating electrons (e_{aq}^-) with reducing properties, whose reactivities have been studied by this method. Pulse radiolysis deals less with the direct effects of the electron pulse on the compounds, rather it observes their reactions with reactive particles. In this way, the reduction of methemoglobin by hydrated electrons and the subsequent binding of oxygen could be studied.

3.5.4
Evaluation of Rapid Kinetic Reactions (Transient Kinetics)

From experiments with rapid methods, especially relaxation methods, relaxation curves are obtained (Fig. 3.56A). The system approaches asymptotically the equilibrium forced upon it by the new conditions (elevated temperature, relaxed pressure). Δ_0 is the maximum achievable deviation from the original equilibrium, Δ_t the deviation at a distinct time t. A characteristic value is the *relaxation time* τ, defined as the time required to achieve 63.2% deviation from Δ_0 (Fig. 3.56A):

$$\Delta_t = \Delta_0 e^{-(t/\tau)} . \tag{3.46}$$

For $t = \tau$ read:

$$\Delta_t = \Delta_0 e^{-1} = \Delta_0 \cdot 0.368 .$$

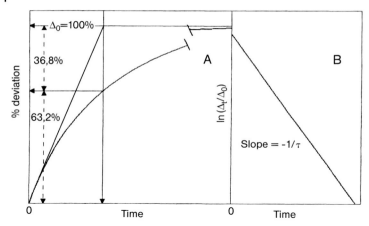

Fig. 3.56 Determination of relaxation time τ from a relaxation curve (A). (B) Semi-logarithmic plot: Δ_0 is the maximum deviation, Δ_t the deviation from time t.

The relaxation time can be obtained from the intercepts of a source tangent with the horizontal straight line for the initial and final states, or from the slope of a semi-logarithmic plot of the deviation $\ln(\Delta_t/\Delta_0)$ against time (Fig. 3.56 B):

$$\ln\frac{\Delta_t}{\Delta_0} = -\frac{t}{\tau}. \tag{3.47}$$

The relaxation time is a complex factor composed of the rate constants and concentration terms of the components of the respective reaction. Here as an example the relaxation time for a bimolecular reaction according to a simple binding equilibrium

$$E + A \underset{k_{-1}}{\overset{k_{+1}}{\rightleftharpoons}} EA \tag{1.18}$$

will be deduced. The perturbation of the equilibrium causes a change of the initial concentrations of the reactants ([A], etc.) by the amount δ into $[\overline{A}]$, etc.:

$$[E] = [\overline{E}] + \delta_E; \quad [A] = [\overline{A}] + \delta_A; \quad [EA] = [\overline{EA}] - \delta_{EA}. \tag{3.48}$$

The total volumes of the reactants remain unchanged by perturbation:

$$[A]_0 = [A] + [EA] = [\overline{A}] + [\overline{EA}], \tag{3.49a}$$

$$[E]_0 = [E] + [EA] = [\overline{E}] + [\overline{EA}]. \tag{3.49b}$$

From this it follows:

$$[\overline{A}] + \delta_A + [\overline{EA}] - \delta_{EA} = [A] + [EA]$$
$$[\overline{E}] + \delta_E + [\overline{EA}] - \delta_{EA} = [E] + [EA]$$

and therefore

$$\delta_A = \delta_E = \delta_{EA} = \delta \ . \tag{3.50}$$

The differential equation for the time-dependent change of the reactants according to the reaction described by Eq. (1.18):

$$-\frac{d[A]}{dt} = k_{+1}[A][E] - k_{-1}[EA]$$

is converted by insertion of Eqs. (3.48) and (3.50) into

$$-\frac{d([\overline{A}] + \delta)}{dt} = k_{+1}([\overline{A}] + \delta)([\overline{E}] + \delta) - k_{-1}([\overline{EA}] - \delta)$$
$$-\frac{d[\overline{A}]}{dt} - \frac{d\delta}{dt} = k_{+1}[\overline{A}][\overline{E}] - k_{-1}[\overline{EA}] + \{k_{+1}([\overline{A}] + [\overline{E}]) + k_{-1}\}\delta + k_{+1}\delta^2 \ . \tag{3.51}$$

In the equilibrium, also after perturbation, the concentration terms are time-independent

$$-\frac{d[\overline{A}]}{dt} = k_{+1}[\overline{A}][\overline{E}] - k_{-1}[\overline{EA}] = 0 \ . \tag{3.52}$$

Thus the first two terms of the left part of Eq. (3.51) can be omitted. For small changes the square term may be disregarded:

$$-\frac{d\delta}{dt} = \{k_{+1}([\overline{A}] + [\overline{E}]) + k_{-1}\}\delta = \frac{\delta}{\tau} \ . \tag{3.53}$$

The bracket term, containing only constant factors (as the concentration terms in the equilibrium can also be regarded as constant), is replaced by $1/\tau$. By integration after time t for a given concentration change Δ:

$$\int_{\Delta_0}^{\Delta_t} \frac{d\delta}{\delta} = -\int_0^t \frac{dt}{\tau} \tag{3.54}$$

the Eqs. (3.46) and (3.47), already mentioned above, are obtained. According to Eq. (3.53), the relaxation time for the present reaction mechanism is:

$$\frac{1}{\tau} = k_{+1}([\overline{A}] + [\overline{E}]) + k_{-1} \ . \tag{3.55}$$

Table 3.3 Significance of relaxation times τ in different reaction mechanisms (after Hiromi 1979)

Reaction mechanism	Reciprocal relaxation time ($1/\tau$)	Remarks
$A \underset{k_{-1}}{\overset{k_1}{\rightleftarrows}} P$	$k_{+1} + k_{-1}$	
$nA \underset{k_{-1}}{\overset{k_1}{\rightleftarrows}} P_n$	$n^2 k_{+1}[\overline{A}]^{n-1} + k_{-1}$	
$A + B \underset{k_{-1}}{\overset{k_1}{\rightleftarrows}} P$	$k_{+1}([\overline{A}] + [\overline{B}]) + k_{-1}$	
$A + B \underset{k_{-1}}{\overset{k_1}{\rightleftarrows}} P$	$k_{+1}[\overline{B}] + k_{-1}$	B in surplus
$A + E \underset{k_{-1}}{\overset{k_1}{\rightleftarrows}} P + E$	$(k_{+1} + k_{-1})[\overline{E}]$	E is catalyst
$A + B \underset{k_{-1}}{\overset{k_1}{\rightleftarrows}} P + Q$	$k_{+1}([\overline{A}] + [\overline{B}]) + k_{-1}([\overline{P}] + [\overline{Q}])$	
$A + B \underset{k_{-1}}{\overset{k_1}{\rightleftarrows}} 2P$	$k_{+1}([\overline{A}] + [\overline{B}]) + 4k_{-1}[\overline{P}]$	
$A + B + C \underset{k_{-1}}{\overset{k_1}{\rightleftarrows}} P$	$k_{+1}([\overline{A}][\overline{B}] + [\overline{A}][\overline{C}] + [\overline{B}][\overline{C}]) + k_{-1}$	

The rate constants can be calculated by the determination of relaxation times at different concentrations of $[\overline{A}]$ and $[\overline{E}]$. From a plot of $1/\tau$ against $[\overline{A}] + [\overline{E}]$, k_{+1} can be obtained from the slope, k_{-1} from the ordinate intercept, and $K_d = k_{-1}/k_{+1}$ from the abscissa intercept. However, concentration values on the abscissa refer to the conditions *after* perturbation, i.e. high temperature, which are obtained by applying Eq. (3.49) if K_d is known. Correspondingly, the constants calculated from these data are valid for conditions after perturbation.

According to Eq. (3.46) the relaxation curve is an exponential function, but only as long as the square term from Eq. (3.51) can be disregarded, i.e. for small changes. Otherwise deviations from the linear progression as in Fig. 3.56 B would occur. Non-linearity may, however, also result from superimposition of different processes. To differentiate between these possibilities, the extent of the perturbation, i.e. the temperature jump, may be reduced. If non-linearity remains, this indicates that several relaxation times superimpose each other. For n states of the reactants $n-1$ relaxation times are to be expected. If they differ by more than one order of magnitude they can be decoupled by variation of the time axis. Otherwise they may be separated by calculation.

The scheme for the derivation of the relaxation time for a bimolecular reaction may also be applied to other reaction mechanisms. Table 3.3 shows a compilation of various frequently occurring mechanisms. The constants are obtained by corresponding plots, as shown in Fig. 3.56 for the bimolecular mechanisms.

References

General Methods and Enzyme Analytics

Bergmeyer, H. U. (1983) *Methods of Enzymatic Analysis*, Verlag Chemie, Weinheim.

Clark, J. M., Switzer, R. L. (1977) *Experimental Biochemistry*, W. H. Freeman, San Francisco.

Eisenthal, R., Danson, J. (1992) *Enzyme Assays. A Practical Approach*, IRL Press, Oxford.

Umbreit, W. W., Burris, R. H., Stauffer, J. F. (1972) *Manometric and Biochemical Techniques*, 5th edn. Burgess, Minneapolis.

Williams, B. L., Wilson, K. (1975) *Principles and Techniques of Practical Biochemistry*, Edward Arnold, London.

Methods to Determine Multiple Equilibria

Ackers, K. G. (1975) Molecular sieve methods of analysis, in Neurath, H., Hill, R. L. (Eds.) *The Proteins*, 3rd edn, Academic Press, New York, Vol. 1, pp. 1–94.

Alberts, R. W., Krishnan, N. (1979) Application of the miniature ultracentrifuge in receptor-binding assays, *Anal. Biochem.* 96, 396–402.

Brumbaugh, E. E., Ackers, K. G. (1974) Molecular sieve studies of interacting protein systems. Direct optical scanning method for ligand-macromolecule binding studies, *Anal. Biochem.* 41, 543–559.

Chanutin, A., Ludewig, S., Masket, A. V. (1942) Studies on the calcium-protein relationship with the aid of the ultracentrifuge, *J. Biol. Chem.* 143, 737–751.

Colowick, S. P., Womack, F. C. (1969) Binding of diffusible molecules by macromolecules: Rapid measurements by rate of dialysis, *J. Biol. Chem.* 244, 774–777.

Draper, D. E., Hippel, P. H. (1979) Measurement of macromolecule binding constants by a sucrose gradient band, *Biochemistry* 18, 753–760.

Englund, P. E., Huberman, J. A., Jovin, T. M., Kornberg, A. (1969) Enzymatic synthesis of deoxyribonucleic acid. Binding of triphosphates to deoxyribonucleic acid polymerase, *J. Biol. Chem.* 244, 3038–3044.

Hummel, J. P., Dreyer, W. J. (1962) Measuring of protein-binding by gel filtration, *Biochim. Biophys. Acta* 63, 530–532.

Martin, R. G., Ames, B. N. (1961) A method for determining the sedimentation behavior of enzymes. Application of protein mixtures, *J. Biol. Chem.* 236, 1372–1379.

Myer, Y. P. & Schellman, J. A. (1962) The interaction of ribonuclease with purine and pyrimidine phosphates. Binding of adenosine-5′-monophosphate to ribonuclease, *Biochim. Biophys. Acta* 55, 361–373.

Paulus, H. (1969) A rapid and sensitive method for measuring the binding of radioactive ligands to proteins, *Anal. Biochem.* 32, 91–100.

Steinberg, I. Z., Schachman, H. K. (1966) Ultracentrifugation studies with absorption optics. Analysis of interacting systems involving macromolecules and small molecules, *Biochemistry* 5, 3728–3747.

Suter, P., Rosenbusch, J. (1976) Determination of ligand binding: Partial and full saturation of aspartate transcarbamylase. Applicability of a filter assay to weakly binding ligands, *J. Biol. Chem.* 251, 5986–5991.

Yamamoto, K. R., Alberts, B. (1974) On the specificity of the binding of estradiol receptor protein to deoxyribonucleic acid, *J. Biol. Chem.* 249, 7076–7086.

Surface Plasmon Resonance

Wilson, W. D. (2002) Analyzing biomolecular interactions, *Science* 295, 2103–2105.

Fägerstam, L. G., Frostell-Karlsson, A., Karlsson, R., Persson, B., Rönnberg, I. (1992) Biospecific interaction analysis using surface plasmon resonance detection applied to kinetic, binding site and concentration analysis, *J. Chromatogr.* 597, 397–410.

Electrochemical Methods

Beechey, R. B., Ribbons, D. W. (1972) Oxygen electrode measurements, *Methods Microbiol.* 6B, 25–53.

Clark, L. C. Jr, Wolf, R., Granger, D., Taylor, Z. (1953) Continuous recording of blood oxygen tensions by polarography, *J. Appl. Physiol.* 6, 189–193.

Degn, H., Lundsgaard, J. S., Peterson, L. C., Ormick, A. (1980) Polarographic measurements of steady-state kinetics of oxygen uptake by biochemical samples, *Methods Biochem. Anal. 26*, 47–77.

Lessler, M. A. (1982) Adaptation to polarographic oxygen sensors for biochemical assays, *Methods Biochem. Anal. 28*, 175–199.

Lessler, M. A., Bierley, G. P. (1969) Oxygen electrode measurements in biochemical analysis. *Methods Biochem. Anal. 17*, 1–29.

Nicolls, D. G., Garland, P. B. (1972) Electrode measurements of carbon dioxide, *Methods Microbiol. 6B*, 55–63.

Weitzman, P. D. J. (1969) Polarographic assay for malate synthase and citrate synthase, *Methods Enzymol. 13*, 365–368.

Calorimetric Methods

Freire, E., van Ospold, W. W. (1990) Calorimetrically determined dynamics of complex unfolding transitions in proteins, *Annu. Rev. Biophys. Biophys. Chem. 19*, 159–188.

Hemminger, W., Höhne, G. (1979) *Grundlagen der Kalorimetrie*, Verlag Chemie, Weinheim.

Jolicoeur, C. (1981) Thermodynamic flow methods in biochemistry, calorimetry, densitometry and dilatometry, *Methods Biochem. Anal. 21*, 171–287.

Spink, C., Wadsö, I. (1976) Calorimetry as an analytical tool in biochemistry and biology, *Methods Biochem. Anal. 23*, 1–159.

Spectroscopic Methods

Adler, A. J., Greenfield, N. J., Fassman, G. D. (1973) Circular dichroism and optical rotation dispersion of proteins and polypeptides, *Methods Enzymol. 27*, 675–735.

Brand, L., Witholt, B. (1961) Fluorescence measurements, *Methods Enzymol. 11*, 776–856.

Brand, L., Gohlke, J. R. (1972) Fluorescence probes for structure, *Annu. Rev. Biochem. 41*, 843–868.

Castor, L. N., Chance, B. (1955) Photochemical action spectra of carbon monoxide-inhibited respiration, *J. Biol. Chem. 217*, 453–465.

Chance, B. (1991) Optical methods, *Annu. Rev. Biophys. Biophys. Chem. 20*, 1–28.

Dewey, T. G. (1991) *Biophysical and Biochemical Aspects of Fluorescence Spectroscopy*, Plenum Press, New York.

Donovan, J. W. (1973) Ultraviolet difference spectroscopy – New techniques and applications, *Methods Enzymol. 27*, 497–525.

Eftink, M. R. (1989) Fluorescence techniques for studying protein structure, *Methods Biochem. Anal. 35*, 127–205.

Fasman, G. D. (1989) *Practical Handbook of Biochemistry and Molecular Biology*, CRC Press, Boca Raton.

Förster, T. (1948) Zwischenmolekulare Energiewanderung und Fluoreszenz, *Ann. Phys. 437*, 6th Series, Vol. 2, pp. 55–75.

Galla, H.-J. (1988) *Spektroskopische Methoden in der Biochemie*, Thieme, Stuttgart.

Graupe, K. (1982) Struktur-Funktionsbeziehung im Pyruvatdehydrogenase-Multienzymkomplex aus *Escherichia coli*, Ph.D. Thesis, University of Tübingen, Germany.

Graupe, K., Abusaud, M., Karfunkel, H., Bisswanger, H. (1982) Reassociation of the pyruvate dehydrogenase complex from *Escherichia coli*: Kinetic measurements and binding studies by resonance energy transfer, *Biochemistry 21*, 1386–1394.

Greenfield, N. J., Fasman, G. D. (1969) Computed circular dichroism spectra for the evaluation of protein conformation, *Biochemistry 8*, 4108–4116.

Guibault, G. G. (1990) *Practical Fluorescence*, Marcel Dekker, New York.

Günzler, H., Gremlich, H.-U. (2002) *IR Spektroscopy*, Wiley-VCH, Weinheim.

Harris, D. A., Bashford, C. L. (1987) *Spectrophotometry and Spectrofluorimetry*, IRL Press, Oxford.

Haugland, R. P. (2002) *Handbook of Fluorescent Probes and Research Products*, 9th edn. Molecular Probes, Eugene, Oregon.

Herskovits, T. T. (1967) Difference spectroscopy, *Methods Enzymol. 11*, 748–775.

Johnson, W. C. (1985) Circular dichroism and its empirical application to biopolymers, *Methods Biochem. Anal. 31*, 61–163.

Lakowicz, J. R. (1999) *Principles of Fluorescence Spectrocopy*, 2nd edn, Kluwer Academic/Plenum Publisher, New York.

Lloyd, D., Scott, R. I. (1983) Photochemical action spectra of CO-liganded terminal oxidases using a liquid dye laser, *Anal. Biochem. 128*, 21–24.

Mendelsohn, R. (1978) Application of infrared and Raman spectroscopy to studies of protein conformation, in Kornberg, H. L. (ed.) *Techniques in Protein and Enzyme Biochemistry.* B109, pp. 1–28. Elsevier/North Holland, County Clare.

Perrin, F. (1926) Polarisation de la Lumire de Fluorescence. Vie Moyenne des Molecules dans l'Etat excit, *J. Phys., Ser. VI 7*, 390–401.

Rosenheck, K., Doty, P. (1961) Far ultraviolet absorption spectra of polypeptide and protein solutions and their dependence on configuration, *Proc. Natl. Acad. Sci. USA 47*, 1775–1785.

Seitz, W. R., Neary, M. P. (1969) Recent advances in bioluminescence and chemoluminescence assay, *Methods Biochem. Anal. 23*, 161–188.

Shikari, M. (1969) *Sci. Pop. Coll. Gen. Educ. Univ. Tokyo 19*, 151.

Strehler, B. L. (1968) Bioluminescence assay: Principles and practice, *Methods Biochem. Anal. 16*, 99–179.

Stryer, L. (1968) Fluorescence spectroscopy of proteins, *Science 162*, 526–533.

Stryer, L. (1978) Fluorescence energy transfer as a spectroscopic ruler, *Annu. Rev. Biochem. 47*, 819–846.

Urry, D. W. (1985) Absorption, circular dichroism and optical rotary dispersion of polypeptides, proteins, prosthetic groups and biomembranes, in Neuberger, A., Van Deenen, L. L. M. (eds.) *Modern Physical Methods in Biochemistry, New Comprehensive Biochemistry 11A*, Elsevier, Amsterdam, pp. 275–346.

Valeur, B. (2002) *Molecular Fluorescence, Principles and Applications*, Wiley-VCH, Weinheim.

Van Holde, K. E. (1985) *Physical Biochemistry*, Prentice Hall, Englewood Cliffs.

Warburg, O. & Negelein, E. (1929) Über das Absorptionsspektrum des Atmungsferments, *Biochem. Z. 214*, 64–100.

Wettlaufer, D. B. (1962) Ultraviolet spectra of proteins and amino acids, *Adv. Protein Chem. 17*, 303–390.

Wharton, C. W. (1986) Infra-red and Raman spectroscopic studies of enzyme structure and function, *Biochem. J. 233*, 25–36.

Measurement of Rapid Kinetic Reactions

Bayley, P. M., Anson, M. (1975) Stopped-flow circular dichroism: A fast-kinetic system, *Biopolymers 13*, 401–405.

Bender, M. L., Kzdy, F. J., Feder, J. (1965) The kinetics of the trypsin-catalyzed hydrolysis of p-nitrophenyl-α-N-benzyloxycarbonyl-L-lysinate hydrochloride, *J. Am. Chem. Soc. 87*, 4953–4954.

Bernasconi, C. F. (1976) *Relaxation Kinetics.* Academic Press, New York.

Bray, R. C. (1961) Sudden freezing as a technique for the study of rapid reactions, *Biochem. J. 81*, 189–193.

Chan, S. S., Austin, R. H. (1984) Laser Photolysis in Biochemistry, *Methods Biochem. Anal. 30*, 105–139.

Chance, B., Eisenhardt, R. M., Gibson, Q. H., Lonberg-Holm, K. K. (1964) *Rapid Mixing and Sampling Techniques in Biochemistry*, Academic Press, New York.

Chance, B. (1974) Rapid-Flow-Methods, in Weissberger, A. (ed.) *Techniques of Chemistry*, Vol. VI, Part II, pp. 5–62. J. Wiley & Sons, New York.

Dorfman, L. M. (1974) Pulse radiolyse, in Weissberger, A. (ed.) *Techniques of Chemistry*, Vol. VI, Part II, pp. 363–419. J. Wiley & Sons, New York.

Eigen, M., De Mayer, L. (1974) Theoretical basis of relaxation spectroscopy, in Weissberger, A. (ed.) *Techniques of Chemistry*, Vol. VI, Part II, pp. 63–146. J. Wiley & Sons, New York.

Eigen, M., De Mayer, L. (1963) Relaxation methods, in Friess, S. L., Lewis, E. S., Weissberger, A. (eds.) *Techniques of Organic Chemistry*, Vol. VIII, Part II, pp. 895–1054. John Wiley & Sons, New York.

Erman, J. E., Hammes, G. G. (1966) Versatile stopped-flow temperature jump apparatus, *Rev. Sci. Instrum. 37*, 746–750.

Fersht, A. R., Jakes, R. (1975) Demonstration of two reaction pathways for the amino acylation of tRNA. Application of the pulsed quenched flow technique, *Biochemistry 14*, 3350–3356.

Flamig, D. P., Parkhurst, L. J. (1977) Kinetics of the alkaline tetramer-dimer dissociation in liganded human hemoglobin: A laser light scattering stopped-flow study, *Proc. Natl. Acad. Sci. USA 74*, 3814–3816.

Gibson, Q. H., Milnes, L. (1964) Apparatus for rapid and sensitive photometry, *Biochem. J.* 91, 161–171.

Grimaldi, J. J., Sykes, B. D. (1975) Concanavalin A: A stopped-flow nuclear magnetic resonance study of conformational changes induced by Mn^{++}, Ca^{++}, and a-methyl-D-mannoside, *J. Biol. Chem.* 250, 1618–1624.

Gutfreund, H. (1969) Rapid mixing: Continuous flow, *Methods Enzymol.* 16, 229–249.

Hartridge, H., Roughton, F. J. W. (1923) The velocity with which carbon monoxide displaces oxygen from combination with haemoglobin, *Proc. R. Soc. London, Ser. B 94*, 336–367.

Hilinski, E. F., Rentzepis, P. M. (1983) Biological applications of picosecond spectroscopy, *Nature 302*, 481–487.

Hiromi, K. (1979) *Kinetics of Fast Enzyme Reactions*, J. Wiley & Sons, New York.

Hoffmann, G. W. (1971) A nanosecond temperature-jump apparatus, *Rev. Sci. Instrum.* 42, 1643–1647.

Hoffmann, H., Yeager, E., Stuehr, J. (1968) Laser temperature-jump apparatus for relaxation studies in electrolytic solutions, *Rev. Sci. Instrum.* 39, 649–653.

Hollaway, M. R., White, H. A. (1975) A double beam rapid-scanning stopped-flow spectrometer, *Biochem. J.* 149, 221–231.

Knoche, W. (1974) Pressure-jump methods, in Weissberger, A. (ed.) *Techniques of Chemistry*, Vol. VI, Part II, pp. 187–210, J. Wiley & Sons, New York.

Kustin, K. (1969) Fast reactions, *Methods Enzymol. 16*.

Macgregor, R. B., Clegg, R. M., Jovin, T. M. (1985) Pressure-jump study of the kinetics of ethidium bromide binding to DNA, *Biochemistry 24*, 5503–5510.

Martin, J. L., Vos, M. H. (1992) Femtosecond biology, *Annu. Rev. Biophys. Biomol. Struct.* 21, 199–222.

Moody, M. F., Vachette, P., Foote, A. M., Tardieu, A., Koch, M. H. J., Bordas, J. (1980) Stopped-flow x-ray scattering: The dissociation of aspartate transcarbamylase, *Proc. Natl. Acad. Sci. USA 77*, 4040–4043.

Norrish, R. G. W., Porter, G. (1949) Chemical reactions produced by very high light intensities, *Nature 164*, 658.

Porter, G., West, M. A. (1974) Flash photolysis, in Weissberger, A. (ed.) *Techniques of Chemistry*, Vol. VI, Part II, pp. 367–462, J. Wiley & Sons, New York.

Roughton, F. J. W., Chance, B. (1963) Rapid reactions, in Friess, S. L., Lewis, E. S., Weissberger, A. (eds.) *Techniques of Organic Chemistry*, Vol. VIII, Part II, pp. 703–792. John Wiley & Sons, New York.

Stehlow, H., Becker, M. (1959) Ein Drucksprungverfahren zur Messung der Geschwindigkeit von Ionenreaktionen, *Z. Elektrochem. 63*, 457–461.

Strittmatter, P. (1964), in Chance, B., Eisenhardt, R. M., Gibson, Q. H., Lonberg-Holm, K. K. (eds.) *Rapid Mixing and Sampling Techniques in Biochemistry*, p. 76. Academic Press, New York.

Subject Index

a

abortive complexes 163
absolute error 186
absorption 228 f.
absorption coefficient 230
absorption measure 230
absorption method 282
absorption photometer 235 f.
acceptor 252
acetylcholine receptor 55
acid base catalysis 152
action spectra 242
active transport 175
Adair, G.S. 34
Adair equation 35 f.
activator, allosteric 38
activation 115
– energy 156
– volume 282
activity 13
– coefficient 13
adiabatic calorimeter 226
AEDANS 249
Alberty-Koerber method 83, 85
Alberty notation 138
alcohol dehydrogenase 47, 121
alkaline phosphatase 47
allosteric centre 38
– effectors 38, 43
– enzymes 38 f.
– – kinetic treatment 147 f.
– inhibition 38, 102
– regulation 50 f.
allostery 39
alternate substrates 122
analysis of dates, see diagrams
anilinonaphthalene-8-sulfonate 248 f.
anisotropy 256
ANS 248, 249
antagonists 110, 182

anti-cooperative behavior 47
anti-Stokes lines 247, 263
apparent dissociation constant 13
– maximum velocity 152
arithmetic mean 189
Arrhenius equation 155 f.
Arrhenius plot 156
– immobilised enzymes 175
aspartate transcarbamoylase 53
aspartokinase 54
association constant, definition 12
– rate constant 12
AUC 183
autoclave 234
automatic burette 224
auto-titrator 281

b

bandwidth 236
batch method 208
bathochromic shift 239
bi 125
binding
– classes 30
– constants 12
– enthalpy 227
– equation 14 f., 30
– – Adair 35 f.
– – concerted model 41
– – general 14 f.
– – Hill 33
– – Pauling 37
– – sequential model 45
– measurements 197 f.
– sites 14
– – identical 14
– – interacting 32, 56
– – non-identical 29, 56
bioavailability 183

Enzyme Kinetics. Principles and Methods. 2nd Ed. Hans Bisswanger
Copyright © 2008 WILEY-VCH Verlag GmbH & Co. KGaA, Weinheim
ISBN: 978-3-527-31957-2

Subject Index

bioluminescence 243
bi-uni-uni-bi ping-pong mechanism 137
Boeker method 83, 85
Bohr effect 53
bolometer 263
Briggs, G. E. 65
Brown, A. J. 19, 65
Brumbaugh-Ackers method 210
burst 63
burst kinetics 274

c

calorimeter 226 f.
calorimeter stopped-flow apparatus 274
catalytic constant 12, 66 f.
– efficiency 66
– quantities 2
– triad 152
CD spectra 260
– spectrometer 262
– spectroscopy 228, 257 f.
– stopped-flow 273
central complexes 125
Cha method 145
Chance, B. 241, 242, 271
Chanutin method 213
chemical potential 202
chlorocruorin 43
chymotrypsin 152
circular dichroism 257 f.
clearance 183
Cleland, W. W. 125
Cleland nomenclature 125, 138
CO_2 electrode 222
coefficient form 141
competition 25, 107
competitive inhibition 25, 107 f.
– constant 109
competitive product inhibition 89
competitive substrates 121
complexes
– abortive 163
– central 125
– Michaelis 65
– transitory 125
concentration-dependent difference spectra 241
concentration quenching 246
concerted model 39 f.
conjugation 183
constant absolute error 186
continuous equilibrium dialysis 203
continuous-flow method 268

continuous test 80
conversion
– external 244
– internal 244
cooperativity 35 ff., 149 f.
– kinetic 51, 149 f.
– negative 34, 47 f., 50, 124
– positive 34, 47, 50
Cornish-Bowden plot 79 f., 83, 84
– determination of initial velocity 83, 84
correlation coefficient 190
Cotton effect 259
coupled assay 80
coupling constant 265
critical micellization concentration 180
CTP-synthase 47
cuvettes 236
– CD 262
– fluorescence 245
– stopped-flow 271
– tandem 238

d

Dalziel coefficient 138
DANSYL chloride 249, 252
dead-end complex 101
dead time 276
density gradient centrifugation 215
deoxyhaemoglobin 53
deoxythymidin kinase 47
diagrams
– Arrhenius 156
– direct 20, 42, 66 f., 104
– direct linear 73 f., 83, 84, 104
– Dixon
– – enzyme inhibition 106, 109, 112, 114, 118
– – K_m determination 71
– – pK determination 152
– double reciprocal 21, 42, 67, 75 f., 104, 113, 115, 128
– Eadie-Hofstee 67, 77, 104, 109, 111, 114, 116, 130, 133
– Guggenheim 149
– Hanes 21, 42, 67, 77, 104, 108, 111, 113, 129
– Hill 34, 42
– Job 24
– Kilroe-Smith 71
– Klotz 21
– Lineweaver-Burk 67, 75, 104, 108, 111, 117, 133
– residual 187

- Scatchard   21, 42, 77
- secondary   105 f., 108, 128, 133
- semi-logarithmic   20, 66, 67, 70, 98
- Stockell   23
dialysis   197
- membrane   197
- time   200
difference spectroscopy   238 f.
diffraction grating   236
diffusion   8 f.
- coefficient   8 f., 170
- controlled reaction   10, 171
- external   169
- facilitated   175
- internal   169, 172 f.
- law   9 f., 172, 178
- limited   10, 169
- non-saturable   176
dimethylaminonaphthalene sulfonate   249
diode array photometer   237, 273
direct plot   20, 42, 66 f.
- linear plot   73 f., 83, 84, 104
dispersion method   282
dissociation constant   12 f., 65
- definition   12
- intrinsic   15
- macroscopic   15 f.
- microscopic   15 f.
distribution volume   184
Dixon plots
- enzyme inhibition   106, 109, 112, 118
- K_m determination   71
- pK determination   152
Donnan effect   202
donor   252
dose response curve   182
dose response relationship   182
double beam spectrophotometer   237
double beam stopped-flow apparatus   272
double reciprocal plot   21, 42, 67, 75 f., 127
- see also Lineweaver-Burk plot
Draper-Hippel method   216
Drude equation   258
dual wavelength spectrophotometer   241

e

Eadie-Hofstee plot   67, 77, 104, 111, 117, 130, 133
EC_{50} value   182
effectors, definition   1, 38
- heterotropic   43
effective concentration   182

effects
- heterotropic   38
- homotropic   38
Eigen, M.   278
Einstein's relationship   8
Einstein-Sutherland equation   10
E jump   283
electric field method   283
electron spin resonance spectroscopy   264 f.
ellipticity   258
elution of broad zones   210
emission   243 f.
- monochromator   245
- spectra   246
end-product inhibition, see feedback inhibition
end-stop system   271
energy
- activation   156
- free   158
- kinetic   8
energy transfer   252 f.
enthalpy   158, 278
- binding   227
entropy   158
enzyme inhibition   91 ff.
- memory   148
- specificity   110
- substrate complex   63 f.
enzyme test   80
- coupled   80
- fluorimetric   248
- stopped   80
eosin iodoacetamide   249
EPR   264
equilibrium   64
- constant   13, 39, 44
- - apparent   13
- dialysis   197 f.
- - apparatus   200
- - continuous   203
error
- constant absolute   76, 186
- relative   76, 186
- scattering   76
- systematic   186
erythrocruorin   43
ESR spectroscopy   264 f.
etheno group   251
exchange rate   159
excimers   246
excitation   228

- monochromator 228
- spectra 228
exciton splitting 260
external conversions 229, 244
extinction 230
Eyring theory 157

f

facilitated diffusion 175
fast reactions 267 f.
feedback inhibition 38, 54, 90
femtosecond spectroscopy 283
Fick's diffusion laws 9 f., 172, 178
filter photometer 236
first reaction order 60
Fischer, Emil 47
flash photolysis 283
flash spectroscopy 284
flow calorimeter 227
flow methods 267, 268 f.
fluorescein 249
fluorescence 228, 243 f.
- apparatus 244 f.
- monochromator 245
- polarisation 254
- quenching 246
- resonance energy transfer 252
- spectra 246
- stopped-flow 273
- T-jump 279
fluorimeter 244 f.
fluorodeoxyuridylate 94
fluorophores 247 f.
Förster, T. 252
Förster equation 252
Foster-Niemann method 89, 119
Franck-Condon principle 230
free energy 158
FRET 252
Fromm method 144
front-stop system 271
fructose-1,6-bisphosphatase 55
FT-IR spectroscopy 263
futile cycle 53

g

gating model 12
gel filtration 207 f.
general binding equation 14
generative flow apparatus 271
g-factor 264
GFP 250
Gibbs-Donnan equilibrium 202

Gibbs free standard energy 158
glass electrode 223 f.
globar 263
glyceraldehyde-3-phosphate dehydrogenase 47
glycogen phosphorylase 55
glycogen synthase 55
Golay cell 263
G-protein-coupled receptors 56
gradient ultracentrifugation 214 f.
graph theory 144
grating 236
Guggenheim plot 149

h

Haldane, J. B. S. 65, 75, 88
Haldane relationship 87 f., 135
half life 61, 62, 184
half-of-the-sites reactivity 47
half-wave potential 225
Hanes plot 21, 31, 42, 77, 104, 108, 111, 117, 129, 133
Hartridge, H. 268
heat conducting calorimeter 226
heat quantity 227
α-helix
- CD spectrum 260
- IR spectrum 264
- UV spectrum 232
hemocyanin 43
hemoglobin 32, 43, 52 f.
Henri, V. 19, 65
heterotropic effectors 43
- effects 38
hexa-uni ping-pong mechanism 137
Hill, A. V. 33
Hill coefficient 34, 42, 48, 52
- equation 33
- plot 34, 42
homoserine dehydrogenase 54
homotropic effects 38
Hummel-Dreyer method 209
hyperbola, right-angle 19, 69
hyperbolic saturation curves 19, 69, 78
hyperfine splitting 265
hysteretic enzymes 148 f.

i

immobilised enzymes 168 f.
induced-fit hypothesis 44
infrared spectroscopy 262
inhibition 91 ff.
- competitive 26, 107 f.

– – product   89
– complete   101
– feedback   38, 54
– immobilised enzymes   173
– in multi-substrate reactions   132
– irreversible   92 f.
– mixed   103
– non-competitive   101 f., 174
– partial   101, 113 f.
– partially competitive   29, 117 f.
– – non-competitive   100, 113 f.
– – uncompetitive   115
– product   88 f., 135
– reversible   98 f.
– substrate   120
– uncompetitive   111 f., 120
inhibition constants   100 f.
– competitive   108
– in multi-substrate reactions   127
– uncompetitive   113
inhibitors   1, 91 f.
– allosteric   38
– feedback   38
– irreversible   93
inhibitory concentration   182
initial velocity   81
– determination   84 f.
inner filter effect   246
integrated Michaelis-Menten equation   78 f.
– determination of initial velocity   84 f.
– enzyme inhibition   107 f., 113
– product inhibition   90, 119
– reversible reactions   87
interaction constant   44
– factor   37
internal conversion   229
intersystem crossing   229
intrinsic dissociation constant   15
invertase   65
ion channel   52
ionisation constants   153
irradiation methods   267
irreversible inhibition   92 f.
IR spectroscopy   262 f.
isoenzymes   47, 123
iso mechanism   130
iso ordered mechanism   132
isoperibolic calorimeter   226
iso ping-pong *bi-bi* mechanism   132
– – mechanism   134
isopycnic centrifugation   215
isosbestic point   242
isothermic calorimeter   226

isotope effects   163 f.
– apparent   165
– primary kinetic   163 f.
– secondary kinetic   165
– isotope exchange kinetics   158 f.
isozymes   123

j
Job plot   24
Joule heating   278

k
Kerr cell   257
Kilroe-Smith plot   71
kinetic constants   3, 65, 142
kinetic cooperativity   51, 149 f.
kinetic isotope effect
– primary   163 f.
– secondary   165
kinetically controlled reaction   169
King-Altman method   138 f.
Klotz plot   21
klystron   265
Koshland, D. E.   44
Krönig-Kramers transformation   260
K systems   48, 147

l
lactate dehydrogenase   123, 135
lag phase   148
Lambert-Beer law   230
Langmuir, I.   19
lanthanides   247
laser   263, 279, 284
Le Chatelier-Braun principle   282
ligands, definition   2
limited diffusion   10, 169
LINAC   285
line spectrum   236
linearisation methods
– binding measurements   21 f.
– enzyme inhibition   102 f.
– first order reaction   60
– irreversible enzyme inhibition   98
– Michaelis-Menten equation   75 f.
– – integrated   78 f.
– multi-substrate reactions   128 f.
linear regression   189
Lineweaver-Burk plot   67, 75, 104, 108, 111, 117, 133
lock-and-key model   44
luciferase   243
luciferin   243

m

luminescence 243
luminometer 243

macromolecule, definition 2
macroscopic dissociation constant 15
malate dehydrogenase 47
manometric method 219
mass action law 12
maximum velocity 65 f.
– apparent 152
Maxwell, J.C. 277
mean value 189
median 73, 189
Menten, M. 19, 65
Michaelis, L. 19, 65
Michaelis complex 65
Michaelis constant 12, 66 f., 127
Michaelis-Menten equation 19, 63 f., 127
– derivation 63 f.
– diagram 67
– hyperbolic function 69 f.
– immobilised enzymes 169 f.
– integrated 78 f., 107 f., 113, 119
– non-linear adaptation 190
– reversible reactions 86
microcalorimeter 226
microscopic dissociation constant 15
microwave generator 265, 279
mixed inhibition 103
mixing chamber 270
mnemonic enzymes 148
mode 189
monochromator 236 f.
Monod, J. 39
multifunctional enzyme 54
multi-mixing system 274
multiple equilibria 2, 7 f.
multisite ping-pong mechanism 133
multi-substrate reactions 144 ff., 151
– nomenclature 124, 138
– product inhibition 135
myoglobin 32, 43

n

NBD 249
negative cooperativity 34, 47 f., 50
Nernst stick 263
nicotine receptor 52, 55
nitroxyl radicals 266
NMR spectroscopy 264
NMR stopped-flow apparatus 273
nomenclature 4

– multi-substrate reactions 124, 138, 161
non-competitive binding 27
– inhibition 101 f., 174
– product inhibition 90, 119
non-identical binding sites 29
non-linear regression 190

o

opacity 230
optical activity 257
optical density 230
optical rotation dispersion 257
optical test 231
ORD spectroscopy 228, 257
ORD spectrometer 261
order of reaction, see reaction order
ordered mechanism 125, 131
ordered bi-bi mechanism 125, 131, 139, 161
ordered ter-ter mechanism 136
osmotic pressure 201
oxidation-reduction potential 223
oxygen electrode 220 f.

p

paramagnetic compounds 266
partial inhibitions 101, 113 f.
partially competitive inhibition 29, 111, 117 f.
partially non-competitive inhibition 101, 113 f.
partially uncompetitive inhibition 115
Pauling, L. 37
Pauling model 37
Perrin equation 255
Perutz, M. 52
pharmacokinetics 182 f.
phenylalanine
– absorption spectrum 232, 248
– CD spectrum 261
– fluorescence spectrum 250
pH-dependence of fluorescence 248
pH-dependent difference spectra 241
pH-jump stopped-flow method 276
pH optimum 151 f.
– immobilised enzymes 174
pH stability 153
pH-stat 223
pH stopped-flow apparatus 274
phosphofructokinase 55
phosphorescence 228, 246
photochemical action spectra 242
photolysis 283

photometer   235 f.
photomultiplier   237, 245
photoselection   255
picosecond spectroscopy   283
ping-pong
– *bi-bi* mechanism   133, 145
– mechanism   125, 132 f., 137, 147
– ordered mechanism   137
– random mechanism   137
p-jump method   281
pK values   151
plot, see diagram
PMSF   94
polarisation   254, 257
polarography   225
polymer substrates   167
positive cooperativity   34, 47, 50
potentiometry   223
pressure jump method   281
pre-steady-state phase   63, 274 f.
product inhibition   88 f., 102
– competitive   88, 90
– immobilised enzymes   174
– in multi-substrate reactions   132, 135
– non-competitive   90, 119 f.
– uncompetitive   90, 119 f.
progress curves   60, 77 f., 80
protein spectra
– absorption   232 f.
– CD   260
– fluorescence   250
protomers   39
proxyl   266
pseudo-first order   61, 276
pulse-flow method   270
pulse fluorimetry   254
pulse radiolysis   285
pyren iodoacetamide   249
pyruvate dehydrogenase complex   95, 148, 151
pyruvate-formiate lyase   266

q

quad mechanism   125, 138
quantum yield   243
quenched-flow technique   274
quenching   246

r

radiationless energy transfer   252
radicals   285
Raman effect   228, 247
Raman spectroscopy   262 f.

random coil
– CD spectrum   260
– UV spectrum   232
random mechanism   125, 126 f.
– *bi-bi*   126, 131, 161
– *bi-uni*   139
– rapid equilibrium   127
rapid freezing method   274
rapid-scan stopped-flow apparatus   273
rate constants   10
– determination   288
– first order   60 f.
– pseudo first order   61
– second order   61
– zero order   62
rate equation
– coefficient form   141
– derivation   63 f., 85, 99 f., 127, 138
Rayleigh scattering   247, 263
reaction enthalpy   158, 278
reaction controlled reactions   10
reaction entropy   158
reaction order
– definition   59
– first   60 f.
– pseudo-first   61
– second   61
– zero   62
receptor   47, 52, 55, 182
redox indicator   223
redox potential   223
regression coefficient   189
regression methods   188 f.
– linear   76 f.
– non-linear   70
– Michaelis-Menten equation   190
– weighting factors   76
relative error   186
relaxation curves   285
relaxation methods   267, 277 f.
relaxation time   285 f.
relaxed state   39
replot, see secondary plot
residual plot   187 f.
resonance Raman method   263
reversible enzyme reactions   85 f.
reversible inhibition   98 f.
rhodamine   249
ribonuclease   51, 150
ribozymes   166
RNA enzymes   166
RNase   150
RNase P   166

Subject Index

Rosenthal method   30
rotation relaxation time   256
Roughton, F.J.R.   268
R_s value   49
R state   39, 44

s

saturation function   20, 65
scanning calorimeter   226
Scatchard plot   21, 31, 42
secant method   82
second reaction order   61 f.
secondary plot   105 f., 128
sequential mechanisms   125
sequential model   44 f.
serine protease   94, 152
β-sheet, spectra
– CD   260
– IR   264
– UV   232
sigmoidal saturation curves   32 f., 48, 150
sigmoidicity   41
single point measurement   81, 83
sliding model   12
slow transition model   51, 150 f.
Smoluchowski limit   10
specificity constant   66
spectra
– CD   260
– corrected   264
– ESR   266
– fluorescence   250
– IR   264
– line   236
– Raman   264
– UV   232 f., 240
spectrofluorimeter   244 f.
spectrophotometer   235 f.
– diode array   237
– double beam   237 f.
– dual wavelength   241
– single beam   237
spectroscopic titrations   22 f.
spin labels   266
spin quantum number   264
SPR   218
– angle   219
standard deviation   189
standard reaction enthalpy   158
standard energy   158
statistical methods   185 f.
statistical terms   189
steady-state   64 f.

steady-state phase   65
Steinberg-Schachman method   213
Stockell plot   23
Stokes lines   247, 263
stopped-flow method   271 f.
– calorimeter   274
– CD   273
– dead time   276
– double beam   272
– dual wavelength   272
– fluorescence   273
– generative   271
– light scattering   273
– multi-mixing   274
– NMR   273
– pH   274
– quenched flow   274
– rapid freezing   274
– rapid scan   227
– temperature jump   273, 280
– UV-Vis   272
stopped test   80
Strittmatter, P.   272
student test   186
substrate   1
– analoga   93 f.
– inhibition   120 f.
– module   170, 173
– surplus inhibition   120
sucrose gradient centrifugation   214
suicide substrate   93
surface plasmon resonance   218
$S_{0.5}$ value   48
swinging bucket rotor   214
symmetry model   39 f.
systematic error   186

t

tandem cuvettes   238
tangent method   81
temperature behaviour   154 f.
– immobilised enzymes   174
temperature-dependent difference spectra   241
temperature jump method   278 f.
temperature jump stopped-flow apparatus   273, 280
temperature maximum   154
temperature optimum   154
temperature stability   155
tempo   266
tense state   39
ter   125

term scheme 228
Theorell-Chance mechanism 131
thermodynamic constants 3
thermostability 156
Thiele module 173
thymidylate synthase 94
T jump method 278 f.
TNS 249, 251
toluidinonaphthalene-6-sulfonate 248, 251
transient kinetics 285
transition state 156
– analogues 95 f., 157
transitory complexes 125
transmittance 230
transport
– active 175
– coefficient 133
– processes 175
trigger 271 f.
tryptophan spectra
– CD 261
– fluorescence 247, 250
– UV 232, 240
tryptophan synthase 29
T state 39, 44
t-test 186
turnover rate 63
Tyndall scattering 247
tyrosine spectra
– CD 261
– fluorescence 247, 250
– UV 232, 241

u

ultracentrifugation methods 211 f.
ultrafiltration methods 206 f.
ultrafiltration apparatus 206
ultrafiltration membranes 206
ultrasonic method 282
uncompetitive inhibition 103, 111 f., 120
uncompetitive product inhibition 90, 119
uni 125
uni-bi-bi-uni mechanism 137
uni-uni-bi-bi mechanism 137
UV spectra 232 f.
UV spectroscopy 232, 235 f.

v

van't Hoff reaction isobar 157, 277
variance 189
vibrational level 229, 262
vibration spectra 262
Volkenstein-Goldstein method 144
V systems 48, 147

w

Warburg, O. 219, 242
Warburg manometer 219
W-test 186
Woolf diagrams 75

y

Yamamoto-Alberts method 218

z

zero point oscillations 163
zero reaction order 62, 64
zonal ultracentrifugation 215

MOLECULES THAT CHANGED THE WORLD

K. C. Nicolaou *The Scripps Research Institute and UC San Diego, La Jolla, USA*
& **Tamsyn Montagnon** *Department of Chemistry, University of Crete, Heraklion, Crete, Greece*

ISBN: 978-3-527-30983-2
February 2008
Hardback • 385 Pages
Price £24.95 / $50.00 / €34.90

In this attractively designed book, K.C. Nicolaou introduces some of the world's **most important molecules** and engagingly demonstrates the role certain compounds play in our everyday lives as drugs, flavours or vitamins. Printed in full colour throughout and with its oversize format, this book is **a must for every chemist**, natural scientist and anyone interested in the sciences.

Get into the fascinating world of substances like Aspirin, Taxol and many more:

- Present interesting and entertaining facts, stories and information about the people behind the scenes
- Covers 40 natural products, each of them with an enormous impact on our everyday life
- Includes a plethora of fascinating pictures